Topics in Applied Physics   Volume 73

W0245799

Springer-Verlag Berlin Heidelberg GmbH

Topics in Applied Physics   Founded by Helmut K. V. Lotsch

Hydrogen in Metals III

Properties and Applications

Edited by H. Wipf

With Contributions by
R.G. Barnes   P. Dantzer   H. Grabert   D.K. Ross
H.R. Schober   H. Vehoff   H. Wipf

With 117 Figures

 Springer

Professor Dr. *Helmut Wipf*
Institut für Festkörperphysik
Technische Hochschule Darmstadt
Hochschulstasse 6
D-64289 Darmstadt
Germany

ISBN 978-3-662-30950-6   ISBN 978-3-540-69988-0 (eBook)
DOI 10.1007/978-3-540-69988-0

Library of Congress Cataloging-in-Publication Data. Hydrogen in metals III: properties and applications/edited by H. Wipf: with contributions by R.G. Barnes ... [et al.].   p.   cm. – (Topics in applied physics: v. 73) Includes bibliographical references and index. ISBN 978-3-662-30950-6 (alk. paper) 1. Metals–Hydrogen content. I. Wipf, H. (Helmut), 1941– II. Barnes, R.G. (Richard G.) III. Title: Hydrogen in metals 3. IV. Title: Hydrogen in metals three. V. Series. TN690.H9694 1997 669'.94–DC20 96–38496

© Springer-Verlag Berlin Heidelberg 1997
Originally published by Springer-Verlag Berlin Heidelberg New York in 1997
Softcover reprint of the hardcover 1st edition 1997

Cover design: Design & Production GmbH, Heidelberg

Typesetting: Scientific Publishing Services (P) Ltd, Madras

SPIN: 10068864   54/3144/SPS – 5 4 3 2 1 0 – Printed on acid-free paper

Preface

In 1978, the two volumes *Hydrogen in Metals I* and *II*, edited by G. Alefeld and J. Völkl, appeared in the series *Topics in Applied Physics*, vols. 28, 29. These volumes reviewed the experimental results and the theoretical understanding of the properties of metal-hydrogen systems at that time. Ten years later, L. Schlapbach edited the two volumes *Hydrogen in Intermetallic Compounds I* and *II* in the same series (vols. 63, 67). The contributions of these volumes focused on aspects not discussed in *Hydrogen in Metals I* and *II*, and in particular on the behavior of hydrogen in intermetallic compounds.

The above volumes of *Topics in Applied Physics* are an important source of information. They are well accepted by the scientific community of physicists, chemists, material scientists, metallurgists and engineers, and they are frequently cited in research papers. They will also continue to be of great value and significance in the years to come.

The number of new and important results that were published since these volumes came into print is large. In fact, there are several topics where, beyond a sole increase of the data, the progress was so significant that our present understanding differs principally from what was known a decade before. Such topics are low-temperature hydrogen diffusion and tunneling, where the hydrogen was found to interact directly with the conduction electrons, and the results from nuclear magnetic resonance and neutron scattering. Further, there are two important application-oriented topics which are only partially addressed in the previous volumes of *Topics in Applied Physics*. This holds for the material problems that are caused by the hydrogen, and for the use of metal hydrides for technical applications such as hydrogen storage and purification, chemical engines, sensors, detectors, batteries and fuel cells.

Accordingly, the intention for publishing the new volume *Hydrogen in Metals III* is to provide a thorough discussion of the above topics which were only partially presented before, or for which the recent progress was so large that a new review seemed necessary.

I would like to thank the authors for their willingness to write their contributions and for their patience in the publication process. I would also like to thank *H. Lotsch* of the Springer Verlag for his continuous support.

Darmstadt, September 1995 *Helmut Wipf*

Contents

Contributors

Barnes, Richard G.

Solid State Physics Division, Ames Laboratory, Iowa State University,
Ames, Iowa 50011, USA

Dantzer, Pierre

CNRS - URA 0446 - ISMa, Université Paris-Sud, Bât. 415,
F-91405 Orsay-Cedex, France

Grabert, Hermann

Fakultät für Physik, Albert-Ludwigs-Universität,
Hermann-Herder-Straße 3, D-79104 Freiburg, Germany

Ross, D. Keith

Department of Physics, Joule Laboratory, University of Salford,
Salford M5 4WT, England

Schober, Herbert R.

Institut für Festkörperforschung, Forschungszentrum Jülich,
D-52425 Jülich, Germany

Vehoff, Horst

Technische Fakultät, Universität des Saarlandes, Im Stadtwald,
D-66123 Saarbrücken, Germany

Wipf, Helmut

Technische Hochschule Darmstadt, Hochschulstraße 6,
D-64289 Darmstadt, Germany

1. Introduction

H. Wipf

With 1 Figure

In 1866, *Graham* [1.1] reported that the metal Pd can absorb large amounts of hydrogen gas. In fact, the ability to absorb hydrogen is common to all metals, the only question is how much hydrogen actually is absorbed in a given situation. Figure 1.1 shows, for six representative metals, the hydrogen concentration (hydrogen-to-metal atom ratio x) that is found under *equilibrium conditions* for a temperature T in the presence of a hydrogen gas atmosphere of 1 bar. Obviously, the six metals react quite differently. Thermodynamics tells us that the most decisive quantity for the resulting hydrogen concentration is the heat of solution, i.e., the amount of heat set free per absorbed hydrogen atom. Figure 1.1 indicates the values of this quantity for the respective metals, valid in the range of low hydrogen concentrations. It can be seen that large positive heat of solutions favor high hydrogen concentrations, particularly at low temperatures, whereas the hydrogen concentrations in the case of a negative heat of solution are very low.

As a matter of fact, the actual hydrogen concentrations in a metal differ in general from the equilibrium value since the absorption of a hydrogen molecule, as well as the reverse desorption process, is usually seriously impeded at the metal surface. The (catalytic) surface reactions that impede (or favor) hydrogen absorption and desorption are of importance for many technological applications.

Within the metal, the absorbed hydrogen molecules are dissociated and the hydrogen atoms occupy interstitial sites in the metal lattice. The hydrogen atoms jump from one interstitial site to a neighboring one, causing extremely high diffusion rates even at low temperatures (in some systems, the room-temperature jump rates exceed 10^{12} Hz). Because of the small hydrogen mass, quantum and tunneling effects are of importance for the diffusion process. Further, the hydrogen interstitials change drastically the mechanical, electronic and magnetic properties of the metal, they modify the phonon spectra, and they induce structural changes that may lead to complex phase diagrams. Because of the high jump rates of the hydrogen, structural changes that require diffusion (for instance in the presence of a miscibility gap) will be observed even far below room temperature, in contrast to ordinary metallic alloy systems where such phase transformations will no longer take place since the diffusivity is too low. On the other hand, the hydrogen concentrations of the nonstoichiometric metal-hydrogen systems can also vary in

Topics in Applied Physics, Vol. 73
Wipf (Ed.)
© Springer-Verlag Berlin Heidelberg 1997

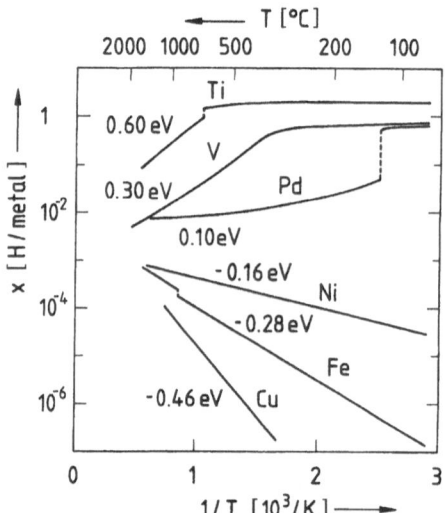

Fig. 1.1. The hydrogen-to-metal atom ratio x found in the six indicated metals under equilibrium conditions at the temperature T in the presence of a hydrogen gas atmosphere of 1 bar. The figure shows also the heat (or enthalpy) of solution for the respective metals, valid in the limit of low hydrogen concentrations

wide ranges, without a structural change, so that these systems can indeed be considered as the realization of a lattice gas.

The properties of metal-hydrogen systems were the subject of intensive and continuous research activities over the last decades. These activities were motivated by both basic science and technological application. In fact, it characterizes the interested community of physicists, chemists and material scientists that the gap between those studying chiefly fundamental aspects and those who are more interested in potential applications never became so deep that there was no vivid mutual communication.

The results of the on-going research on metal-hydrogen systems are reported in a great number of review articles or books so that it is only possible here to mention some of them. As early as 1968, *Mueller* et al. [1.2] published, as editors, a comprehensive book on metal hydrides that still provides a compilation of important basic information. In 1978, *Alefeld* and *Völkl* [1.3] edited the two volumes *Hydrogen in Metals I* and *II*, summarizing the fundamental experimental and theoretical aspects of the behavior of metal-hydrogen systems. These volumes are the leading presentation of the field at that time. Ten years later, they were followed by the two volumes *Hydrogen in Intermetallic Compounds I* and *II*, edited by *Schlapbach* [1.4]. These two volumes accounted for the progress that was made in the field since *Hydrogen in Metals I* and *II* appeared, and they discuss in particular the behavior of hydrogen in a great number of alloy systems which were not specifically addressed in the books edited by *Alefeld* and *Völkl*. The properties of the hydrides of rare earth metals and alloys are summarized by review articles of *Buschow* [1.5], *Arons* [1.6] and *Vajda* [1.12]. Further, the special application of metal hydrides for hydrogen and energy storage was discussed in detail by *Buchner* [1.7] and in a book edited by *Winter* and *Nitsch* [1.8]. The latter book does not only consider the potential application of metal hydrides for hydrogen storage, but it gives a very general presentation of the technological aspects of the use of hydrogen gas as an energy carrier. The basic bulk properties of metal-hydrogen systems were also very recently discussed by *Fukai* [1.9] in a book that focuses on the fundamental points of view. Finally, it seems worth mentioning that metal-hydrogen systems are the topic of a biennial conference series (Intern. Symp. on Metal-Hydrogen Systems). The proceedings of the last two of these conferences in Sweden (1992), Japan (1994) and Les Diableich (1996) are published in special volumes of the *Zeitschrift für Physikalische Chemie* [1.10] and the *Journal of Alloys and Compounds* [1.11], respectively. In these proceedings the interested reader will find a compilation of papers that characterize the state of the research in the field, together with references to previous work.

The present volume *Hydrogen in Metals III* was published in order to provide a thorough discussion of those topics where recent progress was so large that a new review seemed timely and necessary (Chaps. 2–5), and secondly for a discussion of those research fields which were not adequately addressed in the previous volumes of *Hydrogen in Metals* and of *Hydrogen in Intermetallic Compounds* (Chaps. 6, 7). Chapter 2, written by Grabert and Schober on the theory of the tunneling and diffusion of light interstitials in metals, reflects the substantial increase in our understanding of this topic in the preceding years. This holds particularly for hydrogen tunneling at low temperatures and the direct (nonadiabatic) interaction of the tunneling hydrogen with the conduction electrons, which were not comprehensively discussed in a theoretical review paper until now. Chapter 3, which presents a compilation of diffusion data, is the experimental counterpart to Chap. 2. Nuclear magnetic resonance and neutron scattering, both powerful experimental techniques for the study of metal-hydrogen systems, are discussed by Barnes and Ross in Chaps. 4, 5, respectively. The results obtained with these experimental techniques in the last decade were so important that a new review seemed necessary, in spite of a detailed presentation in *Hydrogen in Metals I*. The final two sections of the present volume address application oriented topics. Chapter 6, written by Vehoff, discusses hydrogen related material problems such as hydrogen embrittlement, and Chap. 7, by Dantzer, reviews the technological applications of metal-hydrogen systems, including topics like hydrogen storage and purification, hydride chemical engines, hydrogen sensors and detectors, and hydride batteries and fuel cells. Both topics are important and the motivation for extensive research efforts worldwide so that they certainly deserve a comprehensive presentation in *Hydrogen in Metals III*.

References

1.1   T. Graham: Phil. Trans. Roy. Soc. (London), **156,** 399 (1866)
1.2   W.M. Mueller, J.P. Blackledge, G.G. Libowitz (eds.): *Metal Hydrides* (Academic Press, New York 1968)
1.3   G. Alefeld, J. Völkl (eds.): *Hydrogen in Metals I* and *II*, Topics Appl. Phys., Vols. **28** and **29** (Springer, Berlin, Heidelberg 1978)

1.4 L. Schlapbach (ed.): *Hydrogen in Intermetallic Compounds I* and *II*, Topics Appl. Phys., Vols. **63** and **67** (Springer, Berlin, Heidelberg 1988, 1992)

1.5 K.H.J. Buschow: In *Handbook on the Physics and Properties of Rare Earths*, ed. by K.A. Gschneidner, Jr. and L. Eyring (North Holland, Amsterdam 1984) Vol. **6**, Chap. 47, p. 1

1.6 R.R. Arons: In *Landolt-Börnstein,* New Series III/19d1 (Springer, Berlin, Heidelberg 1991) p. 280

1.7 H. Buchner: *Energiespeicherung in Metallhydriden* (Springer, Wien 1982)

1.8 C.-J. Winter, J. Nitsch (eds.): *Wasserstoff als Energieträger* (Springer, Berlin, Heidelberg 1989)

1.9 Y. Fukai: *The Metal-Hydrogen System. Basic Bulk Properties*, Springer Ser. Mater. Sci., Vol. **21** (Springer, Berlin, Heidelberg 1993)

1.10 Z. Phys. Chemie **179, 181** and **183** (1993, 1994)

1.11 J. Alloys Comp. **231** (1995) and to be published

1.12 P. Vajda: In *Handbook on the Physics and Chemistry of Rare Earths*, ed. by K.A. Gschneidner, Jr. and L. Eyring (Elsevier Science, Amsterdam 1995) Vol. **20**, Chap. 137, p. 207

2. Theory of Tunneling and Diffusion of Light Interstitials in Metals

H. Grabert and H.R. Schober

With 7 Figures

The diffusion and tunneling of hydrogen in metals has been the subject of intensive studies since several decades. Apart from its technological importance hydrogen in metals is also a unique system for basic research in solid state physics. This is especially due to the hydrogen mass which is on one hand small enough to make quantum effects, such as tunneling, observable and on the other hand large enough for the hydrogen to be localized in the host lattice as opposed to conduction electrons. A particular advantage is the large mass ratio of up to three between the isotopes hydrogen, deuterium and tritium or even thirty if one includes the muon μ^+ as an isotope. In this chapter we review the progress made in the theory of tunneling and diffusion of hydrogen over the last years. The relevant experiments are discussed in the following chapter of this book. We restrict ourselves to the limit of low concentration (α-phase) where the hydrogen atoms occupy random interstitial sites. Diffusion is then by random hopping from one site to a neighboring site and interaction with other hydrogen atoms is negligible.

2.1 Background

At low temperatures hydrogen motion arises only by means of quantum tunneling through the intervening potential barrier. The basic theories from the seventies for quantum diffusion of light particles in metals start out from the Born-Oppenheimer approximation. It is assumed that the electrons of the crystal adiabatically follow the nuclear coordinates, i.e. the positions of the metallic ions and the coordinates of the interstitial particle. After the adiabatic elimination of the electronic degrees of freedom one is left with a tunneling particle coupled to the phonons of the host crystal. The defect-phonon problem was treated in the well-known papers on phonon assisted tunneling by *Flynn* and *Stoneham* [2.1], *Kagan* and *Klinger* [2.2] and by *Teichler* and *Seeger* [2.3]. This work originated from the small polaron theory of *Yamashita* and *Kurosawa* [2.4] and of *Holstein* [2.5], which, in turn, is closely related to the familiar treatments of radiationless transitions by *Huang* and *Rys* [2.6] and by *Pekar* [2.7].

These earlier theories break down on both sides of a temperature scale. Conventionally, the dynamics of the system is decoupled into the fast dynamics of the hydrogen, determining the "naked" tunneling probability, and

Topics in Applied Physics, Vol. 73
Wipf (Ed.)
© Springer-Verlag Berlin Heidelberg 1997

the slower dynamics of the host atoms, represented by a phonon heat bath which "dresses" the tunneling probability. The hydrogen is assumed to follow the host lattice adiabatically. As a second, more stringent approximation, the naked tunneling probability is taken to be independent of the state of the phonons, i.e., the temperature (Condon approximation). Both these approximations hold as long as only long wavelength phonon states are populated. They break down at temperatures above $\approx$ 50 K. At higher temperatures the strong coupling between hydrogen and host-lattice vibrations [2.8, 9] gains importance. The Condon approximation is the reason for the failure of these theories to reproduce the experimentally observed kinks in the Arrhenius plots for the diffusion constants in Nb and Ta. Improved theories for the diffusion at elevated temperatures avoid these approximations and will be reviewed in Sect. 2.4.

At still higher temperatures ($T > 300$ K) quantum effects will eventually lose their importance. One can then use standard molecular dynamics to simulate hydrogen diffusion. Such calculations have recently been done for Pd:H [2.66] and for Nb:H [2.67]. In the latter system one finds e.g. two distinct contributions to the residence time on the stable sites. There is a rapid movement on a ring of four neighbouring tetrahedral sites, constituting a so called $4T$ configuration, and a much slower one leading to diffusion. This behaviour had earlier been observed in experiment [2.68]. Substantial contributions of second neighbour jumps to diffusion were observed.

Further complications in hydrogen diffusion arise at finite hydrogen concentrations due to the strong H-H interaction. Experimental and some theoretical results can be found in the book by Fukai [2.69].

Also, the standard small polaron theory is unable to describe the low-temperature limit correctly. In recent years, it has become clear that at low temperatures the tunneling dynamics of light interstitials is strongly affected by a nonadiabatic interaction with the conduction electrons of the host metal. These nonadiabatic electron effects are ignored in the conventional Born-Oppenheimer approximation. The behavior of light particles in metals cannot be explained solely by considering interatomic lattice potentials. The nonadiabatic influence of conduction electrons on the motion of light interstitials was already demonstrated in 1984 in low-temperature ultrasonic experiments on hydrogen in niobium by *Wang* et al. [2.10]. In the same year, nonadiabatic electronic effects were also proposed by *Kondo* [2.11] as a mechanism to understand the previously unexplained temperature dependence observed for the diffusion rate of muons in low-temperature host metals [2.12].

Since a description in terms of lattice potentials is the basis of all conventional treatments, it may be appropriate to elaborate on this point. The Born-Oppenheimer approximation tacitly assumes that the characteristic energy scale of the metal electrons is of the order of the Fermi energy ε_F. However, a tunneling particle causes a slow dynamical perturbation of the conduction electrons screening the interstitial defect. As pointed out by *Kondo* [2.11], the defect motion couples to the low-energy excitation modes of the metallic electrons. Since electron-hole pair excitations can have arbitrarily small energies, parts of the screening cloud cannot follow adiabatically even slow defect motions. In view of the small spectral density of low-frequency phonons, the coupling to the conduction electrons dominates at low temperatures. In a normal metal the lattice vibrations may be eliminated adiabatically at sufficiently low temperatures and one has to deal primarily with a defect-electron problem.

Like in the Anderson orthogonality catastrophe [2.13], an infrared divergence arises in the electronic overlap integral of two screened defect states at different interstitial sites. At finite temperatures the overlap integral becomes proportional to a power of temperature [2.14] which should be reflected in a power-law dependence of the hopping rate on temperature with a negative exponent [2.11, 15]. Hence, the nonadiabatic effects of the conduction electrons lead to an increase of the hopping rate with decreasing temperature, a behavior first observed for the muon hopping rate in aluminum and copper below 10 K [2.12]. In this region the defect hops at random to neighboring sites with an incoherent tunneling rate which is affected by the retarded response of the screening cloud. As the temperature is decreased further, the hopping rate does not continue to grow, rather the defect becomes delocalized with a wave function extending over several interstitial sites. In this case, corresponding to coherent tunneling of the defect, the dynamics can no longer be described in terms of a hopping rate. The same situation arises at somewhat higher temperatures when the metal is in the superconducting state.

The simplest case of coherent tunneling is the delocalization of a particle in a double well potential describing, for instance, two neighboring interstitial sites. The two energetically split tunneling eigenstates of the particle form a two-level system. In 1972, *Anderson* et al. [2.16] and *Phillips* [2.17] postulated the existence of such two-level systems in glasses in order to explain the low-temperature specific heat anomalies in these materials. They attributed the anomalies, which are specific to the amorphous state, to coherent tunneling of atoms or groups of atoms between two mutually accessible potential minima. Ultrasonic experiments revealed that the lifetime of the tunneling eigenstates in amorphous metals is drastically shorter than in nonmetallic glasses. To explain this behavior, *Golding* et al. [2.18] argued that the two-level systems couple nonadiabatically to the conduction electrons in a metallic environment. Following a suggestion by *Black* and *Fulde* [2.19], this electronic influence was proven by *Weiss* et al. [2.20] in ultrasonic studies on a superconducting amorphous metal. Apart from the fact that the microscopic nature of the tunneling entities in amorphous materials is not known, they also exhibit an extremely broad distribution in their physical properties. This tends to smear out experimentally observable effects that specifically reflect properties of the two-level systems. Hence, it is difficult to develop a truly quantitative tunneling theory for amorphous materials and to test its validity through a comparison with experimental data.[1]

---

[1]Very recently individual two-level systems have been observed via conductance fluctuations in a mesoscopic disordered metal [2.21]. This new technique allows for a more detailed comparison with theoretical predictions.

The situation is different for hydrogen interstitials in metals. A large number of low temperature measurements carried out on various metals have indicated the existence of tunneling eigenstates (Chap. 3). In particular, in niobium containing interstitially dissolved oxygen, nitrogen or carbon impurities, the hydrogen is trapped by the (immobile) impurities below ~ 160 K [2.22] and occupies a double well potential allowing tunneling between two neighboring tetrahedral interstitial sites. For these two-level systems, the tunneling entities are clearly identifiable and, in a given sample, they do not exhibit a broad distribution in their physical properties. The nonadiabatic influence of conduction electrons on hydrogen tunneling in niobium becomes manifest both for the coherent tunneling of trapped hydrogen at very low temperatures $T < 5$ K, and for the incoherent tunneling observed at higher temperatures 10 K $< T < 50$ K [2.23]. In the first case the interstitial defect makes transitions between two delocalized tunneling eigenstates, while in the latter case the particle hops randomly from one interstitial site to an adjacent site. Of course, these two seemingly different situations are just two facets of dissipative tunneling in a double well potential [2.24], and one can pass from one situation to the other by changing the temperature.

In this article we first introduce the quantities characterizing the hydrogen metal system and give estimates for the various energies involved. From this qualitative picture we proceed to derive the Hamiltonians needed to calculate the dynamics of the tunneling process. From a general Hamiltonian we abstract through a number of approximations the Hamiltonian for the tunneling system coupled to an electronic and phononic heat bath. We show how standard small polaron theory can be extended to higher temperatures by avoiding low temperature approximations. This can be done conceptually fairly easily using Fermi's golden rule but involves heavy numerical calculations [2.25] and will be discussed in Sects. 2.2–2.4 before the theoretically more involved low temperature behavior including the crossover from coherent to incoherent tunneling in a metallic environment [2.26] will be addressed in Sects. 2.5–2.8. Finally, Sect. 2.9 contains our conclusions.

2.2 Statics and Dynamics of Hydrogen in Metals

Standard theories of diffusion and tunneling of hydrogen in metals start from two interrelated ideas. The first is that the hydrogen is, on a relevant timescale, localized at interstitial sites of the host lattice and that tunneling is from one such site to the next. The second idea is that of self-trapping, meaning that an interstitial hydrogen will distort the host lattice and thus lower the total energy. To get an idea of the various energies involved let us take here and in the following Nb:H as an example. The hydrogen occupies tetrahedral sites and the four nearest-neighbor atoms are displaced by about 0.1 Å [2.27]. From model calculations [2.25, 28] one estimates the difference between the total energies of the undistorted and the relaxed lattices, respectively, to $E_{ST} \approx 450$ meV. This so called self-trapping energy is the sum of the change in potential energy (300 meV) and the change in zero-point vibrational energy. Since the displacements of the Nb atoms are not too large, a harmonic model for the Nb-Nb interaction should be reasonably accurate. The potential energy part (elastic energy) of the energy gain by self-trapping at site i can then be related to the force $\boldsymbol{F}^{(i),m}$ exerted on the Nb atom m due to the presence of the H. This can be a direct force from H on Nb but could also include some induced Nb-Nb forces. The energy gain by this relaxation, the elastic self-trapping energy, is

$$E_{\text{ST}}^{\text{elastic}} = \frac{1}{2}\sum_{m,n} \boldsymbol{F}^{(i),m} G^{mn} \boldsymbol{F}^{(i),n} , \tag{2.1}$$

where $\boldsymbol{G}$ is the harmonic lattice Green's function [2.29], the inverse of the matrix of coupling constants which can be calculated from the measured phonon dispersion [2.30]. The forces have to be calculated self-consistently by minimizing the total energy including its vibrational part. The static displacements $s^{(i),m}$ of the Nb atoms due to the hydrogen are now given by

$$s^{(i),m} = \sum_{n} G^{mn} \boldsymbol{F}^{(i),n} . \tag{2.2}$$

Later on we will need the displacements far away from the hydrogen. The lattice Green's function converges towards its continuum counterpart, the elastic Green's function $G(\boldsymbol{R})$, which is given by the elastic constants. For a description of the long-range displacement field the force pattern may be expanded into multipoles. The zero moment (the sum over all forces) vanishes due to translational invariance and the first moment, the dipole contribution, is the leading term

$$s^{(i)}(\boldsymbol{R}^m) \approx \sum_{n} G(\boldsymbol{R}^n - \boldsymbol{R}^m) \boldsymbol{F}^{(i),n} \approx \boldsymbol{P}^{(i)}\, \nabla G(\boldsymbol{R}^m) \tag{2.3}$$

for $R^m \gg R^n$ with the dipole tensor (α, β cartesian indices)

$$P_{\alpha\beta}^{(i)} = \sum_{n} R_{\beta}^{n} F_{\alpha}^{(i),n} . \tag{2.4}$$

The symmetry of $\boldsymbol{P}^{(i)}$ is determined by the symmetry of the hydrogen site i. For Nb this would imply tetragonal symmetry. In experiment one finds higher symmetry namely an isotropic dipole tensor $P_{\alpha\beta} = 2.8\,\delta_{\alpha\beta}$ [eV] [2.27]. The reason for this isotropy is not quite clear. The dipole tensor is related to the volume expansion ΔV due to the hydrogen via

$$\Delta V = \sum_{\alpha} P_{\alpha\alpha}/3\kappa \tag{2.5}$$

with κ the bulk modulus. Since the Green's function falls of as $1/R$ for large distances, the displacements decay as $1/R^2$. They are in general anisotropic due to the anisotropy of the host lattice and additionally due to a possible anisotropy of the dipole tensor.

The dynamics of the hydrogen on its site is dominated by the localized vibration modes with frequencies of $\hbar\omega_0 =100$ and 170 meV, respectively, where the latter mode is double degenerate. The frequencies of the localized vibrations, which are strongly affected by lattice distortions, would be nearly

50% higher in the unrelaxed lattice. The localized modes account for nearly 99% of the spectral intensity of the hydrogen. Additionally, the H participates in the host lattice vibrations (band modes). These modes are at much lower frequencies, the relevant energy scale is given by the Debye frequency of Nb, $\hbar\omega_D = 22$ meV. The amplitude of the H is strongly enhanced for some of these vibrations and shows a resonant like behavior [2.8, 9]. These latter vibrations determine, because of their lower frequencies, the temperature dependence of the thermal mean square displacements of the H below 1000 K, which is a strong indication that they will also be important for the hydrogen mobility. The resonant like vibrations of the H result from the near-neighbor geometry of the tetrahedral sites and consequently couple to short-wavelength phonons. The coupling of the hydrogen to the long-wavelength phonons is much weaker, the vibrational amplitude of the hydrogen is equal to that of the host atoms for these modes. It is important to distinguish between the three different groups of vibrations: The frequencies of the localized modes determine the order of magnitude of the tunneling matrix element between sites; see (2.8), the weakly coupled low-frequency phonons give the temperature dependence in the standard small polaron theory, and the resonant and localized vibrations cause the breakdown of this description towards higher energies as will be shown later.

In addition to these intrasite energies a number of intersite energies are relevant for diffusion. First, if the hydrogen is moved from its site i to a nearest-neighbor site f site without allowing the lattice to relax, the potential energy will be raised by

$$E_{NN}^{\text{elastic}} = \sum_{m,n} \left(F^{(i),m} - F^{(f),m} \right) G^{mn} F^{(f),n}. \tag{2.6}$$

In model calculations one finds for this term typical values of 100 to 150 meV. The change of the localized modes will add an additional contribution.

This asymmetry between sites prevents tunneling between them. To allow tunneling between sites i and f either the hydrogen has to tunnel together with its deformation field (dressed tunneling, small polaron) or the lattice has to be deformed such that sites i and f become equivalent. There is of course an infinite number of such deformations. The minimal energy needed is called coincidence energy. It can be estimated by hypothetically putting half the hydrogen into each of the two sites. The elastic energy is then determined by the average of the two forces $F^{(i)}$ and $F^{(f)}$.

Relative to the self-trapped state the coincidence energy is then

$$E_c^{\text{elastic}} = \frac{1}{4} \sum_{m,n} \left(F^{(i),m} - F^{(f),m} \right) G^{mn} F^{(i),n} = \frac{1}{4} E_{NN}^{\text{elastic}}. \tag{2.7}$$

To this elastic energy, one has to add again a contribution due to the change in zero-point motion. Models give values for the total coincidence energy E_c of between 35 and 50 meV, of which approximately 10 meV is the vibrational part. This coincidence energy is a measure for the strength of the phonon dressing effects.

Finally, one can define a classical activation energy $E_M^{classical}$. This is the energy needed to move the hydrogen by an adiabatic deformation of the lattice from one site to the next. The saddle-point configuration is then the lowest coincidence configuration for which the distance between the two sites vanishes. The model calculations give values between 70 and 130 meV as the classical activation energy, of which again typically 10 meV is the vibrational part. This energy is of the same order or even smaller than the energies of the localized vibrations and the usual condition $E_M^{classical} \gg \hbar\omega_0$ does not hold for the hydrogen vibrations but only for the lattice vibrations. Hence, deviations from a classical hopping behavior have to be expected in the Nb:H system even at high temperatures. Figure 2.1 depicts schematically the potential energy curves for the hydrogen atom for fixed lattice configurations as discussed above.

The similarity between classical activation energy and vibrational energy has an important consequence for the vibrational states of the hydrogen. If one takes a hydrogen in its relaxed position and excites one of the localized modes, energy can be gained by deforming the lattice towards that classical saddle-point configuration where the particular localized mode is unstable. The elastic energy needed is gained from the vibrational energy. In the model calculations of *Sugimoto* and *Fukai* [2.31] and of *Klamt* and *Teichler* [2.28] this results in delocalization of the excited states whereas experimentally even the higher excited states are observed [2.32]. However, an important consequence of this strong coupling between relaxations and localized modes for the diffusion of hydrogen will be the breakdown of the Condon approximation of the small polaron theory where just this dependence is neglected.

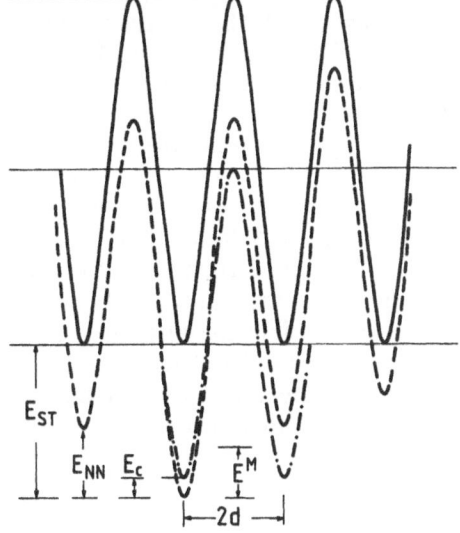

Fig. 2.1. Schematic potential energy curves for a hydrogen interstitial ($2d$: distance between neighboring interstitial sites). *Solid line*: metal host lattice unrelaxed. *Broken line*: host lattice relaxed (self trapped configuration). *Dot-dashed line*: coincidence configuration. E_{ST}: self-trapping energy, E_{NN}: nearest neighbor site energy. E_c: coincidence energy. E_M: classical migration energy

A basic quantity characterizing quantum effects is the tunneling matrix element J in the coincidence configuration. Assuming a sine like potential between the two sites, the tunneling matrix element can be evaluated using the eigenvalues of the Mathieu equation [2.33]

$$J = \hbar\Delta = \hbar\omega \frac{4}{\pi\sqrt{\pi}} \sqrt{\sigma} \exp{(8/\pi^2)\sigma} \tag{2.8}$$

with

$$\sigma = 4m\omega d^2/\hbar , \tag{2.9}$$

where Δ is the tunneling frequency, m is the mass of the tunneling particle, ω the effective vibrational frequency and $2d$ is the distance between the two equilibrium sites. σ scales with $\sqrt{m}$ and due to the exponential dependence the tunneling matrix element shows a very strong isotope effect. *Kehr* [2.34] has evaluated (2.8) for Nb ($2d = 1.17$ Å) and gets the estimates $J_H = 1.5$ meV, $J_D = 0.14$ meV and $J_\mu = 85$ meV for hydrogen, deuterium, and μ^+, respectively. These values are large enough to make tunneling observable and for diffusion to be dominated by quantum effects at least at low temperatures. The observed tunneling matrix element is of course reduced from this naked one by the effects of phonon and electron dressing which are the main subject of this chapter.

In a typical fcc metal such as Pd the situation is different. Due to the larger distance between the octahedral sites occupied by the hydrogen ($2d = 2.7$ Å), the tunneling matrix element is much smaller. Kehr's estimates are $J_H = 4 \times 10^{-6}$ meV, $J_D = 2 \times 10^{-9}$ meV, and $J_\mu = 1.3$ meV, respectively. Quantum effects will only be observable for the light μ^+. The localized mode frequency is much lower than in Nb. For hydrogen one finds $\hbar\omega_0 = 64$ meV [2.35], which is much less than the observed activation energy for diffusion $E_M = 230$ meV [2.36]. Diffusion will be classical apart from small effects due to the vibrational zero point energy.

In hcp metals the situation is more complex as discussed in the following chapter of this book. The hydrogen is thought to occupy tetrahedral interstitial sites. The diffusion involves a jump between nearest-neighbor tetrahedral sites and another jump to octahedral sites. For the first jump, one has a similar situation as in the bcc metals and quantum effects will be important [2.37]. The diffusion itself is limited by the longer second jump which is mainly classical, similar to the fcc case, with a relatively large activation energy of $E_M = 570$ meV for Y.

So far we have treated the hydrogen metal system quasi-statically, merely talking about forces and resulting displacements. Our aim is, however, not so much to calculate the absolute magnitude of the tunneling matrix element but to understand the temperature dependence. Therefore, the elementary excitations, the phonons and the electron-hole pairs have to be included explicitly. For this purpose we will derive the Hamiltonian for our system.

2.3 Hamiltonian of the Hydrogen-Phonon System

As a first approximation, the dynamics of interstitials in metals can be treated within the Born-Oppenheimer approximation which assumes that the electron distribution adjusts instantaneously to the ionic configuration, i.e., the electrons always occupy the momentary groundstate and there is a smooth transition between these states[2]. The only elementary excitations are then the phonons whose energies can be calculated well in this approximation. In a metal there are of course additionally electronic excitations with arbitrarily low energies and these will affect the dynamics at very low temperatures where the phonons are frozen out due to the ω^2 dependence of their spectrum. We will first derive the Hamiltonian for the phonon part and later augment it by an electronic term.

In the Born-Oppenheimer approximation the dynamics of the hydrogen metal system is determined by the kinetic energy and an effective interionic potential. The Hamiltonian can be written as

$$H = T_H + T_M + V_M + V_{HM}.$$ (2.10)

Here T_H and T_M are the kinetic energies of the hydrogen and the metal ions, respectively, V_M is the potential energy of the ideal metal host lattice and V_{HM} is the change due to the hydrogen. Since the displacements of the metal ions by the hydrogen are not too large, a reasonable, often used model for V_M is given by the harmonic approximation

$$V_M = \frac{1}{2} \sum_{\substack{m,n \\ \alpha,\beta}} s_\alpha^m \, \Phi_{\alpha\beta}^{mn} \, s_\beta^n,$$ (2.11)

where $s^m = R^m - R^{om}$ is the displacement of the metal ion m from its ideal position R^{om}. In second quantization the displacements are expressed by phonon creation and annihilation operators $b_{q\lambda}^+$ and $b_{q\lambda}$

$$s^n = \left(\frac{\hbar}{2NM} \right)^{1/2} \sum_{q,\lambda} \frac{e(q,\lambda)}{\sqrt{\omega_{q\lambda}}} \, e^{iq \cdot R^m} \left(b_{q\lambda} + b_{-q\lambda}^+ \right)$$ (2.12)

and

$$H_{ph} = T_M + V_M = \sum_{q\lambda} \hbar \omega_{q\lambda} \, b_{q\lambda}^+ \, b_{q\lambda},$$ (2.13)

here q is the wavevector, λ the polarization and $\omega_{q\lambda}$ the frequency of the phonon. The change in potential energy due to the hydrogen will, in general, be a many-body potential depending on the hydrogen coordinate R^H and on the lattice coordinates R^m. The simplest approximation is to take pairwise potentials between the hydrogen and its near neighbors, yielding

---

[2]We note that strictly speaking, due to the infrared singularities mentioned in the introduction, there is only an effective Born-Oppenheimer approximation for a given temperature. This point will be discussed in Sect. 2.7.3.

$$V_{\mathrm{HM}} = \sum_m V(|\boldsymbol{R}^{\mathrm{H}} - \boldsymbol{R}^m|) \,. \tag{2.14}$$

The pair potential is then fitted to the experimental values of the lattice expansion and of the localized hydrogen vibrations [2.28, 31]. This Hamiltonian has the full translational symmetry of the lattice. Assuming self-trapping, i.e., that the hydrogen atom is localized at one site sufficiently long for the lattice to relax, the various energies given in the previous section can be calculated. For a truly quantum mechanical calculation this Hamiltonian is still too complex and one therefore resorts to an expansion in terms of displacements from a suitably chosen reference configuration.

Due to the exponential distance dependence of the tunnel splitting Δ (2.8), direct transitions to all but the nearest-neighbor sites can be neglected. The tunneling problem can be reduced to the tunneling between two sites only. We introduce approximate total wavefunctions $\Psi_{n_i}^{(i)}(\boldsymbol{R}^{\mathrm{H}}, s)$ and $\Psi_{n_{(f)}}(\boldsymbol{R}^{\mathrm{H}}, s)$ for the hydrogen localized at its initial and final sites, respectively. These wavefunctions are many-body wavefunctions depending both on the hydrogen coordinate $\boldsymbol{R}^{\mathrm{H}}$ and on the lattice displacements s^n. We enumerate the states of the system such that for $n_i = n_f$ the wavefunctions belong to equivalent states related to each other by the symmetry of the lattice of interstitial sites. The wavefunctions $\Psi_{n_i}^{(i)}$ and $\Psi_{n_f}^{(f)}$ are eigenfunctions of the Hamiltonians $H^{(i)}$ and $H^{(f)}$, respectively, where the potential energy in (2.10) was modified such as to produce stable potential energy wells for the hydrogen at sites (i) and (f), respectively.

The solution of the tunneling problem involves the calculation of the matrix element

$$
\begin{aligned}
M_{n_i n_f} &= \left\langle \Psi_{n_i}^{(i)}(\boldsymbol{R}^{\mathrm{H}}, s) \left| H - \tfrac{1}{2}(H^{(i)} + H^{(f)}) \right| \Psi_{n_f}^{(f)}(\boldsymbol{R}^{\mathrm{H}}, s) \right\rangle \\
&= \left\langle \Psi_{n_i}^{(i)}(\boldsymbol{R}^{\mathrm{H}}, s) \left| V_{\mathrm{T}}(\boldsymbol{R}^{\mathrm{H}}, s) \right| \Psi_{n_f}^{(f)}(\boldsymbol{R}^{\mathrm{H}}, s) \right\rangle.
\end{aligned}
\tag{2.15}
$$

The kinetic energy terms cancel in this expression and the transition operator V_{T} is the change in the potential energies only. To calculate the matrix elements one has to integrate the potential energy surface over all coordinates.

The standard solution in the framework of small polaron theory [2.1–3] involves four additional approximations, all of them valid only at low temperatures. First the *adiabatic approximation* for the hydrogen motion: As in the Born-Oppenheimer approximation for the electrons, it is assumed that the hydrogen wavefunction follows the metal ions instantaneously. At any given time the hydrogen tunneling is determined by the frozen lattice configuration at that time. The total wavefunction is then approximated by

$$\Psi_n^{(i)}(\boldsymbol{R}^{\mathrm{H}}, s) = \Phi_n^{(i)}(\boldsymbol{R}^{\mathrm{H}}; s)\, \chi_n^{(i)}(s) \,, \tag{2.16}$$

where $\Phi_n^{(i)}$ represents the hydrogen wavefunction which depends parametrically on the lattice configuration. Second, since the adiabatic approximation is only valid at low temperatures, the hydrogen wavefunction is restricted to the ground state $\Phi^{(i)}$ (*truncation approximation*).

The wavefunction $\Phi^{(i)}$ satisfies

$$\left[T_H + V_{HM}^{(i)}(\boldsymbol{R}^H, s)\right]\Phi^{(i)}(\boldsymbol{R}^H; s) = \epsilon^{(i)}(s)\Phi^{(i)}(\boldsymbol{R}^H; s) \tag{2.17}$$

and the lattice wavefunctions $\chi_n^{(i)}$ satisfy

$$\left[T_M + V_M + \epsilon^{(i)}(s)\right]\chi_n^{(i)}(s) = E_n\,\chi_n^{(i)}(s). \tag{2.18}$$

Inserting these into (2.15) we can define a "naked" tunnel splitting

$$J(s) = 2\Big\langle \Phi^{(i)}(\boldsymbol{R}^H; s)\big|V_T(\boldsymbol{R}^H, s)\big|\Phi^{(f)}(\boldsymbol{R}^H; s)\Big\rangle, \tag{2.19}$$

where the integration is over the hydrogen coordinates only. This tunnel splitting depends still on the lattice configuration. At low temperatures when only long wavelength phonons are excited this dependence is weak and the tunnel splitting is taken as a constant (*Condon approximation*)[3].

The matrix element (2.15) is obtained by averaging $J(s)$ over the lattice wavefunctions $\chi_{n_i}^{(i)}(s)$ and $\chi_{n_f}^{(f)}(s)$

$$M_{n_i n_f} = \frac{1}{2}\Big\langle \chi_{n_i}^{(i)}(s)\big|J(s)\big|\chi_{n_f}^{(f)}(s)\Big\rangle \tag{2.20}$$

or simpler in the Condon approximation

$$M_{n_i n_f} = \frac{1}{2}J\Big\langle \chi_{n_i}^{(i)}(s)\big|\chi_{n_f}^{(f)}(s)\Big\rangle. \tag{2.21}$$

The naked tunneling term is dressed by the overlap of the initial- and final-state lattice wavefunctions, the polaron term.

As a next step, in accordance with the harmonic Ansatz for the inter-metallic coupling V^M (2.11), we expand $\epsilon^{(i)}(s)$ in powers of the displacements

$$\epsilon^{(i)}(s) = \epsilon^{(i)} - \sum_{m\alpha}F_\alpha^{(i)m}\,s_\alpha^m + \frac{1}{2}\sum_{\substack{mn\\\alpha\beta}}s_\alpha^m\,V_{\alpha\beta}^{(i)mn}\,s_\beta^n + \cdots. \tag{2.22}$$

The first term is a constant "chemical" binding energy of the hydrogen, the second is the relaxation energy in the lowest order and the third term is a change of the vibrational coupling parameters to lowest order. Taking the displacements s^m from the coincidence configuration instead of the ideal lattice position, the expansion (2.22) converges much more rapidly. The force term is reduced to

$$\tilde{F}^{(i)m} = F^{(i)m} - F^{(c)m} \approx \frac{1}{2}\left(F^{(i)m} - F^{(f)m}\right) \tag{2.23}$$

---

[3]Some theories have already discussed the configuration dependence of $J(s)$ formally in linear approximation in s. This gives rise to the so-called "barrier fluctuation effect" [2.2, 38].

and similarly the force constant change

$$\tilde{V}^{(i)mn} = V^{(i)mn} - V^{(c)mn} \approx \frac{1}{2}\left(V^{(i)mn} - V^{(f)mn}\right). \tag{2.24}$$

The bulk of the force constant change is now absorbed in $V^{(c)mn}$ which is no longer translationally invariant. Expressing s^n by phonon creation and annihilation operators, one sees that (2.22) is an expansion in powers of phonon operators. Normally, only the first order term is taken (*linear coupling approximation*). Two phonon terms are important in the low-temperature limit of nonmetallic crystals [2.39] whereas in metals the electronic terms, not yet included, prevail in this limit.

At low temperatures, one is interested in the coupling to long-wavelength phonons, which can still be described by plane waves that are slightly shifted between the initial and final states. Inserting (2.4, 12) into the linear term one has in the long wavelength limit

$$\sum_{n\alpha} \tilde{F}_\alpha^{(i)n} s_\alpha^n = \left(\frac{\hbar}{2NM}\right)^{1/2} \frac{1}{2} \sum_{q\lambda} \left[i\left(P_{\alpha\beta}^{(i)} - P_{\alpha\beta}^{(f)}\right) \frac{e(q,\lambda)_\alpha}{\sqrt{\omega_{q\lambda}}} q_\beta \cos(2qd) \right]$$
$$+ \left[\left(P_{\alpha\beta}^{(i)} + P_{\alpha\beta}^{(f)}\right) \frac{e(q,\lambda)_\alpha}{\sqrt{\omega_{q\lambda}}} q_\beta \sin(2qd) \right] (b_{q\lambda} + b_{-q\lambda}^+).$$

$$\tag{2.25}$$

To lowest order in q this gives in view of $\omega_{q\lambda} \sim q$

$$\sum_{n,\alpha} \tilde{F}_\alpha^n s_\alpha^n = \sum_{q\lambda} c_{q\lambda}(b_{q\lambda} + b_{-q\lambda}^+), \tag{2.26}$$

where

$$c_{q\lambda} = u_\lambda \, \omega_{q\lambda}^{s-\frac{1}{2}}. \tag{2.27}$$

Here $s = 1$ applies for tunneling between sites with different orientation and therefore changing the dipole tensor P, e.g., tetrahedral sites in bcc lattices, and $s = 2$ applies if the site symmetry does not change, e.g., octahedral sites in fcc lattices. It should be noted, however, that in Nb and Ta an isotropic dipole tensor is found so that the $s = 2$ term dominates despite the site symmetry.

Under the above approximations, which are reasonable at low temperatures, the hydrogen metal system can be described by a much simpler Hamiltonian. The hydrogen can be considered to move in a double-well potential of a form as depicted in Fig. 2.2. In view of the large single well excitation energy $\hbar\omega_0$, the hydrogen localized in either well can occupy only the vibrational ground state. This may be described in terms of a pseudo-spin by assigning eigenvalues of a Pauli spin matrix to these states, say, $\sigma_z = -1$ to the initial state (i) and $\sigma_z = 1$ to the final state (f). Of course, the two states are coupled by a tunneling matrix element J. The estimates for J presented above show that typically $J \ll \hbar\omega_0, \hbar\omega_D$. Furthermore, at low temperatures only long-wavelength phonons are excited. At sufficiently low temperatures we can thus introduce a phonon cutoff frequency ω_c such that

Fig. 2.2. A double-well potential

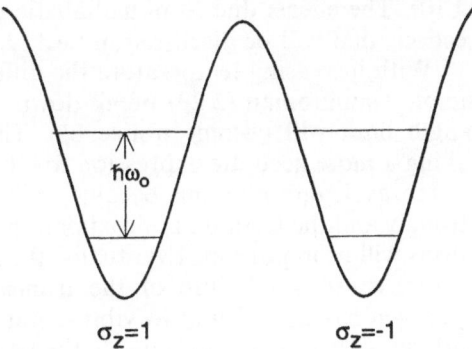

$\sigma_z=1$ $\sigma_z=-1$

$$J, k_B T \ll \hbar\omega_c \ll \hbar\omega_0, \hbar\omega_D . \qquad (2.28)$$

Then, if the tunneling matrix element J is already dressed by the polaron term of all phonons with frequencies above ω_c, and if the linear coupling approximation suffices for the remaining phonons with frequencies $\omega_{q\lambda} < \omega_c$, the original Hamiltonian (2.10) can be replaced by a much simpler one

$$H = H_{hy} + H_{ph} + H_{int} , \qquad (2.29)$$

where

$$H_{hy} = \frac{1}{2} J \sigma_x = \frac{1}{2} \hbar \, \Delta_c \sigma_x \qquad (2.30)$$

is the Hamiltonian of the two-level system describing the hydrogen in the double-well potential. Here Δ_c is an effective tunneling frequency which is reduced compared to the "naked" one by the dressing factor of the high frequency phonons $\omega > \omega_c$. H_{ph} is given in (2.13) and

$$H_{int} = \sigma_z \sum_{q\lambda} c_{q\lambda} (b_{q\lambda}^+ + b_{-q\lambda}^+) \qquad (2.31)$$

describes the interaction (2.26) between the hydrogen and the crystal modes.

So far we have eliminated electronic excitations through the Born-Oppenheimer approximation. There are however electron-hole pairs with arbitrarily low energies and for these the adiabatic approximation breaks down. Similar to the case of phonons the hydrogen will tunnel in a frozen configuration of these excitations. A proper treatment of the effect of the conduction electrons requires two further terms in our Hamiltonian which will be presented in Sect. 2.5.

2.4 Hydrogen Diffusion at Elevated Temperatures

At sufficiently high temperatures ($T \geq 50$ K) hydrogen tunneling is dominated by phonon effects and can therefore be described by the Hamiltonian

(2.10). The effects due to nonadiabatic electronic excitations are small corrections that will be discussed in Sect. 2.7.3.

With increasing temperature the different approximations leading to the simple Hamiltonian (2.29) break down. First corrections to the long wavelength limit will become noticeable. These can easily be incorporated by taking a more accurate expression for the linear coupling term (2.26). When short wavelength phonons become excited, the local geometry will fluctuate strongly and the Condon approximation will break down, barrier fluctuation effects will be important. Eventually, the localized modes of the hydrogen are excited causing a failure of the truncation approximation. Furthermore, hydrogen has been found to vibrate not only with localized modes but also with quasi-resonant modes inside the band of lattice modes. Such vibrations are not properly accounted for by the adiabatic approximation. The aim of the theory of diffusion of hydrogen at higher temperatures is therefore to circumvent the approximations, most importantly the Condon approximation, which lead to the simple Hamiltonian (2.29) describing a two-state system weakly coupled to the phonon bath.

We retain the picture that the hydrogen is in well defined states in the initial and final interstitial sites, i.e., we assume that the hydrogen occupies each site sufficiently long for self-trapping to be effective. The transition probability then can be found by the use of the Fermi golden rule expression for quantal transition rates:

$$\Gamma = \langle\langle w^{if} \rangle\rangle_{n_i} \tag{2.32}$$

with the partial rates

$$w^{if} = \frac{2\pi}{\hbar} \left| \left\langle \Psi_{n_i}^{(i)}(\mathbf{R}^H, s) \middle| V_T(\mathbf{R}^H, s) \middle| \Psi_{n_f}^{(f)}(\mathbf{R}^H, s) \right\rangle \right|^2 \delta[E^{(i)} - E^{(f)}] \tag{2.33}$$

which follows directly from (2.15). Here $E^{(i)}$ and $E^{(f)}$ are the total energies of the initial and final states, respectively and $\langle\langle \; \rangle\rangle_{n_i}$ denotes thermal averaging over initial states n_i and summation over final states n_f. The validity of this expression is restricted to the purely incoherent regime. To study the transition from coherent to incoherent tunneling, different methods have to be employed which are presented in Sect. 2.8. Also correlations between single jumps which become important at high temperatures are not included.

Since in the region of incoherent tunneling, the hopping rate Γ is the same whether the interstitial moves in a double-well potential or in a periodic crystal potential, the rate (2.32) also determines the quantum diffusion coefficient D. For a crystal with cubic symmetry one has

$$D = \frac{z}{6}(2d)^2\Gamma, \tag{2.34}$$

where z is the number of neighboring interstitial sites and $2d$ is the distance between two sites.

Employing the adiabatic, Condon and linear coupling approximations of Sect. 2.3, one gets the simple expression

$$\Gamma = \frac{\pi}{2\hbar} J_0^2 \langle\langle | \langle \chi^{(i)}(s) | \chi^{(f)}(s) \rangle |^2 \, \delta(E^{(i)} - E^{(f)}) \rangle\rangle_{n_i} ,$$ (2.35)

where J_0 is the tunnel splitting of the naked hydrogen in the fixed potential of the host ions and the temperature dependence is given by the averaged overlap of the total vibrational wavefunctions of the host ions with the hydrogen in its initial and final position, respectively. Even in this approximation the evaluation of the transition rate is nontrivial since any vibrational mode in the initial configuration will have a nonvanishing overlap with many modes in the final configuration. Only in the long-wavelength limit will the modes be only weakly affected by the local defect geometry and it is reasonable to approximate them by identical normal modes with only the mean positions of the atoms changed. The total overlap integrals (2.35) then factorizes into products of the single mode overlaps which can easily be calculated. The transition rate follows as [2.1]

$$\Gamma = \frac{J_0^2}{4\hbar^2} \int_{-\infty}^{+\infty} dt \exp \left\{ -\sum_q S_q \Big[\coth(\hbar\omega_q/2k_BT)[1 - \cos(\omega_q t)] \right.$$

$$\left. + i \sin(\omega_q t) \Big] \right\} .$$ (2.36)

Here we have used the index q to denote the phonon q-vector as well as the polarisation. Each mode contributes with a weight given by its Huang-Rhys factor, which is given in terms of the difference of the mode displacements of the initial and final configurations

$$S_q = \frac{1}{2} \frac{m\omega_q}{\hbar} | d_q^{(i)} - d_q^{(f)} |^2$$ (2.37)

with m the host atom mass. The d_q denote here the projections of the total displacement vector onto the eigenvector of the phonon q. These Huang-Rhys factors can be evaluated from (2.25). They are $\propto q$ in the isotropic case and $\propto$ const in the anisotropic case. The above result for the transition probability contains a divergent part due to diagonal transitions, i.e., transitions without phonon excitation. These transitions are not dealt with properly in this model. In fact, we shall show in Sect. 2.7.3 that the main effect of the interaction with the conduction electrons present at sufficiently high temperatures is a suppression of the diagonal transitions. They may, therefore, be subtracted from (2.36) which after some transformations then takes the form [2.5]

$$\Gamma = \frac{J_0^2}{4\hbar^2} e^{-S(T)} \int_{-\infty}^{\infty} dt \left[\exp\left(\sum_q S_q \frac{\cos(\omega_q t)}{\sinh(\hbar\omega_q/2k_BT)} \right) - 1 \right] .$$ (2.38)

The average Huang-Rhys factor $S(T)$ describes the average phonon dressing of the tunneling matrix element

$$S(T) = \sum_q S_q \coth\left(\frac{\hbar\omega_q}{2k_BT} \right) .$$ (2.39)

For low temperatures $S(T)$ is of the form

$$S(T) = S(0) + W(T), \tag{2.40}$$

where the temperature dependent part $W(T)$ is proportional to T^{2s}, with the exponent s introduced in (2.27). The zero-temperature dressing factor may be absorbed into a renormalized tunneling frequency

$$\Delta = \frac{J_0}{\hbar} e^{-S(0)/2} \tag{2.41}$$

and the transition rate then takes the form

$$\Gamma = \frac{\Delta^2}{4} e^{-W(T)} \int_{-\infty}^{+\infty} dt \left(e^{-\Xi(T,t)} - 1 \right) \tag{2.42}$$

in which

$$\Xi(T,t) = -\sum_q S_q \frac{\cos(\omega_q t)}{\sinh(\hbar\omega_q/2k_B T)} . \tag{2.43}$$

For low temperatures the exponential functions can be expanded. The lowest-order contribution is then by two phonon processes giving a transition rate $\propto T^7$ in the isotropic case [2.1]. For nonsymmetric transitions, i.e., for anisotropic defects in an external field, one can also get contributions from one phonon processes giving transition rates $\propto T$ [2.3]. However, these low-temperature results have to be addressed with caution, since the electronic influence may already be substantial in this range of temperatures. The high-temperature limit is treated in the Debye approximation of the phonon spectrum and Γ takes the simple form [2.1]

$$\Gamma = (\pi/16\hbar^2 E_c k_B T)^{1/2} J^2 \exp(-E_c/k_B T). \tag{2.44}$$

This describes a near Arrhenius behavior with the coincidence energy E_c (2.7) as activation energy. Even by fitting the values of J and E_c this result cannot reproduce the measured diffusion constants, in particular the observed change in slope of the measured diffusion constant in Nb and Ta. One remedy would be to introduce transitions between excited states or different sites with a second set of parameters J and E_c [2.40].

A more satisfactory approach is to avoid the serious approximation made, in particular the Condon approximation. In the "occurrence probability" method [2.41, 28] the adiabatic approximation is partially retained but one averages over possible coincidence configurations. Also transitions between differently excited localized H states are included. Since different coincidence configurations have different tunneling probabilities for the "naked" H, an additional temperature dependence is found which can explain the change in slope of the Arrhenius plot of diffusion of H in Nb. *Gillan* [2.42] uses quantum molecular dynamics to identify the symmetric coincidence configurations. A direct calculation of the transition rate is not yet possible and the influence of asymmetric coincidence configurations is difficult to estimate. Both methods allow a comparison with classical diffusion.

A more rigorous approach is the embedded cluster method [2.43]. There one makes use of the spatial localization of the resonant and localized modes. We do not separate the hydrogen and host degrees of freedom but treat a cluster of atoms (typically 6–21 atoms) explicitly and use the above approximations only for the embedding of the cluster into the rest of the host crystal. The limitations of this method are mainly due to computer capacity. The number of degrees of freedom one can treat explicitly is limited owing to the number of integrations involved in calculating a single transition element w^{if} (2.33). On the temperature side one is limited by the rapid increase in terms w^{if} contributing to the total transition rate (2.32).

To see the temperature range where the polaron expression (2.38) breaks down, we first calculate an average coherent tunneling rate given by thermally averaging the diagonal elements ($n_i = n_f$) of (2.15)

$$\hbar\Delta(T) \approx 2 \left\langle \Psi^{(i)}_{n_i} | V_T | \Psi^{(f)}_{n_i} \right\rangle_{n_i} . \tag{2.45}$$

Employing the same simplifications leading to (2.38) for the transition rate in the incoherent case, (2.45) simplifies to

$$\hbar\Delta(T) = J e^{-S(T)/2} , \tag{2.46}$$

where $S(T)$ is again the average Huang-Rhys factor. The total dressing $\exp[-S(T)/2]$ of the tunneling matrix element is of order 0.1 for Nb:H at low temperatures.

In Fig. 2.3 we compare the temperature dependencies of $\Delta(T)$ in Nb:H for both the full expression (2.45) evaluated in the embedded cluster approach and for the weak coupling formula (2.46). The calculations were done

Fig. 2.3. Temperature dependence of the average tunneling rate for hydrogen and deuterium in Nb calculated from (2.45) (*solid line*) and in the weak coupling approximation (2.46) (*dashed line*), respectively. The error bar gives the variance [2.37]

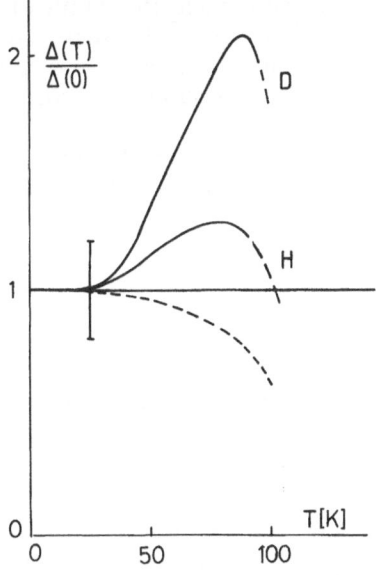

using the interaction potential of *Sugimoto* and *Fukai* [2.31] with harmonic wavefunctions as basis. For low temperatures we find agreement between the two expressions but for $T > 40$ K the full expression (2.45) predicts a strong isotope dependent structure absent in the simplified form (2.46). This indicates a breakdown of the adiabatic and Condon approximations at temperatures where the resonant-like (in-band) vibrations of the H are excited. The same effects are apparent in the variance of $\Delta(T)$ which increases rapidly with temperature. At $T \approx 50$ K it reaches 50%. In passing we want to note that an estimate of $S(T)$ from isotropic elasticity theory seriously underestimates $S(T)$ by about a factor of 5, which again shows that the bulk of the dressing stems from short wavelength phonons.

For the same model we have evaluated the incoherent transition rate. In the embedded cluster approach the partial rates (2.33) take the form

$$w^{if} = \frac{2\pi}{\hbar} \left| \left\langle \Psi^{(i)}_{n_i} | V_T | \Psi^{(f)}_{n_f} \right\rangle \right|^2 G(E^{(i)} - E^{(f)}) . \tag{2.47}$$

Here the wavefunctions Ψ are wavefunctions of the cluster only (or of selected important modes) and the δ function is replaced by a shape function which consists of two distinct factors. One comes from the line broadening, i.e., it accounts for the broadening of the sharp modes of the atomic cluster by the embedding lattice. We take a form

$$G_v(\omega) = (1/\sqrt{\pi})W_v \exp[-(\omega - \Omega)^2/W_v^2] , \tag{2.48}$$

where $W_v = \sum_\sigma w^\sigma (m_\sigma + n_\sigma)$ is the total width and $\hbar\Omega = \sum_\sigma \hbar\omega^\sigma (m_\sigma - n_\sigma)$ is the difference in vibrational energy between the initial and final states. As widths, w^σ, of the individual modes σ we took zero and 1.0 THz for the localized and band modes, respectively. The final result shifts only slightly when we change the width of the band modes from 0.5 to 2.0 THz. The second factor in the shape function stems from the change in mean displacement of the weakly coupled modes outside of the cluster as the hydrogen atom jumps. It corresponds to the dressing terms in (2.35). Its magnitude is, however, strongly reduced since we have included the bulk of the dressing explicitly in (2.47). It accounts mainly for the long wavelength phonons. To simplify the evaluation we may take the high-temperature limit of this term

$$G_R(\omega) = (16\pi \, E_R k_B T)^{-1/2} \exp[-(\hbar\omega + 4E_R)^2/16E_R k_B T] , \tag{2.49}$$

where we estimate from the overlap at $T = 0$ K that $E_R = 6$ meV.

In doing the integrals in (2.47) V_T was expanded up to quadratic terms. We thus take both two phonon processes and barrier shaking into account. Excitation levels were included up to $n = 4$ for the single modes and transitions up to $n_i = n_f \pm 2$ for each mode were included. Figure 2.4 shows the results for the diffusion coefficients of H and D, together with the experimental values. From $T \simeq 50$ K upwards we find an Arrhenius behavior with a sudden change in activation energy at $T \simeq 250$ K. This kink can be un-

Fig. 2.4. Diffusion constants of hydrogen and deuterium in Nb. *Solid line*: experimental results [2.36]. *Broken line*: calculation with harmonic wavefunctions. *Dash-dotted line*: calculation including anharmonic corrections. *Dotted line*: Vineyard approximation for high temperatures [2.45]

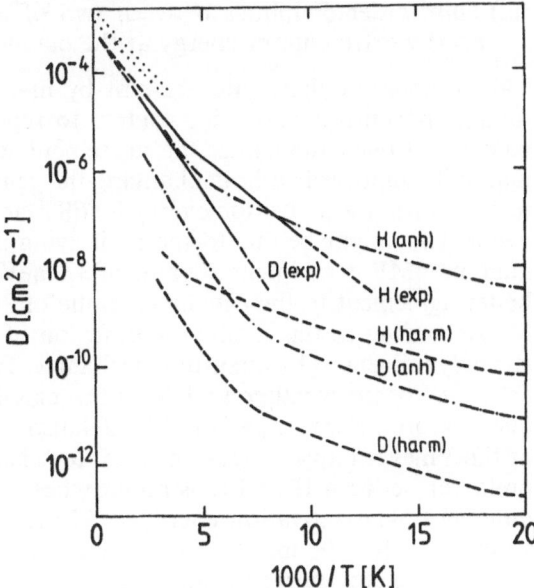

derstood from the partial rates, w^{if}. At low temperatures transitions are only possible without exciting the localized vibrations of the H. The transition rate is therefore determined by the in-band excitations only and can be described by a formula as (2.44) with an activation energy of ≈ 30 meV close to the coincidence energy of our model. At higher temperatures scattering from the band states into the localized vibrational states of the H becomes possible, $m_{loc} = 0 \rightarrow n_{loc} = 1$. These transitions with an excited localized state have much larger tunneling matrix elements and hence gain importance. At still higher temperatures transitions between excited H-vibrations become dominant. This general behavior is in qualitative agreement with the results of the occurrence probability approach [2.28].

The model so far describes the general behavior. It severely underestimates the diffusion constant, however. This has to be expected, since using harmonic wavefunctions, one underestimates the overlap between final and initial states. So far the actual potential shape was included in the Hamiltonian but not in the wavefunctions. To get a better quantitative agreement one has to improve on the latter. For this purpose the anharmonic expansion parameters of the potential energy at the initial and final equilibrium sites were studied. Three types of anharmonic terms were found to be important:

(i) Fourth- and higher-order terms in the localized H vibrations $[\phi^{(4)} s_{loc}^4]$ to bring the tunneling frequency of the naked hydrogen up to the value of WKB-type expressions.

(ii) Third-order couplings between localized and band modes $[\phi^{(3)} s_{band} s_{loc}^2]$. These terms account for a reduction of the localized mode frequency when the lattice is deformed.

(iii) Third-order couplings between two of the localized modes due to the nearby extremum of energy at the octahedral site.

Taking anharmonicity into account by first-order perturbation theory and taking appropriate correction factors to reproduce the WKB result for the naked hydrogen tunneling, the agreement with the experimental results is markedly improved without changing the general shape of the curves. At low temperatures the activation energy is still much smaller than the experimental value. This might be due to the underlying interaction mode. On the other hand in NMR experiments a value of 37 meV was found [2.44] to be in much better agreement to the calculated value of 30 meV.

According to this model the diffusion of H at room temperature is still strongly influenced by quantum mechanics. The calculation is not sufficiently accurate to see whether and how the classical limit is reached at higher temperatures. For comparison Fig. 2.4 also shows the results of a calculation of the Vineyard approximation [2.45] for classical diffusion. The shift in the prefactor between H and D is about what one would expect from the mass difference. The activation energy is 97 meV and 6 meV for H and D, respectively. Here 90 meV is the potential energy part. The discrepancy between our quantum results and these classical ones is not sufficient to draw any definite conclusion. *Gillan* [2.42] puts the classical to quantum transition already at the kink. Since there might be a smooth transition and due to differences in the model this does not need to be in contradiction to our results.

A similar behavior was found for jumps between nearest neighbor tetrahedral sites in hcp metals [2.37]. The diffusion, however, is dominated by the much slower jump to the octahedral site where no strong quantum effects are expected. Since the kink in the diffusion constant in Nb is caused by a subtle interplay between lattice activation and localized mode excitation, small perturbations should be sufficient to shift this kink. Our model predicts strong anomalies of the pressure dependence of the diffusion constant [2.46]. Experimentally these predicted anomalies have not been fully verified so far [2.46, 47].

2.5 Hamiltonian for Two-Level Systems in Metals

We shall consider the simplest case of a tunneling system in a metal, namely a particle that can tunnel between two equivalent interstitial sites of the host crystal. Let us briefly recall some of the relevant energy scales of this problem introduced in Sect. 2.2. At the interstitial site the particle may occupy the ground state in the potential well or excited states, the lowest one being separated from the ground state by the excitation energy $\hbar\omega_0$ where ω_0 is the vibrational frequency of the defect. The degeneracy between the two ground states is lifted by a tunnel splitting $\hbar\Delta$ where Δ is the tunneling frequency. Usually $\hbar\Delta$ is much smaller than $\hbar\omega_0$. The host crystal adds some further energy scales, most importantly, the thermal energy $k_B T$, and energy scales characterizing the electrons and phonons of the metal such as the Fermi energy ε_F and $\hbar\omega_D$ where ω_D is the Debye frequency.

In the following we shall assume that $\hbar\Delta$ and $k_B T$ are both much smaller than the other relevant energy scales, i.e.,

$$\hbar\Delta, k_B T \ll \hbar\omega_0, \hbar\omega_D, \varepsilon_F . \qquad (2.50)$$

Typically, the energy scales on the right-hand side are at least of the order of 0.1 eV so that these inequalities are very well obeyed provided the tunnel splitting is in the meV region or smaller and the temperature does not exceed a few times 10 K. Since $k_B T$ is now much smaller than the excitation energy $\hbar\omega_0$, the particle can occupy only the ground states in the two potential wells and it forms a two-level system which may be described in terms of a pseudo-spin.

The coupling of the two-level system to phonons and electrons arises for simple physical reasons. In Sect. 2.4 we have considered that the hydrogen carries the lattice deformation along when it tunnels to the neighboring site. Furthermore, interstitial particles such as protons and muons are surrounded by an electron cloud screening the Coulomb potential of the defect. Since tunneling of the particle gives rise to a displacement of this screening cloud, the defect motion also couples to the metallic electrons.

It is natural to consider the spectral densities of the excitations of the host crystal that couple to the two-level system. Of course, the characteristic frequency of our problem is the tunneling frequency Δ. Since Δ is much less than the Debye frequency ω_D, we deal in this frequency range with acoustic phonons the spectral density of which vanishes as ω^2 for small frequencies. Hence, there are few phonons that are slower than the tunneling system. These low-frequency modes give only a very small contribution to the phonon overlap integral. The low-energy excitations of the conduction electrons are the electron-hole pairs. They have a large effective density of states in the low-frequency range of interest; in fact, the electronic overlap integral has an infrared divergence. Hence, the slow parts of the electronic screening cloud that cannot follow instantaneously the tunneling motion will strongly affect the tunneling dynamics.

As described earlier, crystal modes with frequencies much larger than Δ and $k_B T/\hbar$ may be eliminated adiabatically. Thereby modes with frequencies well above ω_0 lead primarily to a potential and mass renormalization of the interstitial, while modes with frequencies below ω_0 rather modify the overlap of the two defect states. However, the outcome of either effect is a reduction of the effective tunneling frequency Δ. Most of the electronic excitations of the metal may also be eliminated, since $\varepsilon_F \gg \hbar\omega_0$. We introduce an arbitrary energy scale $\hbar\omega_c$ in between the small and large energy scales in (2.50) and assume that all crystal excitations with energies above $\hbar\omega_c$ have been eliminated adiabatically. We are then left with an effective Hamiltonian

$$H = H_{hy} + H_{el} + H_{ph} + H_{int} , \qquad (2.51)$$

where H_{hy} and H_{ph} have been introduced in (2.13, 30).

$$H_{el} = \sum_{ks} \varepsilon_k a_{ks}^\dagger a_{ks} \tag{2.52}$$

describes the quasi-particles near the Fermi surface with momentum $\hbar k$ and spin s. The energy $\varepsilon_k < \hbar\omega_c$ is measured from the Fermi level. Finally,

$$H_{int} = \sigma_z(R_{el} + R_{ph}) \tag{2.53}$$

describes the interaction between the hydrogen interstitial and the crystal modes. In view of (2.31), the coupling to the phonon modes may be written as

$$R_{ph} = \sum_{q\lambda} c_{q\lambda}(b_{q\lambda}^\dagger + b_{q\lambda}) . \tag{2.54}$$

The metal electrons are scattered by the hydrogen interstitial. Since the Coulomb potential of the tunneling particle is screened by the high frequency modes that have been eliminated, the remaining pseudopotential can be assumed to be of short range and we have

$$R_{el} = \sum_{kk's} V_{kk'} a_{ks}^\dagger a_{k's} \tag{2.55}$$

where

$$V_{kk'} = iV \sin[(k - k')d] , \tag{2.56}$$

contains the strength V of the pseudopotential of the interstitial, and the sine describes the difference in phase factors $\exp[i(k - k')r]$ between the two interstitial sites at $r = d$ and $-d$, respectively [2.14]. In (2.53) we have neglected the coupling of the electrons to Pauli matrices other than σ_z. The matrix element for these so-called electron-assisted tunneling processes is reduced by an overlap factor of order Δ/ω_0. Electron-assisted tunneling only becomes relevant at very low temperatures [2.48].

2.5.1 Reduced Density Matrix

Let us first consider the reduced density matrix ρ_{hy} of the interstitial embedded in the host crystal. Decomposing the trace according to

$$tr = tr_{hy} tr_{el} tr_{ph} \tag{2.57}$$

we find for the unnormalized density matrix

$$\rho_{hy} = \frac{tr_{el} tr_{ph} \exp(-\beta H)}{tr_{el} \exp(-\beta H_{el}) tr_{ph} \exp(-\beta H_{ph})} . \tag{2.58}$$

When this is expanded in powers of H_{int} one readily obtains

$$\rho_{hy} = \exp(-\beta H_{hy})$$

$$\left[1 + \frac{1}{\hbar^2} \int_0^{\hbar\beta} d\tau \int_0^\tau d\tau' \sigma_z(\tau)\sigma_z(\tau')(\langle R_{el}(\tau)R_{el}(\tau')\rangle_{el} \right. \tag{2.59}$$

$$\left. + \langle R_{ph}(\tau)R_{ph}(\tau')\rangle_{ph}) + \dots \right] ,$$

where we have restricted ourselves to the lowest order nontrivial terms. Here, the time dependence of operators is calculated for the non-interacting system. For instance, $R_{el}(\tau) = \exp[(\tau/\hbar)H_{el}] R_{el} \exp[-(\tau/\hbar)H_{el}]$, and $\langle\ \rangle_{el}$ denotes an average over the unperturbed electron density matrix $\exp(-\beta H_{el})/\mathrm{tr}_{el}\exp(-\beta H_{el})$.

2.5.2 Effective Electronic Density of States in a Normalconducting Metal

The imaginary time correlation functions appearing in (2.59) are easily evaluated. Using (2.52, 55) we find

$$
\begin{aligned}
M_{el}(\tau) &= \langle R_{el}(\tau)R_{el}(0)\rangle_{el} \\
&= \sum_{kk's} |V_{kk'}|^2 f(\varepsilon_k)[1 - f(\varepsilon_{k'})] \exp\left[\frac{1}{\hbar}(\varepsilon_k - \varepsilon_{k'})\tau\right],
\end{aligned}
\tag{2.60}
$$

where $f(\varepsilon) = 1/[1 + \exp(\beta\varepsilon)]$ is the Fermi function. Inserting now (2.56) and performing the angular average for wavevectors k and k' of length of the Fermi wavevector k_F, one obtains

$$
M_{el}(\tau) = \frac{K}{2} \int d\varepsilon \int d\varepsilon' n(\varepsilon)n(\varepsilon')f(\varepsilon)[1 - f(\varepsilon')]\exp\left[\frac{1}{\hbar}(\varepsilon - \varepsilon')\tau\right],
\tag{2.61}
$$

where

$$
K = 2V^2\rho^2\left(1 - \frac{\sin^2(2k_Fd)}{(2k_Fd)^2}\right)
\tag{2.62}
$$

is the dimensionless Kondo parameter characterizing the coupling to the fermionic bath [2.11]. ρ is the electronic density of states at the Fermi surface and $n(\varepsilon)$ the relative density of states for energies ε above the Fermi level. Finally, $2d$ is the distance between the two interstices.

We may assume $n(\varepsilon) = \exp(-|\varepsilon|/\hbar\omega_c)$, since excitations with energies above $\hbar\omega_c$ have been eliminated, and the microscopic electronic density of states changes significantly on the scale ε_F only. For $k_BT \ll \hbar\omega_c$, we then obtain from (2.61)

$$
M_{el}(\tau) = \frac{\hbar^2}{2}\int_0^\infty d\omega\, J_{el}(\omega)[\coth(\hbar\beta\omega/2)\cosh(\omega\tau) - \sinh(\omega\tau)],
\tag{2.63}
$$

where

$$
J_{el}(\omega) = K\omega\exp(-\omega/\omega_c)
\tag{2.64}
$$

is an effective density of modes of the fermionic environment. This linear increase of the spectral density for small frequencies is a characteristic feature of so-called Ohmic dissipation [2.24].

2.5.3 Effective Electronic Density of States in a Superconducting Metal

It is straightforward, to derive a corresponding result for the case in which the metal is superconducting. In the presence of a BCS energy gap Δ_{gap} the result (2.61) is generalized to read [2.19]

$$M_{\text{el}}(\tau) = \frac{K}{2} \int d\varepsilon \int d\varepsilon' n_{\text{qp}}(\varepsilon) n_{\text{qp}}(\varepsilon') \left(1 - \frac{\Delta_{\text{gap}}^2}{\varepsilon\varepsilon'}\right)$$
$$f(\varepsilon)[1 - f(\varepsilon')] \exp\left[\frac{1}{\hbar}(\varepsilon - \varepsilon')\tau\right],$$

(2.65)

where

$$n_{\text{qp}}(\varepsilon) = \Theta(|\varepsilon| - \Delta_{\text{gap}}) \frac{|\varepsilon|}{\sqrt{\varepsilon^2 - \Delta_{\text{gap}}^2}} \exp(-|\varepsilon|/\hbar\omega_{\text{c}})$$

(2.66)

is the reduced quasi-particle density of states in which $\Theta(x)$ is the unit step function. The correlation (2.65) may be written in the form of (2.63) with $J_{\text{el}}(\omega)$ replaced by

$$J_{\text{qp}}(\omega) = \frac{K}{\hbar} (1 - e^{-\hbar\beta\omega}) \int d\varepsilon \, n_{\text{qp}}(\varepsilon - \hbar\omega) n_{\text{qp}}(\varepsilon) \left[1 - \frac{\Delta_{\text{gap}}^2}{\varepsilon(\varepsilon - \hbar\omega)}\right]$$
$$f(\varepsilon - \hbar\omega) f(-\varepsilon).$$

(2.67)

This effective density of modes depends on temperature. At $T = 0$ we find

$$J_{\text{qp}}(\omega) = K\Theta(\hbar\omega - 2\Delta_{\text{gap}})\omega \exp(-\omega/\omega_{\text{c}})$$
$$E\left(\sqrt{\hbar^2\omega^2 - 4\Delta_{\text{gap}}^2}/\hbar\omega\right),$$

(2.68)

where $E(k)$ is the complete elliptic integral of the second kind with modulus k. Hence, for $T = 0$ K the effective spectral density $J_{\text{qp}}(\omega)$ vanishes for all frequencies below $2\Delta_{\text{gap}}/\hbar$ where it jumps to the value $J_{\text{qp}}(2\Delta_{\text{gap}}/\hbar) = \pi K\Delta_{\text{gap}}/\hbar$. For $\hbar\omega \gg \Delta_{\text{gap}}$ the normal state density (2.64) is approached. At finite temperatures $J_{\text{qp}}(\omega)$ is finite for frequencies below $2\Delta_{\text{gap}}/\hbar$. For small frequencies we have

$$J_{\text{qp}}(\omega) = \frac{2K\omega}{1 + \exp(\beta\Delta_{\text{gap}})}[1 + O(\omega)].$$

(2.69)

Hence, for $T > 0$ K the effective spectral density has an Ohmic term linear in ω like the spectral density $J_{\text{el}}(\omega)$ in the normal state, however, with a reduced effective coupling parameter $2K/[1 + \exp(\beta\Delta_{\text{gap}})]$ which vanishes at $T = 0$ K.

2.5.4 Effective Phononic Density of States

In a similar way we may evaluate the corresponding correlation function of the phonon operators. Using (2.13, 54) we find

$$M_{ph}(\tau) = \langle R_{ph}(\tau)R_{ph}(0)\rangle_{ph}$$
$$= \sum_{q\lambda} c_{q\lambda}^2[n_{q\lambda}\exp(\omega_{q\lambda}\tau) + (n_{q\lambda} + 1)\exp(-\omega_{q\lambda}\tau)], \qquad (2.70)$$

where $n_{q\lambda} = 1/[\exp(\hbar\beta\omega_{q\lambda}) - 1]$ is the equilibrium phonon occupancy. The result (2.70) can again be written in a form corresponding to (2.63). We have

$$M_{ph}(\tau) = \frac{\hbar^2}{2}\int_0^\infty d\omega\, J_{ph}(\omega)[\coth(\hbar\beta\omega/2)\cosh(\omega\tau) - \sinh(\omega\tau)], \qquad (2.71)$$

where

$$J_{ph}(\omega) = \frac{2}{\hbar^2}\sum_{q\lambda} c_{q\lambda}^2\delta(\omega - \omega_{q\lambda}) \qquad (2.72)$$

is an effective phononic density of states. Assuming now that the cutoff frequency ω_c is sufficiently below the Debye frequency ω_D so that the coefficients $c_{q\lambda}$ are well described by their low-frequency form (2.27), we find for a Debye density of phonon states an effective spectral density of the form

$$J_{ph}(\omega) = U\omega^{2s+1}\exp(-\omega/\omega_c) \quad (s = 1, 2), \qquad (2.73)$$

where all frequency independent factors have been absorbed in the effective coupling parameter U, and the exponent s depends on the crystal structure of the host metal as explained in Sect. 2.3.

2.5.5 Effective Spectral Density

The reduced density matrix (2.59) of the interstitial now takes the form

$$\rho_{hy} = \exp(-\beta H_{hy})$$
$$\left[1 + \frac{1}{\hbar^2}\int_0^{\hbar\beta} d\tau \int_0^\tau d\tau' K(\tau - \tau')\sigma_z(\tau)\sigma_z(\tau') + \dots\right], \qquad (2.74)$$

where

$$K(\tau) = \frac{\hbar^2}{2}\int_0^\infty d\omega\, J(\omega)\frac{\cosh[\omega(\hbar\beta/2 - \tau)]}{\sinh(\hbar\beta\omega/2)}. \qquad (2.75)$$

Here

$$J(\omega) = J_{el}(\omega) + J_{ph}(\omega) \qquad (2.76)$$

is an effective spectral density describing the influence of the host crystal on the two-level system. For a superconducting metal $J_{el}(\omega)$ has to be replaced by $J_{qp}(\omega)$. Within second order perturbation theory the expression (2.74) is equivalent to

$$
\rho_{hy} = \mathbf{T} \exp \left[-\frac{1}{\hbar} \int_0^{\hbar\beta} d\tau \, H_{hy,\tau} \right.
$$

$$
\left. + \frac{1}{2\hbar^2} \int_0^{\hbar\beta} d\tau \int_0^{\hbar\beta} d\tau' K(\tau - \tau') \sigma_{z,\tau} \sigma_{z,\tau'} \right], \tag{2.77}
$$

where $\mathbf{T}$ is a time ordering operator which orders the Schrödinger operators $H_{hy,\tau}$ and $\sigma_{z,\tau}$ so that the ordering index τ decreases from left to right. As far as the effect of the phonon modes is concerned, the time ordered exponential (2.77) gives the correct summation of all orders of perturbation theory. For the electronic coupling the expression (2.77) corresponds to a summation of diagrams with decoupled self-energy insertions [2.49]. The terms omitted in (2.77) do not contribute to the linear low-frequency behavior of $J_{el}(\omega)$ and can usually be disregarded at low temperatures.

2.5.6 Equivalent Spin-Boson Hamiltonian

From the above analysis it is clear that (2.77) is the exact reduced density matrix for a system described by the so-called spin-boson Hamiltonian

$$
H_{SB} = H_{hy} + \sigma_z \sum_j G_j(b_j + b_j^\dagger) + \sum_j \hbar\omega_j b_j^\dagger b_j \, . \tag{2.78}
$$

Of course, the mode frequencies ω_j and the coupling parameters G_j have to be chosen in such a way that the effective spectral density (2.76) of the problem under consideration is recovered, i.e.,

$$
J(\omega) = \frac{2}{\hbar^2} \sum_j G_j^2 \delta(\omega - \omega_j) \, . \tag{2.79}
$$

The Hamiltonian (2.78) can now be used as a model Hamiltonian for a two-level system in a metal[4]. In the normal conducting state of the metal the spectral density takes the form

$$
J(\omega) = (K\omega + U\omega^{2s+1}) \exp(-\omega/\omega_c) \, . \tag{2.80}
$$

Henceforth, K and U will be treated as phenomenological constants parametrizing the damping of the hydrogen motion by the electrons and phonons of the host crystal. In the superconducting state of the metal the first term in (2.80), which gives the electronic contribution, has to be replaced by the spectral density (2.67) which contains the BCS energy gap Δ_{gap} as an additional parameter.

---

[4]While we have established the approximate equivalence between the spin-boson Hamiltonian (2.78) and the full Hamiltonian (2.51) only for the reduced density matrix of the interstitial, one can easily show that the equivalence extends to the dynamics of the interstitial without further approximation.

2.6 Time Evolution of the Two-Level System

We have seen that the low-temperature tunneling of trapped hydrogen interstitials can be described by the model of a dissipative two-level system. The tunneling dynamics of the defect between the two interstitial sites at $r = \pm d$ is then related to the time development of the Pauli matrix $\sigma_z(t)$. In this section we study the autocorrelation function of this variable for the effective spin-boson model obtained in (2.78).

2.6.1 Unitary Transformation

It is convenient [2.5, 2.50] to make a unitary transformation of the spin-boson Hamiltonian (2.78)

$$\tilde{H}_{\mathrm{SB}} = S\,H_{\mathrm{SB}}S^{-1} \tag{2.81}$$

where S is a unitary operator defined by

$$S = \exp\left[-\sigma_z \sum_j \frac{G_j}{\hbar\omega_j}(b_j - b_j^\dagger)\right]. \tag{2.82}$$

One readily obtains from (2.30, 78)

$$\tilde{H}_{\mathrm{SB}} = \frac{1}{2}\hbar\Delta_c(B_+\sigma_- + B_-\sigma_+) + \sum_j \hbar\omega_j b_j^\dagger b_j - \sum_j G_j^2/\hbar\omega_j, \tag{2.83}$$

where

$$\sigma_\pm = \sigma_x \pm i\sigma_y \tag{2.84}$$

and

$$B_\pm = \exp\left[\pm\sum_j \frac{2G_j}{\hbar\omega_j}(b_j - b_j^\dagger)\right]. \tag{2.85}$$

The Pauli matrix σ_z is invariant under the transformation (2.82).

2.6.2 Equation of Motion

To determine the time development of the symmetrized correlation function

$$C(t) = \frac{1}{2}\langle\sigma_z(t)\sigma_z(0) + \sigma_z(0)\sigma_z(t)\rangle \tag{2.86}$$

we make use of the projection operator technique [2.51]. Let us introduce the Liouville operator

$$L = \frac{i}{\hbar}[\tilde{H}_{\mathrm{SB}}, X], \tag{2.87}$$

where X is an arbitrary operator, as well as the projection operator

$$PX = \frac{1}{2}\langle\sigma_z X + X\sigma_z\rangle\sigma_z \tag{2.88}$$

which has the property $P^2 = P$. Clearly, we then have

$$Pe^{iLt}P = C(t)P, \tag{2.89}$$

so that the correlation function (2.86) can be seen as a reduction of the time evolution operator $\exp[iLt]$ to the one-dimensional subspace projected out by P.

From (2.89) we find

$$\dot{C}(t)P = Pe^{iLt}iLP \tag{2.90}$$

into which we may insert the operator identity

$$Pe^{iLt} = Pe^{iLt}P + \int_0^t ds Pe^{iLs}PiL(1-P)e^{i(1-P)L(1-P)(t-s)} \tag{2.91}$$

to give

$$\dot{C}(t)P = Pe^{iLt}PiLP$$
$$+ \int_0^t ds\, Pe^{iLs}PiL(1-P)e^{i(1-P)L(1-P)(t-s)}i\, LP. \tag{2.92}$$

Now, from (2.83) we have

$$\dot{\sigma}_z = iL\sigma_z = -i\Delta_c(B_+\sigma_- - B_-\sigma_+) \tag{2.93}$$

which gives

$$P\dot{\sigma}_z = 0, \tag{2.94}$$

since $\sigma_z\sigma_\pm + \sigma_\pm\sigma_z = 0$. Hence, $PLP = 0$ and the first term on the rhs of (2.92) vanishes. Further, using (2.89, 93), the second term may be transformed to yield

$$\dot{C}(t) = -\int_0^t ds\, \Phi(t-s)C(s), \tag{2.95}$$

where we have introduced the memory function

$$\Phi(t) = -\frac{\Delta_c^2}{2}\langle A(t)A(0) + A(0)A(t)\rangle \tag{2.96}$$

which is a symmetrized correlation function of the operator

$$A(t) = e^{i(1-P)L(1-P)t}(B_+\sigma_- - B_-\sigma_+) \tag{2.97}$$

that obeys a modified time evolution law.

2.6.3 Memory Function

The equation of motion (2.95) for $C(t)$ is formally exact, however, the memory function (2.96) has not been determined, as yet. Now, the memory function is explicitly of second order in the tunneling frequency Δ_c which is the smallest characteristic frequency of our problem. We therefore now assume that the time development of $A(t)$ can be calculated for $\Delta_c = 0$. In this approximation we have

$$A(t) = B_+(t)\sigma_- - B_-(t)\sigma_+ , \tag{2.98}$$

where the time development of $B_\pm(t)$ is given by the free motion of the bosonic bath, i.e.,

$$B_\pm(t) = \exp\left[\pm \sum_j \frac{2G_j}{\hbar\omega_j} \left(b_j e^{-i\omega_j t} - b_j^\dagger e^{i\omega_j t} \right) \right]. \tag{2.99}$$

Using well-known properties of a harmonic oscillator system, we find for the bath correlation functions

$$\langle B_+(t)B_- \rangle_0 = \langle B_-(t)B_+ \rangle_0 = \langle B_+B_-(t) \rangle_0^* = \langle B_-B_+(t) \rangle_0^*$$

$$= \exp\left\{ -\sum_j \left(\frac{2G_j}{\hbar\omega_j} \right)^2 [\coth(\hbar\beta\omega_j/2)[1 - \cos(\omega_j t)] \right.$$

$$\left. + i \sin(\omega_j t)] \right\} \tag{2.100}$$

where $\langle \ \rangle_0$ denotes the average over the equilibrium state of the unperturbed bosonic bath. Inserting now (2.98) into (2.96) and using (2.100) and properties of the Pauli matrices, we find for the memory function

$$\Phi(t) = \Delta_c^2 \, \Re \, \exp\left\{ -2 \int_0^\infty d\omega \, \frac{J(\omega)}{\omega^2} \, [\coth(\hbar\beta\omega/2)[1 - \cos(\omega t)] \right.$$

$$\left. + i \sin(\omega t)] \right\}, \tag{2.101}$$

where $\Re$ denotes the real part, and where we have written the bath correlation function (2.100) in terms of the spectral density (2.79). The equation of motion (2.95) with the memory function (2.101) will be the basis of the further analysis.

2.7 Incoherent Tunneling

As we have already mentioned in the introduction, the hydrogen interstitial hops randomly from an interstitial site to a neighboring site except for the region of low temperatures where coherent tunneling may occur. In this section we discuss the behavior in the region of incoherent tunneling where subsequent tunneling transitions are statistically independent so that the dynamics can be described in terms of a rate.

2.7.1 Hopping Rate

To solve the equation of motion (2.95) we introduce the Laplace transform

$$\hat{C}(z) = \int_0^\infty dt \, e^{-zt} C(t) \tag{2.102}$$

of the correlation function $C(t)$ which is found to read

$$\hat{C}(z) = \frac{1}{z + \hat{\Phi}(z)}, \tag{2.103}$$

where $\hat{\Phi}(z)$ is the Laplace transform of the memory function (2.101) and where we have used $C(0) = 1$. Now, in the region of incoherent tunneling of the interstitial (2.103) can be evaluated in the Markovian approximation where it is replaced by

$$\hat{C}(z) = \frac{1}{z + 2\Gamma}, \tag{2.104}$$

where

$$\Gamma = \frac{1}{2}\hat{\Phi}(0) = \frac{1}{2}\int_0^\infty dt \, \Phi(t) \tag{2.105}$$

is the hopping rate. One can easily convince oneself that (2.104) describes the correlation function of an interstitial tunneling with the rate Γ between the two sites corresponding to $\sigma_z = \pm 1$. In the following section we will present a more detailed evaluation of (2.103) and discuss the deviations from (2.104) occurring at low temperatures.

From (2.101, 105) we find for the hopping rate after a shift of the integration contour the result

$$\Gamma = \frac{1}{4}\Delta_c^2 \int_{-\infty}^\infty dt \exp[-\Lambda(t)], \tag{2.106}$$

where we introduced

$$\Lambda(t) = 2\int_0^\infty d\omega \frac{J(\omega)}{\omega^2} \left[\coth(\hbar\beta\omega/2) - \frac{\cos(\omega t)}{\sinh(\hbar\beta\omega/2)}\right]. \tag{2.107}$$

For a normal conducting metal we now insert the spectral density (2.80) into (2.107) which gives after an evaluation of the frequency integrals

$$\Lambda(t) = \Lambda_{el}(t) + \Lambda_{ph}(t), \tag{2.108}$$

where

$$\Lambda_{el}(t) = 2K \ln\left[\theta_c \left|\Gamma\left(1 + \frac{1}{\theta_c}\right) \middle/ \Gamma\left(\frac{1}{2} + \frac{1}{\theta_c} + i\frac{t}{\hbar\beta}\right)\right|^2\right] \tag{2.109}$$

and

$$\Lambda_{\mathrm{ph}}(t) = 2U \left\{ \frac{\Gamma(2s)}{(\hbar\beta)^{2s}} \left[2\zeta\left(2s, 1+\frac{1}{\theta_{\mathrm{c}}}\right) + \theta_{\mathrm{c}}^{2s} \right] \right.$$
$$\left. + (-1)^s \frac{2}{\hbar\beta} \left(\frac{\partial}{\partial t}\right)^{2s-1} \Im\Psi\left(\frac{1}{2} + \frac{1}{\theta_{\mathrm{c}}} + i\frac{t}{\hbar\beta}\right) \right\}. \tag{2.110}$$

Here $\Gamma(z)$ and $\Psi(z)$ are the gamma and digamma functions, respectively, $\zeta(z,q)$ is a Riemann zeta function, and $\theta_{\mathrm{c}} = \hbar\beta\omega_{\mathrm{c}}$. Finally, $\Im$ denotes the imaginary part.

Now, since the model Hamiltonian applies only for temperatures where $k_{\mathrm{B}}T \ll \hbar\omega_{\mathrm{c}}$, we have $\theta_{\mathrm{c}} \gg 1$, and the expressions (2.109, 110) can be simplified to read [2.52]

$$\Lambda_{\mathrm{el}}(t) = 2K \ln\left[\frac{\hbar\beta\omega_{\mathrm{c}}}{\pi} \cosh\left(\frac{\pi t}{\hbar\beta}\right)\right] \tag{2.111}$$

and

$$\Lambda_{\mathrm{ph}}(t) = 2\Gamma(2s)U\omega_{\mathrm{c}}^{2s} + W + \Xi(t), \tag{2.112}$$

where

$$W = \frac{4\Gamma(2s)\zeta(2s)U}{(\hbar\beta)^{2s}}$$

$$\Xi(t) = (-1)^s \frac{2\pi U}{\hbar\beta} \left(\frac{\partial}{\partial t}\right)^{2s-1} \tanh\left(\frac{\pi t}{\hbar\beta}\right). \tag{2.113}$$

In these latter expressions terms of order $1/\theta_{\mathrm{c}}$ were disregarded.

When (2.111, 112) are inserted into (2.106), the rate formula takes the form

$$\Gamma = \left(\frac{\Delta_{\mathrm{c}}}{2}\right)^2 \left(\frac{\pi}{\hbar\beta\omega_{\mathrm{c}}}\right)^{2K}$$
$$\exp(-2\Gamma(2s)U\omega_{\mathrm{c}}^{2s} - W) \int_{-\infty}^{\infty} dt \, \frac{e^{-\Xi(t)}}{\cosh^{2K}\left(\frac{\pi t}{\hbar\beta}\right)} \tag{2.114}$$

which will be evaluated further below.

2.7.2 Electron Dominated Region

For low temperatures the rate formula (2.114) can be simplified. Both W and $\Xi(t)$ are of order $U/(\hbar\beta)^{2s}$. We may relate the coupling constant U to a characteristic phonon coupling temperature

$$T_{\mathrm{ph}} = \frac{\hbar}{\pi k_{\mathrm{B}}} (2U)^{-1/2s}, \tag{2.115}$$

which will be used as a convenient phenomenological parameter, henceforth. Note that T_{ph} decreases as the phonon coupling strength U increases. Typically, T_{ph} is a few tens of Kelvin. Now, for $T \ll T_{ph}$, the rate expression (2.114) takes the form

$$
\begin{aligned}
\Gamma &= \left(\frac{\Delta_c}{2}\right)^2 \left(\frac{\pi}{\hbar\beta\omega_c}\right)^{2K} \exp(-2\Gamma(2s)U\omega_c^{2s}) \int_{-\infty}^{\infty} dt \frac{1}{\cosh^{2K}\left(\frac{\pi t}{\hbar\beta}\right)} \\
&= \frac{\sqrt{\pi}\Delta_c^2}{4} \frac{\Gamma(K)}{\omega_c \Gamma(K+1/2)} \exp(-2\Gamma(2s)U\omega_c^{2s}) \left(\frac{\pi k_B T}{\hbar\omega_c}\right)^{2K-1} .
\end{aligned}
\tag{2.116}
$$

Introducing a cutoff-independent, renormalized tunneling frequency

$$
\Delta = [\cos(\pi K)\Gamma(1-2K)]^{\frac{1}{2(1-K)}} \Delta_c \left(\frac{\Delta_c}{\omega_c}\right)^{\frac{K}{1-K}} \exp\left[-\Gamma(2s)\frac{U\omega_c^{2s}}{1-K}\right]
\tag{2.117}
$$

the result (2.116) may be written as

$$
\Gamma = \frac{\Gamma(K)}{\Gamma(1-K)} \frac{\Delta}{2} \left(\frac{2\pi k_B T}{\hbar\Delta}\right)^{2K-1} ,
\tag{2.118}
$$

where the cutoff frequency ω_c has dropped out. One can show that in the frequency region where the microscopic electronic density of states is constant and the phononic density of states is of the Debye form, a change of the cutoff frequency ω_c modifies the tunneling frequency Δ_c in such a way that the renormalized tunneling frequency (2.117) remains invariant [2.53].

Since the Kondo parameter $K < 1/2$ [2.54], the rate (2.119) shows a power-law increase $\propto T^{2K-1}$ as the temperature decreases [2.11, 15]. This explains the anomalous behavior of the low-temperature hopping rate observed for muons in copper and aluminium [2.12]. When we consider the general rate formula (2.106), the temperature dependence of the hopping rate can be traced back to the time dependence of the electronic part of $\Lambda(t)$. In an adiabatic approximation $\Lambda_{el}(t)$ would be replaced by a time-independent constant. Hence, the power-law behavior of the tunneling rate for temperatures below T_{ph} is intimately connected with a breakdown of the Born-Oppenheimer approximation.

2.7.3 Phonon Dominated Region

For temperatures near and above T_{ph} we have to evaluate the expression (2.114) more precisely. We shall consider here the case $s = 1$. Then, using (2.115) we find from (2.113)

$$W = \frac{1}{3}\left(\frac{T}{T_{\text{ph}}}\right)^2 ,$$

$$\Xi(t) = -\left(\frac{T}{T_{\text{ph}}}\right)^2 \frac{1}{\cosh^2\left(\frac{\pi t}{\hbar\beta}\right)} . \tag{2.119}$$

When this is inserted into the rate formula (2.114) we find

$$\Gamma = \frac{\Delta_c^2}{4\omega_c}\left(\frac{\pi k_B T}{\hbar\omega_c}\right)^{2K-1} \exp\left[-2\Gamma(2s)U\omega_c^{2s} - T^2/3T_{\text{ph}}^2\right] I\left(T^2/T_{\text{ph}}^2\right) , \tag{2.120}$$

where the integral

$$I(z) = \int_{-\infty}^{\infty} dx \, \frac{e^{z/\cosh^2(x)}}{\cosh^{2K}(x)} = \frac{\sqrt{\pi}\,\Gamma(K)}{\Gamma(K+1/2)} M(K, K+1/2, z) . \tag{2.121}$$

Here $M(a,b,z)$ is a Kummer confluent hypergeometric function which is also denoted as ${}_1F_1(a;b;z)$. When (2.120) is written in terms of the renormalized tunneling frequency (2.117) it takes the form [2.26]

$$\Gamma = \frac{\Gamma(K)}{\Gamma(1-K)}\frac{\Delta}{2}\left(\frac{2\pi k_B T}{\hbar\Delta}\right)^{2K-1} \exp(-T^2/3T_{\text{ph}}^2) M(K, K+1/2, T^2/T_{\text{ph}}^2) . \tag{2.122}$$

For $T \ll T_{\text{ph}}$ this reduces to our earlier result (2.118). On the other hand, for $T \gg T_{\text{ph}}$ we find from the asymptotic behavior of the Kummer function

$$\Gamma = \frac{\Gamma(K+1/2)}{\Gamma(1-K)}\frac{\Delta}{2}\left(\frac{2\pi k_B T_{\text{ph}}}{\hbar\Delta}\right)^{2K-1}\left(\frac{T}{T_{\text{ph}}}\right)^{2(K-1)} \exp\left(\frac{2T^2}{3T_{\text{ph}}^2}\right) , \tag{2.123}$$

which leads to an exponentially fast increase of the rate with temperature. We note that the result (2.123) has to be addressed with some caution, since our model is restricted to temperatures $T \ll \hbar\omega_c/k_B$ which basically means to temperatures well below the Debye temperature T_D.

Figure 2.5 shows the temperature dependent rate (2.122) for various values of the Kondo parameter K. The temperature is measured in units of T_{ph} and the rate is given in units of Γ_0 which is the expression (2.118) for $T = T_{\text{ph}}$. In these units the phonon coupling parameter drops out.

As the temperature is increased the long wavelength approximation for the thermally occupied phonon modes, which leads to the form (2.73) of the effective phononic density of states, will break down and the transition rate has to be calculated by means of the methods presented in Sect. 2.4. To see the connection between the results derived in this section, which include the electronic coupling, and the transitions rates in Sect. 2.4, we write the rate expression (2.114) in the form

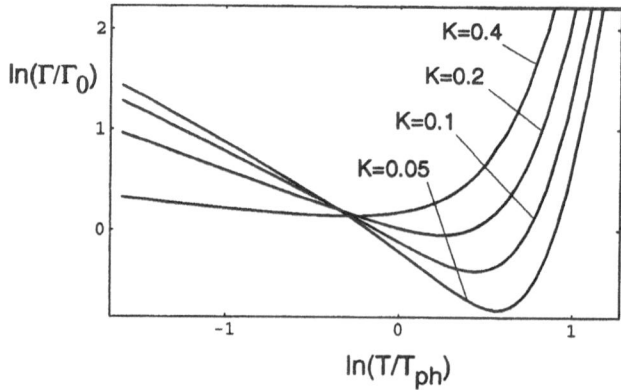

Fig. 2.5. The rate of incoherent tunneling Γ, (2.112) is shown for various values of the Kondo parameter K. The temperature is given in units of T_{ph} and the rate in units of $\Gamma_0 = (\Delta/2)[\Gamma(K)/\Gamma(1-K)](2\pi k_B T_{ph}/\hbar\Delta)^{(2K-1)}$. In these units the graph is independent of the phonon coupling strength. The figure shows the changeover from the electron-dominated region to the phonon-dominated region

$$\Gamma = \Gamma_1 + \Gamma_2 \; , \tag{2.124}$$

where

$$\Gamma_1 = \frac{\Gamma(K)}{\Gamma(1-K)} \frac{\Delta}{2} \left(\frac{2\pi k_B T}{\hbar\Delta}\right)^{2K-1} e^{-W} \tag{2.125}$$

and

$$\Gamma_2 = \frac{\Delta^2}{4} \left(\frac{\pi k_B T}{\hbar\Delta}\right)^{2K} \frac{1}{\cos(\pi K)\Gamma(1-2K)} e^{-W} \int_{-\infty}^{+\infty} dt \frac{e^{-\Xi(t)} - 1}{\cosh^{2K}\left(\frac{\pi t}{\hbar\beta}\right)} \; . \tag{2.126}$$

Here, we have made use of (2.117). Now, the contribution Γ_1 rapidly decreases as the temperature increases and the transition rate at elevated temperatures is dominated by Γ_2. Furthermore, for $T > T_{ph}$ the numerator of the integrand in (2.126)) leads to a rapid convergence of the integral and the denominator can be replaced by 1. We thus obtain for $T > T_{ph}$ [2.55]

$$\Gamma = \frac{\Delta_0^2}{4} \left(\frac{T}{T_{BO}}\right)^{2K} e^{-W} \int_{-\infty}^{+\infty} dt \left[e^{-\Xi(t)} - 1\right] , \tag{2.127}$$

where we have introduced a temperature T_{BO} for which the lattice potentials in the effective Born-Oppenheimer approximation of the conventional small polaron theory are valid and where

$$\Delta_0 = \Delta \left(\frac{\pi k_B T_{BO}}{\hbar\Delta}\right)^K [\cos(\pi K)\Gamma(1-2K)]^{-1/2} \tag{2.128}$$

is an effective tunneling frequency for this temperature. Of course, when (2.128) is inserted into (2.127), the temperature T_{BO} drops out. However, this choice of Δ_0 is necessary to make contact with theories that disregard the electronic coupling. Apart from the factor $(T/T_{\text{BO}})^{2K}$ the rate (2.127) is identical to our earlier result (2.42). In a large temperature range around T_{BO} this latter factor is not very different from 1 for $K \ll 1$, as it is the case for Nb:H. However, in general the influence of conduction electrons weakly affects the transition rate even at elevated temperatures.

In particular, when excited states become occupied, the electronic influence can no longer be described by a $(T/T_{\text{BO}})^{2K}$ correction factor but a more complete theory avoiding the truncation to a two–state model is needed.

2.8 Coherent Tunneling

In the previous section we have discussed the rate of incoherent tunneling of a hydrogen interstitial. As one lowers the temperature, the rate, or the corresponding quantum diffusion coefficient rapidly decreases until one reaches the temperature T_{ph} characterizing the phonon coupling strength. Near T_{ph} the rate goes through a minimum below which it shows a power-law increase with decreasing temperature. Clearly, since $K < 1/2$, the rate formula (2.118) will fail for $T \to 0$ K. As a matter of fact, the Markovian approximation of the equation of motion (2.103) is no longer valid at very low temperatures and we have to evaluate the memory function (2.101) more carefully.

2.8.1 Low-Temperature Memory Function

In the temperature region we are addressing now, $T \ll T_{\text{ph}}$, we may disregard finite temperature effects of phonons as we did in deriving (2.118). For a metal in the normal conducting state (2.64, 101) combine to yield for the memory function

$$\Phi(t) = \Delta_c^2 \Re \exp[-\Lambda'_{\text{el}}(t) - 2\Gamma(2s)U\omega_c^{2s}], \qquad (2.129)$$

where[5]

$$\Lambda'_{\text{el}}(t) = 2K \int_0^\infty d\omega \frac{e^{-\omega/\omega_c}}{\omega} \left\{ \coth(\hbar\beta\omega/2)[1 - \cos(\omega t)] + i\sin(\omega t) \right\}$$

$$= 2K \ln \left[\frac{\Gamma\left(1 + \dfrac{1}{\theta_c}\right)\Gamma\left(\dfrac{1}{\theta_c}\right)}{\Gamma\left(1 + \dfrac{1}{\theta_c} - i\dfrac{t}{\hbar\beta}\right)\Gamma\left(\dfrac{1}{\theta_c} + i\dfrac{t}{\hbar\beta}\right)} \right]. \qquad (2.130)$$

---

[5]We note that the function $\Lambda_{\text{el}}(t)$ introduced in (2.109) is given by $\Lambda_{\text{el}}(t) = \Lambda'_{\text{el}}(t - i\hbar\beta/2)$.

Now, since $\theta_c = \hbar\beta\omega_c \gg 1$, this can be simplified to read

$$\Lambda'_{el}(t) = 2K \ln\left[\frac{i\hbar\beta\omega_c}{\pi}\sinh\left(\frac{\pi t}{\hbar\beta}\right)\right], \tag{2.131}$$

and the memory function takes the form

$$\Phi(t) = \Delta_c^2 \cos(\pi K)\left(\frac{\pi}{\hbar\beta\omega_c}\right)^{2K} \exp(-2\Gamma(2s)U\omega_c^{2s})\frac{1}{\sinh^{2K}\left(\frac{\pi t}{\hbar\beta}\right)}. \tag{2.132}$$

The Laplace transform is found to read

$$\hat{\Phi}(z) = \frac{\Delta_c^2}{\omega_c}\cos(\pi K)\left(\frac{2\pi}{\hbar\beta\omega_c}\right)^{2K-1}$$
$$\exp(-2\Gamma(2s)U\omega_c^{2s})\frac{\Gamma(1-2K)\Gamma(K+z\hbar\beta/2\pi)}{\Gamma(1-K+z\hbar\beta/2\pi)}, \tag{2.133}$$

which can be written in terms of the renormalized tunneling frequency (2.117) as

$$\hat{\Phi}(z) = \Delta\left(\frac{\hbar\beta\Delta}{2\pi}\right)^{1-2K}\frac{\Gamma(K+z\hbar\beta/2\pi)}{\Gamma(1-K+z\hbar\beta/2\pi)}. \tag{2.134}$$

At zero temperature this reduces to

$$\hat{\Phi}(z) = \Delta\left(\frac{\Delta}{z}\right)^{1-2K}, \tag{2.135}$$

which is nonanalytic for $z \to 0$ and clearly indicates the breakdown of the Markovian approximation. The result (2.135) combines with (2.103) to yield for the Laplace transform of the correlation function

$$\hat{C}(z) = \frac{1}{z}\frac{1}{1+(\Delta/z)^{2(1-K)}}, \tag{2.136}$$

We note that for $K \to 0$ this is the Laplace transform of $C(t) = \cos(\Delta t)$, which is the correlation function of a system tunneling with frequency Δ in a coherent, clock-like fashion between two sites. Because of the interaction with the conduction electrons, the tunneling oscillations will be damped. This can be seen from the neutron structure factor discussed in the next subsection.

2.8.2 Structure Factor for Neutron Scattering

The experimentally measured differential cross section for neutron scattering is given in terms of the dynamic structure factor [2.56]. The latter is defined by

$$S(\boldsymbol{k}, \omega) = \frac{1}{2\pi}\int_{-\infty}^{+\infty} dt \, \exp(i\omega t)\langle\exp[-i\boldsymbol{k}\boldsymbol{r}(0)]\exp[i\boldsymbol{k}\boldsymbol{r}(t)]\rangle, \tag{2.137}$$

where $\hbar k$ and $\hbar\omega$ are the momentum and energy transfer for the neutron during the scattering process and $r(t)$ is the position of the scatterer at time t. When the scatterer tunnels between two positions d and $-d$, the operator r may be written $r = d\sigma_z$, and the structure factor takes the form

$$S(k, \omega) = \cos^2(kd)\delta(\omega)$$
$$+ \sin^2(kd)\frac{1}{2\pi}\int_{-\infty}^{+\infty} dt \, \exp(i\omega t)\langle\sigma_z(0)\sigma_z(t)\rangle . \tag{2.138}$$

The first term represents an elastic peak while the second term describes inelastic scattering. This latter part is connected with the tunneling dynamics of the interstitial. Thus, writing

$$S(k, \omega) = \cos^2(kd)\delta(\omega) + \sin^2(kd)S(\omega) \tag{2.139}$$

we see that the quantity of interest here is the Fourier transform

$$S(\omega) = \frac{1}{2\pi}\int_{-\infty}^{+\infty} dt \, \exp(i\omega t)\langle\sigma_z(0)\sigma_z(t)\rangle \tag{2.140}$$

of the correlation function of $\sigma_z(t)$. Now, using the relation between the time-ordered and symmetrized correlation functions, $S(\omega)$ can be expressed as

$$S(\omega) = \frac{1}{1 + \exp(\hbar\beta\omega)}\frac{1}{\pi}\int_{-\infty}^{+\infty} dt \, \exp(i\omega t)C(t), \tag{2.141}$$

where $C(t)$ is the symmetrized correlation function introduced in (2.86). Noting that $C(t) = C(-t)$, the above result can also be written in terms of the Laplace transform (2.102) as

$$S(\omega) = \frac{2}{\pi}\frac{1}{1 + \exp(\hbar\beta\omega)}\Re\hat{C}(i\omega) \tag{2.142}$$

which relates the structure factor $S(\omega)$ with the results obtained in the previous sections.

Now combining (2.103, 142) we obtain [2.26]

$$S(\omega) = \frac{2}{\pi}\frac{1}{1 + \exp(\hbar\omega/k_B T)}\Re\left\{\frac{1}{i\omega + \hat{\Phi}(i\omega)}\right\}, \tag{2.143}$$

where $\hat{\Phi}(z)$ is given by (2.134). This formula determines the cross-section for inelastic neutron scattering. It should be noted that aside from the temperature T the result (2.143) for $S(\omega)$ depends only on two parameters, namely the Kondo parameter K and the renormalized tunneling frequency Δ.

2.8.3 Crossover from Coherent to Incoherent Tunneling

Frequently, the result (2.143) for the structure factor can be simplified further. For instance, for hydrogen in Nb the Kondo parameter $K \ll 1$, and the Laplace transform (2.134) of the memory function may be written

$$\hat{\Phi}(z) = \Delta \left(\frac{\hbar \beta \Delta}{2\pi} \right)^{1-2K} \frac{\Gamma(1 + K + z\hbar\beta/2\pi)}{\Gamma(1 - K + z\hbar\beta/2\pi)} \frac{1}{K + z\hbar\beta/2\pi}$$

$$= \Delta \frac{\hbar \beta \Delta}{2\pi} \frac{1 + 2K[\Psi(1 + z\hbar\beta/2\pi) - \ln(\hbar\beta\Delta/2\pi)]}{K + z\hbar\beta/2\pi}, \qquad (2.144)$$

where we have neglected corrections of order K^2 in the second line. When we insert this into (2.143) and again disregard higher-order terms in K, we find that the structure factor may be written

$$S(\omega) = \frac{1}{\pi} \frac{1}{1 + \exp(\hbar\omega/k_B T)} \left(\frac{\gamma}{(\omega - \tilde{\Delta})^2 + \gamma^2} + \frac{\gamma}{(\omega + \tilde{\Delta})^2 + \gamma^2} \right), \qquad (2.145)$$

where [2.26]

$$\tilde{\Delta} = \Delta \left\{ 1 + K \left[\Re \left\{ \psi(i\hbar\Delta/2\pi k_B T) \right\} - \ln(\hbar\Delta/2\pi k_B T) \right] \right\} \qquad (2.146)$$

is an effective temperature dependent tunnel splitting with $\tilde{\Delta} = \Delta$ for $T = 0$, and

$$\gamma = \frac{\pi K}{2} \Delta \coth(\hbar\Delta/2k_B T) \qquad (2.147)$$

is a Korringa-type relaxation rate of the two-level system. A corresponding temperature dependence of the energy splitting and the relaxation rate is also found for the levels of rare earth impurities in metals [2.57].

The result (2.145) describes two Lorentzians centered at $\omega = \pm\tilde{\Delta}$ with linewidth 2γ. The approximate result (2.145) is only valid if $\gamma \ll \Delta$ which is the case for $k_B T \ll \hbar\Delta/K$. Hence, for $K \ll 1$ and low temperatures the hydrogen interstitial forms delocalized tunneling eigenstates with an energy splitting $\hbar\tilde{\Delta}$. This is very different from the situation discussed in the previous section where the interstitial makes transitions between the pocket states in the potential wells.

The relative intensity of the two peaks of (2.145) is determined by a detailed balance factor. The expressions (2.146, 147) describe the finite temperature shift and broadening of the peaks in the low temperature regime. Since electronic excitations with energies below $k_B T$ are incoherent and do not contribute to the renormalization of the tunneling frequency, $\tilde{\Delta}$ increases weakly at finite temperatures. On the other hand, the linewidth 2γ grows rather strongly with temperature. As soon as γ becomes comparable with Δ the result (2.145) is no longer valid.

On the other hand, for sufficiently high temperatures, the structure factor (2.143) can again be simplified. For $\hbar\beta\Delta/2\pi \ll K$, that is $k_B T \gg \hbar\Delta/2\pi K$, we see from (2.134) that $\hat{\Phi}(i\omega)$ can be replaced by $\hat{\Phi}(0)$, since the relevant frequencies are of order Δ. Now, according to (2.105), $\hat{\Phi}(0) = 2\Gamma$. Inserting this into (2.143) the structure factor takes the form

$$S(\omega) = \frac{1}{\pi} \frac{1}{1 + \exp(\hbar\omega/k_B T)} \frac{4\Gamma}{\omega^2 + 4\Gamma^2},$$ (2.148)

where Γ is the incoherent tunneling rate (2.118). This describes a quasi-elastic Lorentzian peak. In this region the interstitial tunnels incoherently between the two sites. Hence, subsequent tunneling transitions are statistically independent as discussed in the previous section.

We see, the model of a two-level system coupled to conduction electrons leads to a crossover from low temperature coherent tunneling characterized by a tunneling frequency to incoherent tunneling characterized by a tunneling rate at elevated temperatures. The transition between coherent and incoherent quantum dynamics arises near temperatures of the order of $\hbar\Delta/k_B K$. This crossover is seen explicitly in Fig. 2.6 where the structure factor (2.143) is depicted for a system with a tunnel splitting $\hbar\Delta = 0.2$ meV and a Kondo parameter $K = 0.05$. These values roughly correspond to the parameters for trapped hydrogen in niobium, which is the tunneling system discussed in detail elsewhere in this volume. At zero temperature, inelastically scattered neutrons suffer an energy loss of the order of the tunnel splitting $\hbar\Delta$. In view of the interaction with the metallic electrons the scattering peak at $T = 0$ has a finite width. At 2.5 K there is also a peak on the energy gain side. Further, compared with the zero-temperature result, the resonance has shifted and widened. Since K is small, this low-temperature behavior is well described by the approximation (2.145).

We also see from Fig. 2.6 that the tunneling eigenstates are readily destroyed with increasing temperature. At $T = 10$ K we find a very broad, relatively structureless energy distribution of the inelastically scattered neutrons. Upon further raising the temperature the structure factor approaches a

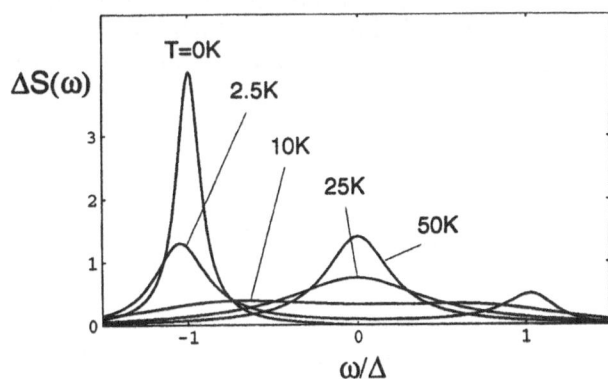

Fig. 2.6. The function $S(\omega)$, (2.143), which determines the energy distribution of inelastically scattered neutrons is shown for a tunneling system with a tunnel splitting $\hbar\Delta = 0.2$ meV and a Kondo parameter $K=0.05$ for various temperatures between 0 and 50 K. In this temperature range the system shows a crossover from coherent to incoherent tunneling

quasi-elastic peak that is a Lorentzian centered about $\omega = 0$. Comparing the results for 25 K and 50 K in Fig. 2.6, we see that the quasi-elastic peak is found to narrow with increasing temperature. In this temperature range we recover the region of incoherent tunneling described by the result (2.148).

To observe the nonadiabatic effects of conduction electrons rather purely in the entire range of temperatures of interest, $\hbar\Delta$ must be sufficiently small so that there is a region where the incoherent quantum dynamics is not affected by temperature dependent phonon effects. On the other hand, $\hbar\Delta$ must be sufficiently large so that the low temperature region of coherent quantum dynamics is experimentally accessible. Also systems with small tunnel splitting are more strongly affected by lattice imperfections leading to an asymmetry between the interstitial sites.

2.8.4 Effect of an Asymmetry Energy

The results presented here may partially be extended to account for an asymmetry energy $\hbar\epsilon$ which is unavoidable in samples with higher defect concentrations in view of the strain field interaction between the tunneling systems. In this case a term $\frac{1}{2}\hbar\epsilon\sigma_z$ has to be added to the Hamiltonian (2.51) and accordingly to the spin-boson Hamiltonian (2.78). Since the asymmetry energy does not matter for $k_B T \gg \hbar\epsilon$, we may restrict ourselves to the low-temperature region where the influence of the phonon excitations may be disregarded.

In the region of incoherent tunneling the extension is straightforward, since then the rate formula (2.106) is simply modified to read

$$\Gamma_\pm = \frac{1}{4}\Delta_c^2 \exp[\pm\hbar\beta\epsilon/2] \int_{-\infty}^{\infty} dt \ \exp[\pm i\epsilon t - \Lambda(t)]. \tag{2.149}$$

Of course, now the rates from left to right and vice versa differ, and in the structure factor (2.148) 2Γ has to be replaced by the sum of Γ_+ and Γ_-. Using in (2.149) the result (2.111) and the zero-temperature limit of (2.112), one finds that the average hopping rate is modified to read [2.58, 59]

$$\Gamma = \frac{\sin(\pi K)}{\pi}\cosh\left(\frac{\hbar\epsilon}{2k_B T}\right)\left|\Gamma\left(K + \frac{i\hbar\epsilon}{2\pi k_B T}\right)\right|^2 \frac{\Delta}{2}\left(\frac{2\pi k_B T}{\hbar\Delta}\right)^{2K-1}, \tag{2.150}$$

which reduces to (2.119) for $\hbar\epsilon \ll k_B T$. For small K (2.150) can be simplified to yield [2.59]

$$\Gamma = \frac{K}{K^2 + (\hbar\epsilon/2\pi k_B T)^2}\frac{\hbar\epsilon}{2k_B T}\coth\left(\frac{\hbar\epsilon}{2k_B T}\right)\frac{\Delta}{2}\left(\frac{2\pi k_B T}{\hbar\Delta}\right)^{2K-1}. \tag{2.151}$$

This shows that asymmetric tunneling systems have their hopping rate reduced roughly by the factor $1 + (\hbar\epsilon/2\pi K k_B T)^2$.

Evaluating (2.149) for a $\Lambda(t)$ comprising the electronic term (2.111) and the phonon terms (2.119) one finds [2.60]

$$\Gamma = \frac{\sin(\pi K)}{\pi} \cosh\left(\frac{\hbar\epsilon}{2k_BT}\right) \left|\Gamma\left(K + \frac{i\hbar\epsilon}{2\pi k_BT}\right)\right|^2 \frac{\Delta}{2} \left(\frac{2\pi k_BT}{\hbar\Delta}\right)^{2K-1}$$

$$\exp\left(-\frac{T^2}{3T_{ph}^2}\right) {}_2F_2\left(K + \frac{i\hbar\varepsilon}{2\pi k_BT}, K - \frac{i\hbar\varepsilon}{2\pi k_BT}; K, K + \frac{1}{2}; \frac{T^2}{T_{ph}^2}\right), \qquad (2.152)$$

where $_2F_2(a,b;c,d;z)$ is a generalized hypergeometric function. The result (2.152) reduces to (2.123) for small asymmetry energy and/or higher temperatures. The effect of the asymmetry energy on the hopping rate is shown in Fig. 2.7.

In the region of coherent tunneling the situation is more complicated [2.59, 2.61–63] and explicit results are only available for small K. Even in the absence of damping, the peak positions are shifted by the asymmetry energy according to $\Delta_\epsilon = \sqrt{\Delta^2 + \epsilon^2}$. This leads to a broadening of the resonance peaks. Furthermore, there is now an additional quasi-elastic peak at low temperatures which is associated with the relaxation into the lower well. For a discussion of the various line positions and widths we refer to the literature [2.61–63]. A thorough study of dissipative, asymmetric double-well systems is given by *Weiss* [2.64].

2.8.5 Coherent Tunneling in a Superconducting Metal

So far, all results presented are for interstitials tunneling in normal state metals. In principle, we can use the same approach as presented above for hydrogen tunneling in superconducting metals. In this case the effective electronic density of states $J_{el}(\omega)$ has to be replaced by the effective quasiparticle density of states $J_{qp}(\omega)$ introduced in (2.67). However, due to the more complicated form of $J_{qp}(\omega)$, one has to evaluate the formulas numerically.

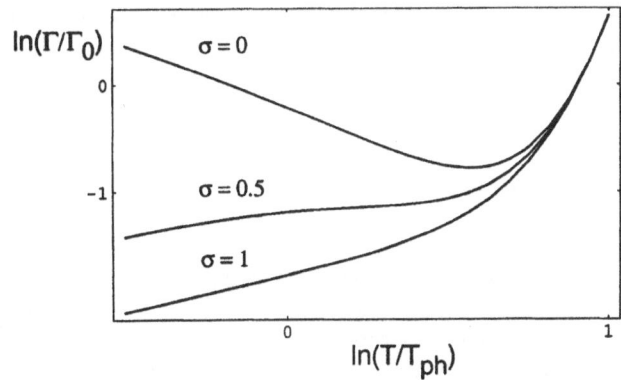

Fig. 2.7. The average rate of incoherent tunneling in the presence of an asymmetry energy, (2.152), is shown for a Kondo parameter $K = 0.05$ and various values of $\sigma = \hbar\varepsilon/k_BT_{ph}$. The units are the same as in Fig. 2.5

Here we shall restrict ourselves to the case $T = 0\,$K where $J_{qp}(\omega)$ is given by (2.68) which vanishes for frequencies below $2\Delta_{gap}/\hbar$. Hence, for $2\Delta_{gap} \gg \hbar\Delta_c$ all quasi-particle excitations can be eliminated adiabatically. In this case the memory function (2.101) takes the form

$$\Phi(t) = \Delta_c^2 \exp[-\Lambda_{qp} - 2\Gamma(2s)U\omega_c^{2s}], \qquad (2.153)$$

where

$$\Lambda_{qp} = 2\int_0^\infty d\omega\, \frac{J_{qp}(\omega)}{\omega^2}. \qquad (2.154)$$

This is the memory function of an undamped system tunneling coherently with the tunneling frequency

$$\Delta_s = \Delta_c \exp[-\Lambda_{qp}/2 - \Gamma(2s)U\omega_c^{2s}]. \qquad (2.155)$$

Now inserting (2.68) into (2.154) we obtain for $\hbar\omega_c \gg \Delta_{gap}$

$$\Lambda_{qp} = -2K\ln(2\Delta_{gap}/\hbar\omega_c) - 2aK. \qquad (2.156)$$

Here $a = 1 - 2\ln(2) + \gamma \sim 0.1909\ldots$ is a numerical factor where γ is the Euler constant. Combining (2.155, 156) we see that Δ_s can be expressed in terms of the renormalized cutoff independent tunneling frequency Δ in the normal-conducting metal. Using (2.117) we find

$$\Delta_s = \Delta\left(\frac{2\Delta_{gap}}{\hbar\Delta}\right)^K [\cos(\pi K)\Gamma(1 - 2K)]^{-1/2} \exp(aK). \qquad (2.157)$$

An approximate version of this result has been derived by *Teichler* [2.65]. For a superconducting metal one may pass to the normal state by applying a magnetic field. In view of (2.157) the Kondo parameter K can then be determined from the shift of the low-temperature resonance line. Hence, both parameters of the low-temperature theory, Δ and K, can be determined in the mK region.

At $T = 0\,$K the linewidth in the superconducting metal is entirely due to inhomogeneous broadening by asymmetry energies. At finite but low temperatures the homogeneous linewidth obeys the formula (2.147) with the effective Kondo parameter $2K/[1 + \exp(\beta\Delta_{gap})]$ obtained in (2.69).

2.9 Conclusions

In this chapter the theory of tunneling and quantum diffusion of hydrogen in metals has been treated over a large temperature range from 0 K to well above room temperature. Depending on temperature different effects dominate (coherent or incoherent tunneling, coupling to electrons or to phonons) and correspondingly different approximations are suitable. We have restricted ourselves to the quantum regime which is found at low temperatures and for hydrogen diffusion at elevated temperatures in cases where the distance between equilibrium sites is small, e.g., between tetrahedral sites in bcc metals.

To put the models for low and high temperatures on a common footing, the appropriate Hamiltonians were derived by successive approximations from a general interaction. Thus as a basic approximation the standard small polaron theory is recovered where a linear coupling of Debye phonons to a tunnel system is assumed. This simple picture breaks down for both low and high temperatures. Conceptually easier is the extension to higher temperatures, however, it can only be done numerically. First, one has to include realistic vibration modes of the metal hydrogen system instead of of Debye phonons. Secondly, for higher temperatures the dominant effect to be accounted for is the excitation of localized hydrogen vibrations with a resulting rapid increase of the tunneling frequency (violation of the Condon approximation). This effect is the reason for the observed change in apparent activation energy for hydrogen diffusion in Nb:H and Ta:H and its strong isotope dependence. To obtain quantitative results rather than a qualitative description also phonon anharmonicities have to be dealt with.

On the other hand, at low temperatures the tunneling dynamics of a light particle in a metallic environment is controlled by its nonadiabatic coupling to the conducting electrons. We have emphasized the connection between the two seemingly different situations of coherent and incoherent tunneling. Coherent tunneling or the existence of well-defined tunneling states and incoherent tunneling or hopping represent in fact only the low- and the high-temperature manifestations of the same tunneling phenomenon. Both, the theoretical calculations presented here and the experimental results discussed elsewhere in this book demonstrate the mutuality of coherent and incoherent tunneling. The theory describes the transition and, consequently, the quantitative relations between the two limiting situations of coherent and incoherent tunneling.

There are of course a number of theoretical aspects not discussed within this chapter. We have only considered the delocalization of interstitial atoms tunneling in a double-well potential. For particles moving in a periodic crystal field one expects the formation of Bloch states at very low temperatures. Such a long-range coherent tunneling dynamics, however, seems not to have been observed so far for light particles in metals. The effects of lattice imperfections and of the strain field interactions between interstitial atoms make this region very hard to access.

Acknowledgements. Part of the work presented here has been performed in collaboration with others and has benefited by discussions with colleagues. We wish to thank all, in particular, S. Dattagupta, R. Jung, A.M. Stoneham, U. Weiss, and H. Wipf. Financial support was provided by the Deutsche Forschungsgemeinschaft through SFB237.

References

2.1   C.P. Flynn, A.M. Stoneham: Phys. Rev. B **1**, 3966 (1970)
2.2   Yu. Kagan, M.I. Klinger: J. Phys. C **7**, 2791 (1974)
2.3   H. Teichler, A. Seeger: Phys. Lett. A **82**, 91 (1981)
2.4   J. Yamashita, T. Kurosawa: J. Phys. Chem. Sol. **5**, 34 (1958)

2.5   T. Holstein: Ann. Phys. (NY) **8**, 325, 343 (1959)
2.6   K. Huang, A. Rys: Proc. Roy. Soc. (London) A **204**, 406 (1951)
2.7   S. Pekar: Zh. Eksp. Theor. Fiz. **20**, 510 (1950)
2.8   V. Lottner, H.R. Schober, W.J. Fitzgerald: Phys. Rev. Lett. **42**, 1162 (1979)
2.9   H.R. Schober, V. Lottner: Z. Phys. Chem. (NF) **114**, 203 (1979)
2.10 J.L. Wang, G. Weiss, H. Wipf, A. Magerl: in *Phonon Scattering*, ed. by
      W. Eisenmenger, K. Labann, S. Döttinger, Springer Ser. Solid-State Sci. Vol. 51,
      (Springer, Berlin 1984) p. 401
2.11 J. Kondo: Physica **125 B**, 279 (1984); ibid **126 B**, 377 (1984); see also J. Kondo: in
      Fermi Surface Effects, ed. by J. Kondo, A. Yoshimiro: Springer Ser. Solid-State Sci.,
      Vol. 77, (Springer, Berlin, Heidelberg 1988) p. 1
2.12 O. Hartmann, E. Karlsson, L.O. Norlin, T.O. Niinikoski, K.W. Kehr, D. Richter,
      J.-M. Welter, A. Yaouanc, J. Le Hericy: Phys. Rev. Lett. **44**, 337 (1980)
      K.W. Kehr, D. Richter, J.-M. Welter, O. Hartmann, E. Karlsson, L.O. Norlin,
      T.O. Niinikoski, A. Yaouanc: Phys. Rev. B **26**, 567 (1982)
2.13 P.W. Anderson: Phys. Rev. Lett. **18**, 1049 (1967)
2.14 J. Kondo: Physica **84 B**, 40 (1976)
2.15 K. Yamada: Prog. Theor. Phys. **72**, 195 (1984)
2.16 P.W. Anderson, B.I. Halperin, C.M. Varma: Phil. Mag. **25**, 1 (1972)
2.17 W.A. Phillips: J. Low Temp. Phys. **7**, 351 (1972)
2.18 B. Golding, J.E. Graebner, A.B. Kane, J.L. Black: Phys. Rev. Lett. **41**, 1487 (1978)
2.19 J.L. Black, P. Fulde: Phys. Rev. Lett. **43**, 453 (1979)
2.20 G. Weiss, W. Arnold, K. Dransfeld, H.J. Güntherodt: Solid State Commun. **33**, 111
      (1980)
2.21 B. Golding, N.M. Zimmerman, S.N. Coppersmith: Phys. Rev. Lett. **68**, 998 (1992)
2.22 C.C. Baker, H.K. Birnbaum: Acta Met. **21**, 865 (1973); G. Pfeiffer, H. Wipf: J. Phys.
      F: Metal Phys. **6**, 167 (1976)
      A. Magerl, J.J. Rush, J.M. Rowe, D. Richter, H. Wipf: Phys. Rev. B **27**, 927 (1983)
2.23 for a review see: H. Grabert, H. Wipf: Festkörperprobleme - Advances in Solid State
      Physics, ed. by U. Rössler (Vieweg, Braunschweig 1990) Vol. 30, p. 1
2.24 A.J. Leggett, S. Chakravarty, M.P.A. Fisher, A.T. Dorsey, A. Garg, W. Zwerger:
      Rev. Mod. Phys. **59**, 1 (1987)
2.25 H.R. Schober, A.M. Stoneham: Phys. Rev. Lett. **60**, 2307 (1988)
2.26 H. Grabert, S. Linkwitz, S. Dattagupta, U. Weiss: Europhys. Lett. **2**, 631 (1986)
      H. Grabert: in *Quantum Aspects of Molecular Motions in Solids*, ed. by A. Heidemann
      et al. (Springer, Berlin 1987) p. 130
2.27 H. Peisl: In *Hydrogen in Metals I*, ed. by G. Alefeld, J. Völkl Topics Appl Phys. Vol.
      28 (Springer, Berlin, Heidelberg 1978) p. 53
2.28 A. Klamt, H. Teichler: Phys. Status. Solidi (b) **134**, 103 (1986); *ibid.* **134**, 533
      (1986)
2.29 G. Leibfried, N. Breuer: *Point Defects in Metals I*, Springer Tracts Mod. Phys., Vol.
      81 (Springer, Berlin 1978)
2.30 H.R. Schober, M. Mostoller, P.H. Dederichs: Phys. Status. Solidi. (b) **64**, 173 (1974)
2.31 H. Sugimoto, Y. Fukai: Phys. Rev. B **22**, 670 (1980)
2.32 R. Hempelmann, D. Richter, A. Kollmar: Z. Phys. B **44**, 159 (1981)
2.33 M. Abramowitz, I.A. Stegun: *Handbook of Mathematical Functions* (National Bureau
      of Standards, Washington DC 1964)
2.34 K. Kehr: In *Hydrogen in Metals I*, ed. by G. Alefeld, J. Völkl Topics Appl. Phys., Vol.
      28, (Springer, Berlin Heidelberg 1978) p. 197
2.35 T. Springer: In *Hydrogen in Metals I*, ed. by G. Alefeld, J. Völkl Topics Appl. Phys.,
      Vol. 28, (Springer, Berlin, Heidelberg 1978) p. 75
2.36 J. Völkl, G. Alefeld: In *Hydrogen in Metals I*, ed. by G. Alefeld, J. Völkl Topics Appl.
      Phys., Vol. 28, (Springer, Berlin, Heidelberg 1978) p. 321
      Z. Qi, J. Völkl, R. Lässer, H. Wenzl: J. Phys. F **13**, 2053 (1983)
2.37 H.R. Schober, A.M. Stoneham: J. Less.- Comm. Met. **172–174**, 538 (1991)
2.38 Yu. Kagan, N.V. Prokof'ev: Sov. Phys. JETP **69** 836 (1990)
      Yu. Kagan, Ber. Bunsenges: Phys. Chem. **95**, 411 (1991)

2.39 Yu. Kagan, N.V. Prokof'ev: Sov. Phys. JETP **69**, 1250 (1990)
2.40 P. Vargas, L. Miranda, M. Lagos: Z. Phys. Chem. (NF) **164**, 975 (1989)
2.41 D. Emin, M.J. Baskes, W.D. Wilson: Phys. Rev. Lett. **42**, 791 (1979)
2.42 M.J. Gillan: Phys. Rev. Lett. **58**, 563 (1987)
2.43 H.R. Schober, A.M. Stoneham: Phys. Rev. B **26**, 1819 (1982)
2.44 R. Messer, A. Blessing, S. Dais, D. Höpfel, G. Majer, C. Schmidt, A. Seeger, W. Zag, R. Lässer: Z. Phys. Chem. (NF) Suppl. **H2**, 62 (1984)
2.45 G.H. Vineyard: J. Phys. Chem. Solids **3**, 121 (1957)
2.46 D. Steinbinder, H. Wipf, H.R. Schober, H. Blank, G. Kearley, C. Vettier, A. Magerl: Europhys. Lett. **8**, 269 (1989)
2.47 U. Stuhr, H. Wipf, H.R. Schober: Solid State Commun. **80**, 987 (1991)
2.48 K. Vladár, A. Zawadowski, G.T. Zimányi: Phys. Rev. B **37**, 2001 (1987)
2.49 F. Guinea: Phys. Rev. Lett. **53**, 1268 (1964)
2.50 W. Zwerger: Z. Phys. B **53**, 53 (1983); C. Aslangul, N. Pottier, D. Saint-James: J. Physique **47**, 1657 (1986)
2.51 H. Grabert: *Projection Operator Techniques in Nonequilibrium Statistical Mechanics*, Springer Tracts Mod. Phys., Vol. 95 (Springer, Berlin, Heidelberg 1982)
2.52 H. Grabert, U. Weiss, H.R. Schober: Hyperfine Interact. **31**, 147 (1986)
2.53 U. Weiss, H. Grabert, P. Hänggi, P. Riseborough: Phys. Rev. B **35**, 9535 (1987)
2.54 K. Yamada, A. Sakurai, S. Miyazima: Prog. Theor. Phys. **73**, 1342 (1985)
2.55 H. Grabert (unpublished 1992)
2.56 D. Richter: *Transport mechanism of light in terstitials in metals*. Springer Tracts Mod. Phys. Vol. 101, 85 (Springer, Berlin, Heidelberg 1983)
      S. Dattagupta: *Relaxation Phenomena in Condensed Matter Physics* (Academic Press, New York 1987)
2.57 P. Fulde, I. Peschel: Adv. Phys. **21**, 1 (1972)
2.58 H. Grabert, U. Weiss: Phys. Rev. Lett. **54**, 1605 (1985)
      M.P.A. Fisher, A.T. Dorsey ibid. 1609
2.59 S. Dattagupta, H. Grabert, R. Jung: J. Phys. C **1**, 1405 (1989)
2.60 H. Grabert: Phys. Rev. B **46**, 12753 (1992)
2.61 U. Weiss, M. Wollensak: Phys. Rev. Lett. **62**, 1663 (1989)
2.62 U. Weiss, H. Grabert, S. Linkwitz: J. Low Temp. Phys **68**, 213 (1987)
2.63 S. Dattagupta, T. Qureshi: Physica B **174**, 262 (1991)
2.64 U. Weiss: *Quantum dissipative systems*, Series Mod. Condensed Matter Phys., Vol. 2 (World Scientific, Singapore 1993)
2.65 H. Teichler: Quantitative theory of the tunnel level splitting for hydrogen trapped at oxygen in mobium, in *Quantum Aspects of Molecular Motions in Solids*, ed. by A. Heidemann, A Magerl, M. Prager, D. Richter, T. Springer. *Springer Proc. Phys.* **17**, 167 (Springer, Berlin, Heidelberg 1987)
2.66 Y. Li and G. Wahnström: Phys. Rev. B **46**, 14528 (1992)
2.67 Y. Li and G. Wahnström: Phys. Rev. B **51**, 12233 (1995)
2.68 H. Dosch, F. Schmid, P. Wiethoff and J. Peisl: Phys. Rev. B **46**, 55 (1992)
2.69 Y. Fukai, The Metal-Hydrogen System, Springer Verlag, Berlin 1993

3. Diffusion of Hydrogen in Metals

H. Wipf

With 24 Figures

The diffusional properties of hydrogen interstitials in metals are unique for several reasons. They are influenced, or even controlled, by quantum and tunneling effects since hydrogen is the lightest of all elements. The fact that hydrogen has the highest possible isotopic mass ratios of all elements leads further to large – and experimentally well observable – isotope effects in diffusion. These isotope effects provide in many cases a sensitive and critical test for theoretical predictions. Hydrogen interstitials are finally extremely mobile, even at low temperatures. This allows the application of a great variety of different and complimentary techniques for diffusion studies, and it allows investigations to be carried out over wide temperature ranges. All these features make hydrogen in metals a perfect model system for diffusion and transport studies on mobile lattice defects.

The diffusion behavior of hydrogen in metals is of importance for a large number of technological problems including, for instance, the use of hydrogen gas as a secondary energy carrier, hydrogen gas storage and purification, fusion reactor technology and material failure processes.

The present state of hydrogen diffusion theory is the subject of Chap. 2 of this volume. The theoretical and experimental aspects of hydrogen diffusion were further discussed in monographs that appeared in the last decade [3.1–6]. The present review addresses chiefly experimental aspects. It presents first an introduction to the theoretical background required for a phenomenological description of hydrogen diffusion. Subsequently, it gives a compilation of data for the diffusion coefficient of hydrogen in pure metals and some representative alloys, and it finally discusses recent experimental results for the local diffusion or tunneling processes of trapped hydrogen interstitials. The experiments on the trapped hydrogen are particularly important from a theoretical point of view since they can be carried out at extremely low temperatures where both the tunneling and the nonadiabatic interaction with the conduction electrons (Sect. 3.3.1) can dominate the transport processes of the hydrogen.

Topics in Applied Physics, Vol. 73
Wipf (Ed.)
© Springer-Verlag Berlin Heidelberg 1997

3.1 Theoretical Background

3.1.1 Chemical Diffusion and Self Diffusion

The diffusivity of hydrogen interstitials exceeds that of the host metal atoms by many orders of magnitude. Therefore, the diffusive transport of the hydrogen can be considered to take place within a fixed reference lattice that is formed by the immobile metal atoms. This fact allows an elementary and transparent phenomenological description of hydrogen diffusion, simpler than that usually required for the description of diffusion processes in alloy systems [3.7–9].

We have to consider two different diffusion coefficients of the hydrogen, the chemical diffusion coefficient D and the self (or tracer) diffusion coefficient D_s. The chemical diffusion coefficient describes the hydrogen flux

$$j(r, t) = -D \cdot \nabla n(r, t) \tag{3.1}$$

in a concentration gradient according to Fick's law, where $n(r, t)$ is the hydrogen density (hydrogen interstitials per unit volume) as a function of the space vector r and the time t. In view of our later discussion, we also consider the diffusion equation

$$\mathring{n}(r, t) = D \cdot \nabla^2 n(r, t) \ , \tag{3.2}$$

which follows from (3.1) and the continuity equation $\mathring{n}(r, t) = -\nabla \cdot j(r, t)$. Equation (3.2) determines, for appropriate boundary conditions, the variation of the hydrogen density $n(r, t)$ as a function of both t and r.

The chemical diffusion coefficient D may be defined by either of the two equations above. According to these definitions, it can experimentally be determined by setting up a nonequilibrium hydrogen distribution and by observing the resulting hydrogen flux or the diffusive decay of this initial distribution. Appropriate experimental techniques for instance, are, electrochemical [3.10] or gas volumetric [3.11] measurements of hydrogen absorption (desorption) or permeation, resistance relaxation [3.12, 13] and the Gorsky effect [3.14]. Coherent neutron scattering, probing time-dependent hydrogen density fluctuations in a metal under equilibrium conditions, does not, in general, directly yield the chemical diffusion coefficient [3.15]. Rather, the results obtained by this technique require a careful correction for the effects of coherency stresses, as discussed for instance in [3.15].

The self diffusion coefficient D_s describes the diffusive behavior of a given, or tagged, hydrogen interstitial, characterized by the probability $p(r, t)\mathrm{d}V$ for finding this interstitial at time t in the volume element $\mathrm{d}V$ at r. The dependence of $p(r, t)$ on both r and t is described by the diffusion equation

$$\mathring{p}(r, t) = D_s \cdot \nabla^2 p(r, t) \ , \tag{3.3}$$

which is identical to (3.2) except for the fact that D is replaced by D_s. A suitable measure for the self-diffusion behavior is the expectation value $\langle [r(t) - r(0)]^2 \rangle$ of the square of the distance the considered hydrogen interstitial diffuses away within time t from its original position $r(0)$ at time $t = 0$. The solution of (3.3) with the appropriate initial condition $p(r, t = 0) = \delta(r - r(0))$ yields

$$\left\langle [r(t) - r(0)]^2 \right\rangle = \int [r(t) - r(0)]^2 p(r, t) \, dV = 6 D_s t \ . \tag{3.4}$$

Accordingly, the self diffusion coefficient may be defined by (3.3) or by the equation

$$D_s = \lim_{t \to \infty} \frac{1}{6t} \left\langle [r(t) - r(0)]^2 \right\rangle = \lim_{t \to \infty} \frac{1}{6t} \frac{1}{nV} \sum_i [r_i(t) - r_i(0)]^2 \ , \tag{3.5}$$

where the summation in the last term extends over all hydrogen interstitials i found at $t = 0$ in a partial volume V. The diffusion equation itself does not require the limit $t \to \infty$ in (3.5). However, this limit is conventionally in cluded in the definition of D_s since it eliminates, for instance, the effects of the initial (thermal) velocity of the hydrogen atoms which, for small t, would yield deviations from a purely diffusive behavior as described by (3.3, 4). In the present situation, the limit $t \to \infty$ is usually well fulfilled if the considered time t exceeds the mean residence time of the hydrogen on its interstitial sites. An exception is the anomalous diffusion in metallic glasses where the required times t can considerably exceed the (average) mean residence time (Sect. 3.2.5).

Experimental techniques directly measuring the self diffusion coefficient D_s are incoherent neutron scattering and the field gradient method in nuclear magnetic resonance. These two techniques are discussed in Chaps. 4, 5. It seems worth mentioning that measurements of $T_1, T_{1\rho}$ and T_2 in nuclear magnetic resonance do not directly probe D_s since they require the assumption of specific diffusion models in order to yield this quantity.

The above equations are strictly valid only for a spatial continuum. However, they also describe the situation in a real metal if we consider $n(r, t)$ or $p(r, t)$ as a local average carried out over lattice parameter distances. Further, the equations assume an isotropic diffusion behavior as found, for instance, in metals with cubic symmetry. For a lower crystal symmetry, the diffusive behavior is characterized by a diffusion coefficient tensor rather than by a scalar diffusion coefficient [3.16]. Experimental data for the special case of a hexagonal crystal will be discussed in Sect. 3.2.3.

It is important to point out that the two diffusion coefficients D and D_s in general have different values. The comparison between (3.2, 3) shows that both quantities are identical only if the hydrogen density $n(r, t)$ varies in the same way with time as does the probability density $p(r, t)$ of a single tagged hydrogen interstitial. This requires the absence of any correlation in the motion of different hydrogen interstitials. An example for such a situation is the limit of low hydrogen concentrations where the interaction, and thus correlation, between different hydrogen atoms can be neglected. The diffusion coefficients D and D_s are therefore identical at sufficiently low hydrogen concentrations.

For arbitrary concentrations, the relationship between the diffusion coefficients D and D_s is given by the ratio [3.17–21]

$$\frac{D}{D_s} = \frac{n\,\partial\mu/\partial n}{k_B T}\,\frac{1}{f_{cor}}\ ,$$ (3.6)

where μ is the chemical potential of the hydrogen, f_{cor} is the correlation factor (or Haven's ratio), and $k_B T$ is the thermal energy. The quantity $(n\,\partial\mu/\partial n)/(k_B T)$, which is called the thermodynamic factor f_{therm}, can experimentally be determined from both hydrogen solubility and Gorsky effect measurements [3.14, 22]. The correlation factor f_{cor} describes the correlation in time between the diffusive jump paths of a given hydrogen, due to interaction effects with neighboring hydrogen interstitials. Its value is 1 in the absence of interaction effects, i.e., for low hydrogen concentrations. Theoretical [3.20, 23–25] calculations carried out for simple lattice gas models yielded, with rising concentration, a continuous decrease of f_{cor} down to values typically between 0.5 and 0.8. These theoretical predictions are also supported by experimental studies [3.15, 18, 19]. Equation (3.6) finally shows that D and D_s are identical in the low-concentration limit where both f_{cor} and $f_{therm} = (n\,\partial\mu/\partial n)/(k_B T)$ assume the value 1.

3.1.2 Correlation Factor and Velocity Correlation Functions

A rigorous derivation of the relation (3.6) between the chemical and the self diffusion coefficient is obtained with the help of velocity correlation functions. It is an immediate consequence of (3.5) that the self diffusion coefficient can also be written as [3.26–28]

$$D_s = \frac{1}{3}\int_0^\infty dt\langle v(0)v(t)\rangle = \frac{1}{3nV}\int_0^\infty dt \sum_i v_i(0)v_i(t)\ ,$$ (3.7)

where $\langle v(0)v(t)\rangle$ is the velocity correlation function of a given hydrogen interstitial, and where the final term expresses this correlation as a summation over all hydrogen interstitials being at $t = 0$ in a partial volume V. The chemical diffusion coefficient can be described by the Kubo-Green formula [3.26–28], yielding the analogous expression

$$D = \frac{n\,\partial\mu/\partial n}{k_B T}\,\frac{1}{3nV}\int_0^\infty dt \sum_{i,j} v_i(0)v_j(t)\ ,$$ (3.8)

which involves also velocity correlations between different hydrogen interstitials i and j. From (3.7, 8) the ratio D/D_s is found to be

$$\frac{D}{D_s} = \frac{n\partial\mu/\partial n}{k_B T} \int\limits_0^\infty dt \sum_{i,j} v_i(0)v_j(t) \Bigg/ \int\limits_0^\infty dt \sum_i v_i(0)v_i(t) \ . \tag{3.9}$$

The comparison with (3.6) demonstrates that the correlation factor can be written as

$$f_{cor} = 1 - \int\limits_0^\infty dt \sum_{i,j(\neq i)} v_i(0)v_j(t) \Bigg/ \int\limits_0^\infty dt \sum_{i,j} v_i(0)v_j(t) \ . \tag{3.10}$$

Equation (3.10) is a rigorously valid result for the correlation factor. It shows that this factor is 1 if correlations between the velocities of different hydrogen interstitials do not exist, such as in the low-concentration limit. For a jump diffusion process, (3.10) can be used [3.29] to obtain the usual expression [3.30]

$$f_{cor} = \frac{1 + \langle\cos\theta\rangle}{1 - \langle\cos\theta\rangle} \tag{3.11}$$

for the correlation factor where $\langle\cos\theta\rangle$ is the expectation value for the angle between the directions of two consecutive jumps. In contrast to (3.10), the validity of (3.11) requires some simplifying assumptions on the types and the symmetries of the diffusive jumps, and the existence of a direct correlation only between two consecutive jumps. It seems also worth mentioning that (3.8) yields the equation [3.26–28]

$$D = \lim_{t\to\infty} \frac{1}{6t} \frac{n\partial\mu/\partial n}{k_B T} \frac{1}{nV} \left[\sum_i r_i(t) - r_i(0)\right]^2 \tag{3.12}$$

for the chemical diffusion coefficient, which is the analog of (3.5) for the self diffusion coefficient.

The chemical diffusion coefficient D in (3.8) or (3.12) is proportional to the thermodynamic factor $f_{therm} = (n\,\partial\mu/\partial n)/(k_B T)$ which can strongly vary with both hydrogen concentration and temperature in the neighborhood of phase transformations (e.g., above a miscibility gap). This fact led some authors to define a reduced diffusion coefficient [3.2, 31]

$$D_{red} = \frac{k_B T}{n\partial\mu/\partial n} D \ , \tag{3.13}$$

which does not exhibit the concentration and temperature variations resulting from f_{therm}.[1] Accordingly, the reduced diffusion coefficient is expected to demonstrate more clearly the atomistic jump behavior of the

---

[1]In [3.2, 6, 31], the chemical diffusion coefficient D and the reduced diffusion coefficient D_{red} were called D^* and D, respectively. Note also that the influence of the correlation factor was not considered in these references so that the reduced diffusion coefficient and the self (or tracer) diffusion coefficient were assumed to be identical.

individual hydrogen atoms. The comparison with (3.6) or (3.9) shows that it differs from the self diffusion coefficient by the correlation factor f_{cor}.

3.1.3 External Forces

The presence of an electric current (electrotransport), a temperature gradient (thermotransport) or a stress gradient causes a force F on the hydrogen interstitials in a metal [3.8, 32–34]. The force in electrotransport is described by an effective charge number Z^*, according to the equation

$$F = -eZ^*\nabla\Phi \ , \tag{3.14}$$

in which e is the (positive) elementary charge and $-\nabla\Phi$ is the applied electric field that causes the electric current. The force in the presence of a temperature gradient is given by

$$F = -Q^*\frac{\nabla T}{T} \ , \tag{3.15}$$

where Q^* is the heat of transport. For a more detailed discussion of the forces in both electrotransport and thermotransport, the reader is referred to [3.8, 32–34] and references therein. In the case of stress gradients, the force F can be written as (Einstein's summation convention)

$$F = P\nabla(s_{iikl}\sigma_{kl}) \ , \tag{3.16}$$

where P is the mean value of the trace components of the double force tensor of the hydrogen interstitials; a typical value for P is ~ 3 eV [3.32, 33, 35]. The quantities s_{iikl} and σ_{kl} are tensor components of the elastic coefficients and of the applied stresses, respectively.

The presence of a force F induces a hydrogen flux in addition to that caused by concentration gradients. The total flux is given by

$$j(r, t) = -D\nabla n(r, t) + nMF \ , \tag{3.17}$$

where M is the mobility of the hydrogen interstitials. This quantity is related to the chemical diffusion coefficient D according to [3.2, 6, 8, 26, 27, 32, 33]

$$D = n\frac{\partial\mu}{\partial n}M = k_B T f_{therm} M \ . \tag{3.18}$$

Equation (3.18) can be considered to represent a generalized Einstein relation. In the limit of small hydrogen concentrations, it can be written $D = k_B T M$ since the thermodynamic factor f_{therm} assumes the value 1.

All three of the above forces are of technological importance, see [3.3, 34] and Chap. 6. It was discussed in Sect. 3.1.1 that they can be used for experimental studies of hydrogen diffusion, and for a determination of the diffusion coefficient, by creating a hydrogen flux and, thus, a nonequilibrium hydrogen distribution within a sample [3.12–14]. For instance, the force in the presence of stress gradients, as given in (3.16), is the basis of the Gorsky effect [3.14].

3.1.4 Jump Diffusion

The diffusion of hydrogen interstitials is a leading example for a jump diffusion process [3.36–39]. It is characterized by a mean residence time τ of the hydrogen on a given interstitial site or, correspondingly, by a jump rate $1/\tau$ of hydrogen atoms located on this site. The self diffusion coefficient D_s can be calculated from the mean residence time under certain assumptions about jump paths and correlation effects between subsequent jumps.

Figure 3.1 shows octahedral interstitial sites in a fcc lattice and tetrahedral sites in a bcc lattice. These sites are usually occupied by the hydrogen in the respective lattices [3.22, 40]. The simplest, and most common, assumptions for a jump diffusion process are that correlations in the directions and probabilities of subsequent jumps do not exist, and that the hydrogen jumps only to nearest-neighbor interstitials sites. Under these assumptions, the occupation of the sites in Fig. 3.1 yields the relation

$$D_s = \frac{d^2}{6\tau} \, , \qquad (3.19)$$

where d is the distance between two nearest-neighbor interstitial sites.

For more complex lattices and jump geometries, relations that correspond to that in (3.19) can be derived from a Fourier transformation of the self correlation function [3.36, 38]. This method applies also to noncubic lattices. A second method consists in a calculation of the square of the diffusive paths, as performed for tetrahedral interstitial sites in a hexagonal crystal [3.39]. It is finally worth mentioning that the simple equation (3.19) applies also for interstitial sites forming any of the cubic Bravais lattices, and also for octahedral interstitial sites in a bcc lattice.

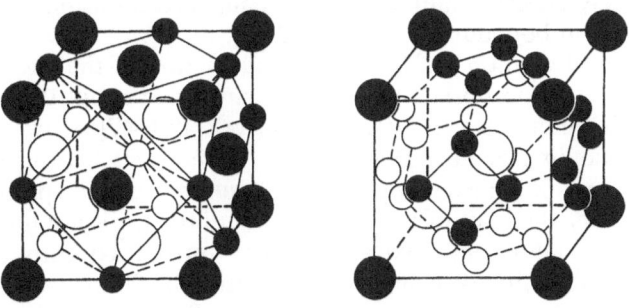

Fig. 3.1. Octahedral interstitial sites in a fcc lattice (*left*) and tetrahedral interstitial sites in a bcc lattice (*right*). The host lattice atoms and the interstitial sites are indicated by large and small circles, respectively. For both lattice structures, the respective interstitial sites have the largest possible separation from the host lattice atoms. The distance between nearest neighbor interstitials sites is $d = a/\sqrt{2}$ (*left*) or $d = a/\sqrt{8}$ (*right*), where a is the cubic lattice parameter

3.1.5 The Influence of Traps

The diffusion behavior of hydrogen interstitials in metals is severely influenced by traps [3.41–47]. The traps are formed by lattice defects such as substitutional or interstitial impurity atoms, vacancies, dislocations and grain boundaries. The traps reduce, in general, the diffusivity. Their influence is, at least qualitatively, exemplified by the widely used model of *Oriani* [3.41] according to which the total hydrogen density $n(r, t)$ can be divided into the density $n_t(r, t)$ of the trapped and the density $n_f(r, t)$ of the free (untrapped) hydrogen. The model assumes that the trapped hydrogen is completely immobile, whereas the free hydrogen can diffuse unimpededly through the metal lattice. Therefore, only the free hydrogen contributes to a diffusive hydrogen flux , j according to the equation

$$j(r, t) = -D\nabla n_f(r, t) = -D(\partial n_f/\partial n)\nabla n(r, t) \ , \qquad (3.20)$$

where D is the chemical diffusion coefficient in the absence of the traps. The second part of (3.20) presupposes a local equilibrium between the trapped and free hydrogen, so that $n_f(r, t)$ is a unique function of the total hydrogen density $n(r, t)$. We can define an effective diffusion coefficient

$$D_{\text{eff}} = D \, \partial n_f/\partial n \qquad (3.21)$$

in the presence of the traps if we write Fick's law in its usual form according to which the hydrogen flux is proportional to the gradient of the total hydrogen density $n(r, t)$. This definition also leads, together with the continuity equation $\dot{n}(r, t) = -\nabla \cdot j(r, t)$, to a diffusion equation for $n(r, t)$ identical to that of (3.2) except for the fact that D is replaced by D_{eff}.

It can be anticipated that the above model provides a realistic description for small trap concentrations. In this case, only a very minor fraction of the entire metal volume is expected to be directly influenced by the traps so that the hydrogen, which is not trapped, can essentially freely diffuse through the metal lattice. However, it is pointed out that the model does not account for any effects resulting from an enhanced diffusion along dislocation pipes or grain boundaries, see, e.g., [3.48].

Equation (3.21) allows a calculation of the effective diffusion coefficient D_{eff} if appropriate assumptions are made about the (temperature dependent) relation between $n_f(r, t)$ and $n(r, t)$ [3.41–47]. However, a qualitative understanding of how D_{eff} depends on the total hydrogen density $n(r, t)$ follows from (3.21) even without any assumption of a specific model. We first consider the limit of small $n(r, t)$ where the traps are not saturated. In this case, any additional hydrogen atoms introduced into the metal have a high probability for being trapped so that the factor $\partial n_f/\partial n$ in (3.21) is small. This means a low effective diffusion coefficient D_{eff}. The situation is different for large $n(r, t)$ where most of the traps are saturated. In this situation, an additional hydrogen atom is unlikely to be trapped. Therefore, the factor $\partial n_f/\partial n$ is close to 1 so that D_{eff} will not differ very much from the chemical diffusion coefficient in the absence of traps. This qualitative argumentation shows that the influence of traps is most severe for low hydrogen concentrations. It is also evident that this influence is larger the lower the temperature, since a hydrogen atom has a higher probability of becoming trapped.

The number of experimental studies investigating the influence of traps on hydrogen diffusion is extremely large. The interested reader is referred to [3.41–47] and references therein. However, it is important to point out that defects, which act as traps, will be found in any metal, even if their concentration may be very small in some fortunate cases. This means that the measured diffusion coefficient will generally be found to be reduced by trapping effects if the hydrogen concentration becomes sufficiently small. Therefore, in order to avoid this influence of the traps on the measured diffusion coefficient, it is necessary to carry out experiments with hydrogen concentrations that considerably exceed the presumable trap concentration.

3.2 The Diffusion Coefficient of Hydrogen in Pure Metals and Some Representative Alloys

This section compiles experimental results for the diffusion coefficient of hydrogen interstitials in pure metals and alloys. With a few exceptions that are especially mentioned, the results were obtained for low hydrogen concentrations. This means in particular that interaction effects between the hydrogen interstitials were small so that the chemical and the self diffusion coefficient are expected to be identical.

3.2.1 Diffusion in fcc Metals

It is generally assumed that hydrogen in fcc metals occupies octahedral interstitial sites (Fig. 3.1). This assumption is based on that an octahedral occupancy was found in those cases for which the interstitial position of the hydrogen was experimentally established (e.g., for Pd [3.22]). The distance between two nearest neighbor octahedral sites ranges between 2.5 and 2.9 Å for the metals discussed in the present section.

(A) Copper, Silver, Gold, and Platinum

Figure 3.2 presents Arrhenius plots of results for the diffusion coefficient of hydrogen in Cu, Ag, Au, and Pt. The plots also show the reference number of the data, omitting the number 3 for the chapter. A common characteristic of the metals in Fig. 3.2 is that their hydrogen solubility, and, therefore, the hydrogen concentrations of the studies in Fig. 3.2 are very small [3.50–54, 56, 57, 60–64, 66, 67]. This implies that the accuracy, and thus the congruity, of the data in general is poor since the measurements required the handling, or even the quantitative determination, of very small amounts of hydrogen. A

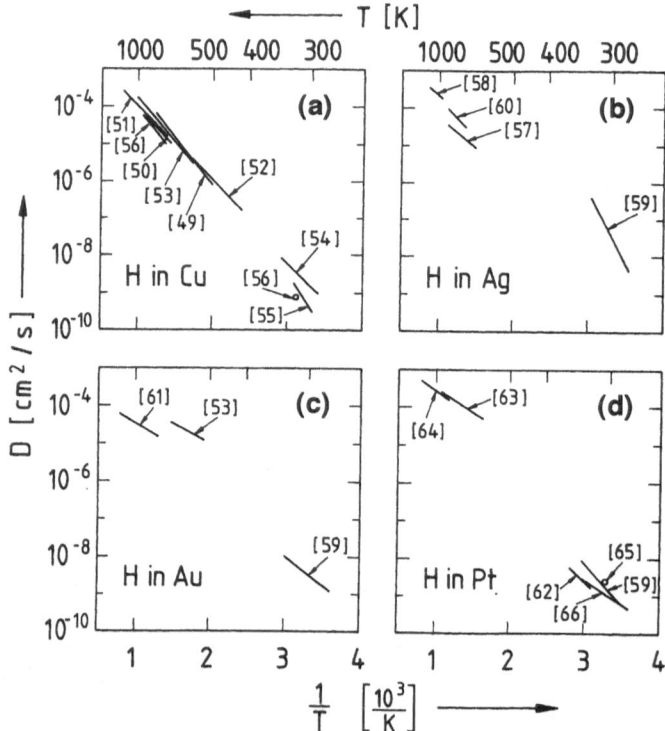

Fig. 3.2. Arrhenius plots of results for the diffusion coefficient of hydrogen in Cu (**a**), Ag (**b**), Au (**c**) and Pt (**d**). The diffusion coefficients are logarithmically plotted versus the reciprocal temperature. The figure also shows the references from which the data were taken

second problem of the small hydrogen concentrations is that traps tend to have a significant influence on the measured diffusivity, particularly at lower temperatures.

It seems that the quality of the results in Fig. 3.2 is not sufficient to firmly establish any deviations from an Arrhenius temperature behavior as, possibly, suggested by some of the data, see also [3.59]. The diffusivity of deuterium and tritium was investigated in Cu [3.50, 51], and that of deuterium in Pt [3.63]. In the investigated temperature range (700–1250 K), the data for Cu show a decrease of both the diffusivity and the activation energy with rising isotopic mass. In Pt, hydrogen is reported to diffuse at 750 K about 16% faster than deuterium [3.63].

(B) Aluminum

Figure 3.3 shows a compilation of literature data for the diffusion coefficient of hydrogen in Al. The scatter in the data reported from different authors is large. Similarly as in the case of Cu, Ag, Au and Pt, this is likely to result

Fig. 3.3. Arrhenius plots of results for the diffusion coefficient of hydrogen in Al. The figure also shows the references for the data

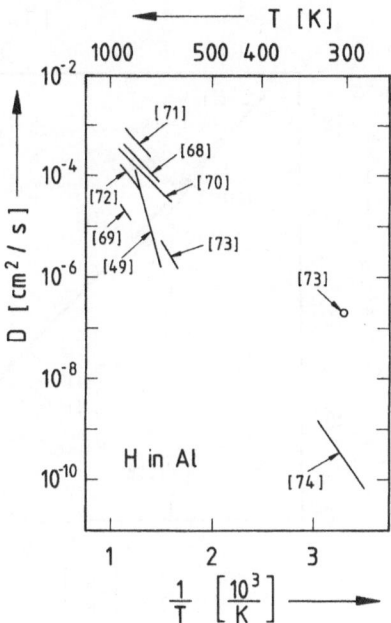

from the extremely small hydrogen solubility of Al [3.49, 67, 68]. Specifically in the case of Al, it also cannot be excluded that surface oxide layers may have influenced the hydrogen permeation during the measurements [3.75].

(C) Nickel

Figure 3.4 presents results for the diffusion coefficient of hydrogen in Ni. The data published before 1978 were reviewed by *Völkl* and *Alefeld* [3.2], and the thick solid lines in Fig. 3.4 represent a best representation of these data as given by these authors. Later data are shown in Fig. 3.4 together with their respective reference where, however, the results of three studies [3.78, 83, 85] are omitted since they are practically identical to the best representation of *Völkl* and *Alefeld* [3.2].

The results from different laboratories for the diffusion coefficient of hydrogen in Ni demonstrate a surprisingly good agreement. The good agreement may reflect the fact that Ni has a somewhat higher hydrogen solubility than the metals discussed before [3.50, 67, 78, 81, 83, 87]. It can also be assumed that surface barrier effects are of lesser importance in the case of Ni, possibly because of a catalytic surface reaction. The data analysis of *Völkl* and *Alefeld* [3.2] yielded a best representation which exhibited a small steplike increase above the Curie point at 631 K. However, the size of this increase is at the border of the experimental accuracy so that the actual existence of such an increase may be questioned. The isotope effect in the

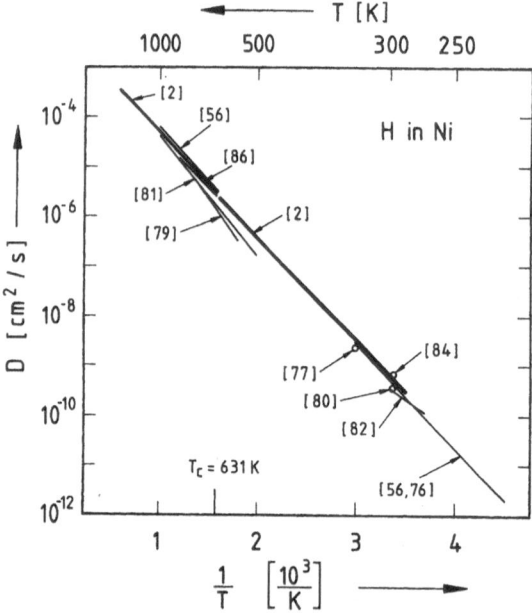

Fig. 3.4. Arrhenius plots of results for the diffusion coefficient of hydrogen in Ni. The thick solid lines are a best representation of the literature data before 1978, according to the analysis of *Völkl* and *Alefeld* [3.2]. Data published later are presented together with their reference. The figure does not show results of three further studies [3.78, 83, 85] which are practically identical with the best representation of *Völkl* and *Alefeld* [3.2]. The abscissa indicates also the Curie temperature $T_C = 631$ K of Ni

diffusion coefficient of hydrogen in Ni was repeatedly investigated [3.50, 51, 76, 85, 87]. The measurements carried out at higher temperatures (500–1270 K) [3.50, 51, 85, 87] yielded – in agreement with the isotope effect in the other fcc metals above – a decrease (i) for the absolute value of the diffusion coefficient and (ii) the activation energy, with rising isotopic mass. In contrast to this, the data reported in [3.76] for low temperatures (220–340 K) demonstrate a reversed isotopic behavior for the activation energies of hydrogen and deuterium.

(D) Palladium

The diffusion of hydrogen in Pd has been investigated in numerous studies, and the data reported by different groups generally agree excellently with one another [3.2, 22, 88–91]. The agreement can be explained by an appreciable hydrogen solubility which tends to reduce the influence of trapping effects in experimental studies. Further, the surface of the noble metal Pd is particularly permeable for hydrogen so that surface barrier effects are very small.

Figure 3.5 presents data reported for the diffusion coefficients of hydrogen, deuterium and tritium in Pd. *Völkl* and *Alefeld* [3.2] reviewed the results published before 1978 and showed that the diffusion coefficient of the isotope hydrogen is, between 230 and 900 K, well described by an Arrhenius law $D = D_0 \exp(-E/k_B T)$ with an activation energy $E = 0.23$ eV and a pre-exponential factor $D_0 = 2.9 \times 10^{-3}$ cm^2/s. The diffusion coefficient resulting from these values is indicated in Fig. 3.5. A later study [3.89] demonstrated

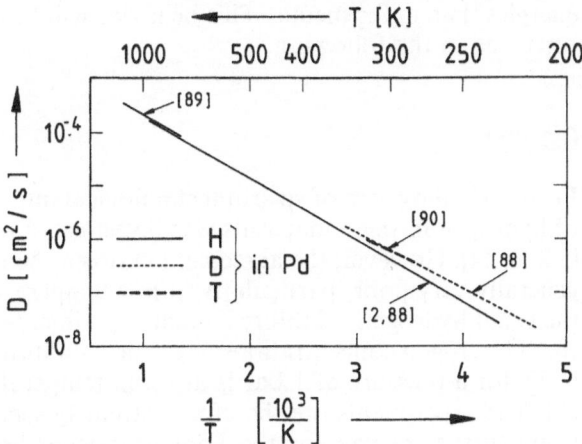

Fig. 3.5. Arrhenius plots of results for the diffusion coefficient of hydrogen, deuterium and tritium in Pd. The figure also shows the references for the data

that the Arrhenius behavior extends up to a temperature of 1220 K. The deuterium data in Fig. 3.5 are the results of Gorsky effect measurements [3.88], yielding a lower activation energy $E = 0.206$ eV and a lower pre-exponential factor $D_0 = 1.7 \times 10^{-3}$ cm^2/s. The lower activation energy of deuterium agrees with the isotopic behavior in the case of the fcc metals discussed above (with exception of one out of five studies on Ni [3.76]). Figure 3.5 finally shows results for the diffusion coefficient of tritium [3.90], described by $E = 0.185$ eV and $D_0 = 7.3 \times 10^{-4}$cm^2/s. Within the investigated temperature range, the tritium data are nearly identical to those for deuterium.

3.2.2 Diffusion in bcc Metals

Experimental studies on bcc metals demonstrated that the hydrogen was located on tetrahedral interstitial sites (Fig. 3.1) [3.22, 40], at least in those cases in which the hydrogen concentration did not exceed the range of solid solution (α phase). It is therefore suggestive to assume that the hydrogen occupies quite generally this type of sites in metals with a bcc structure. For the bcc metals discussed in this section, the distance between two nearest-neighbor tetrahedral sites is in the range between 1.01 and 1.17 Å. These distances are smaller by a factor of two than the distances between nearest-neighbor octahedral sites in fcc metals. One important consequence of the small distances is a large matrix element for hydrogen tunneling between two neighboring interstitial sites so that hydrogen diffusion in bcc metals is strongly influenced by tunneling effects (Chap. 2 and [3.1, 5, 6]). A further consequence is (i) that hydrogen diffusion in bcc metals is characterized by higher absolute values of the diffusion coefficient and (ii) by lower activation energies than in fcc metals. This behavior will be confirmed by the data to be discussed in the following.

(A) Iron

Fe and Fe alloys are of special technological importance so that the diffusion of hydrogen in these materials was investigated in a large number of studies [3.2, 6, 44]. However, the agreement between the results of different studies is generally very poor, particularly at low temperatures. One reason for this is the small hydrogen solubility so that experiments have to be performed with small hydrogen concentrations (e.g., the equilibrium hydrogen concentration in Fe for a pressure of 1 bar is at room temperature below 1 at-ppm) [3.44, 67]. The experiments are therefore extremely sensitive to trapping effects. A likely further reason for the poor agreement between the published data, particularly those at lower temperatures, is surface barrier effects.

The literature data reported for hydrogen diffusion in (pure) Fe were reviewed in 1978 by *Völkl* and *Alefeld* [3.2] and in 1983 by *Kiuchi* and *McLellan* [3.44]. The three thick solid lines in Fig. 3.6 show best representations as suggested by these authors for the diffusion coefficient of the hydrogen. In fact, *Völkl* and *Alefeld* [3.2] discuss two representations differing by almost an order of magnitude at room temperature. One of these representations (D_1 in Fig. 3.6) was selected according to the highest values reported for the diffusion coefficient, with the reasoning that trapping effects

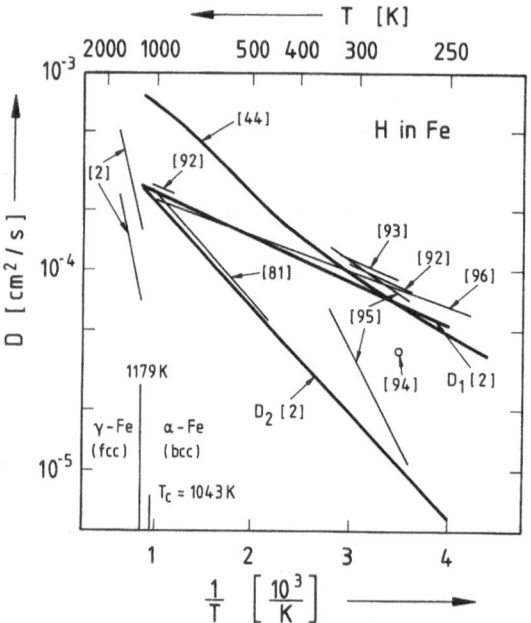

Fig. 3.6. Arrhenius plots of results for the diffusion coefficient of hydrogen in Fe. The thick solid lines represent two different best representations (D_1 and D_2) of literature data as suggested by *Völkl* and *Alefeld* [3.2], and a best representation given by *Kiuchi* and *McLellan* [3.44]. The other data are shown together with their references. The abscissa indicates the Curie temperature $T_C = 1043$ K and the structural transition between α-Fe and γ-Fe at 1179 K

reduce the measured diffusivity so that the highest measured diffusivities are most likely to describe diffusion in the absence of traps. The second representation (D_2) agreed with the large majority of the diffusion data below ~ 600 K. The best representation according to *Kiuchi* and *McLellan* [3.44] exceeds, at higher temperatures, considerably both of the representations suggested by *Völkl* and *Alefeld* [3.2]. However, it is essentially identical with the D_1 representation of these authors around room temperature. Figure 3.6 shows further literature data for the diffusion coefficient of hydrogen in Fe which were published after the review of *Kiuchi* and *McLellan* [3.44] was written. It can be seen that the majority of these most recent data support the D_1 representation of *Völkl* and *Alefeld* [3.2]. This holds in particular for the data of *Hayashi* et al. [3.96] which cover the wide temperature range between 240 and 970 K. It is therefore suggestive to assume that the result of Hayashi et al. [3.96) for the diffusion coefficient, an Arrhenius relation with an activation energy $E = 0.035$ eV and a pre-exponential factor $D_0 = 3.35 \times 10^{-4}$ cm^2/s, provides a reasonably good description of hydrogen diffusion in pure Fe.

It is obvious that the accuracy of the presently available data does not allow to establish any influence of the Curie point at 1043 K on the measured hydrogen diffusivities. However, the structural transformation of bcc α-Fe into fcc γ-Fe at 1179 K causes a well-established decrease of the diffusion coefficient. Figure 3.6 shows the upper and the lower limit of data reported in literature according to [3.2] for the diffusion coefficient of hydrogen in γ-Fe (i.e., above 1179 K). The data demonstrate clearly that the activation energy for hydrogen diffusion is smaller in the bcc phase than in the fcc phase, as discussed in the introductory part of this section.

The isotope effect of hydrogen diffusion in Fe was repeatedly investigated. Figure 3.7 presents a compilation of *Hayashi* et al. [3.96] which shows experimental ratios between the diffusion coefficients for hydrogen and deuterium in a semilogarithmic plot versus reciprocal temperature. The compilation demonstrates that the isotope hydrogen diffuses faster than deuterium, and that the activation energy for hydrogen is smaller than that for deuterium. Both of these findings are characteristic for hydrogen diffusion in bcc metals. It is specifically worth pointing out that the isotopic behavior of the activation energy is reversed as compared to that found in Sect. 3.2.1 for fcc metals in that the activation energies there were, with one questionable exception, found to be larger for hydrogen than for deuterium. A recent study [3.94] finally reports a ratio of ~ 4 between the diffusion coefficients of hydrogen and tritium, thus indicating that tritium diffuses slower than both hydrogen and deuterium.

(B) Vanadium, Niobium and Tantalum

Figure 3.8 presents a compilation of Gorsky effect results for the diffusion coefficient of hydrogen, deuterium and tritium in V, Nb and Ta (please not the different ordinate scales) [3.102]. Because of their high hydrogen solu-

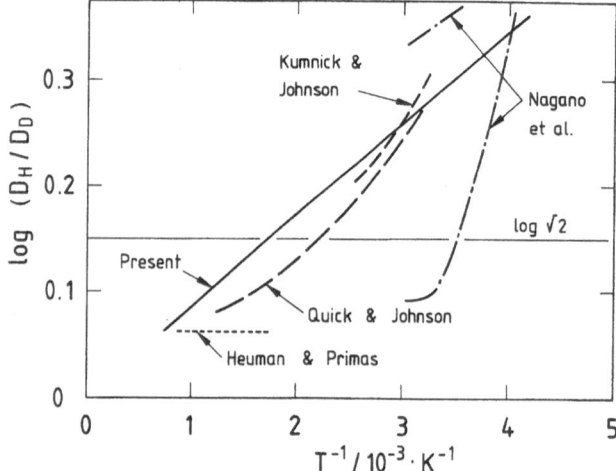

Fig. 3.7. Isotope dependence of hydrogen diffusion in Fe, as compiled by *Hayashi* et al. [3.96]. The figure shows experimental ratios between the diffusion coefficients of hydrogen and deuterium in a semilogarithmic plot versus reciprocal temperature. The individual experiments were carried out by *Hayashi* et al. [3.96] (present), *Heumann* and *Primas* [3.97], *Kumnick* and *Johnson* [3.98], *Quick* and *Johnson* [3.99], and *Nagano* et al. [3.92, 100, 101]. The figure is taken from [3.96]

bility [3.40, 67], hydrogen diffusion in these metals can be studied with higher concentrations than in most of the metals discussed before so that trapping effects are of lesser importance. The possibility to investigate higher hydrogen concentrations further allows the application of the Gorsky effect technique [3.14] which is not impeded by any surface barriers. The results in Fig. 3.8 describe therefore the diffusivity of the three hydrogen isotopes with high accuracy.

The diffusion coefficients in Fig. 3.8 are very large. In fact, V exhibits – together with Fe – the highest hydrogen diffusivity of all metals reported so far. This may be attributed to the particularly small lattice parameters of V and Fe, which lead to a small distance between the tetrahedral sites and, therefore, to a large matrix element for hydrogen tunneling. Figure 3.8 shows large isotopic differences in the diffusion coefficients, exceeding an order of magnitude at low temperatures, and it shows a decrease of the (apparent) activation energies for hydrogen diffusion in Nb and Ta at low temperatures (in the case of V, the data may also indicate a similar small decrease of the activation energy below ~ 300 K). All these effects reflect the quantum nature of hydrogen diffusion in these metals. It should be pointed out that numerical theoretical calculations, based on the polaron concept, were able to reproduce surprisingly well the experimental data for the diffusion coefficients in the case of Nb and Ta ([3.103, 104], see also Chap. 2).

At high temperatures (900–1370 K), the diffusion coefficient of hydrogen in V, Nb and Ta was determined in absorption measurements [3.105]. The results of these measurements are only slightly higher than an extrapolation of

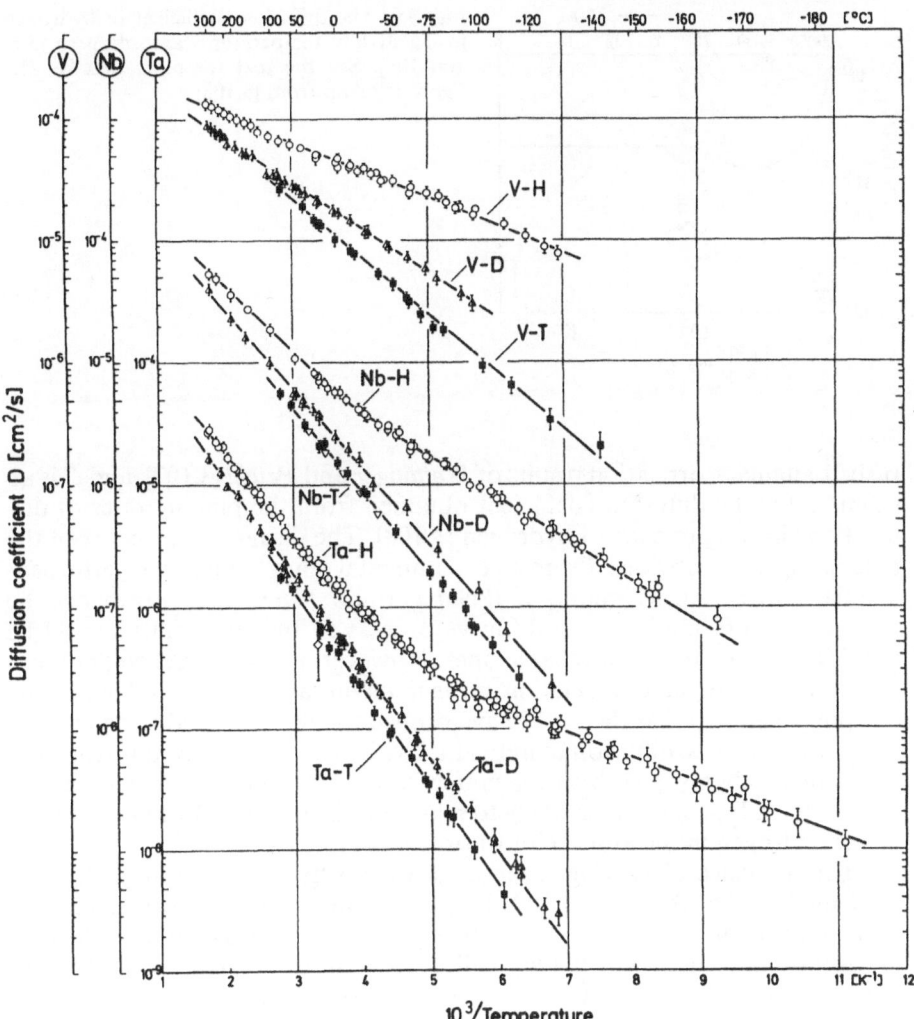

Fig. 3.8. Arrhenius plots of the diffusion coefficients of hydrogen, deuterium and tritium in V, Nb, and Ta, according to *Qi* et al. [3.102]. Note the different ordinate scales for the three metals. The figure is taken from [3.102]

the Gorsky effect data in Fig. 3.8 yields. For temperatures below those covered by the data in Fig. 3.8, the present experimental information on the diffusivity of the hydrogen is conflicting. This holds in particular for Ta where the most detailed information is available. Figure 3.9 shows a compilation of low-temperature values for the diffusion coefficient of hydrogen in Ta, as suggested by different groups [3.6, 106–108]. Curve *A* was derived by *Fukai* [3.5, 6, 106] from an analysis of quenching experiments (the data above 90 K were taken from the Gorsky effect measurements of Fig. 3.8). Curve *B* represents per-

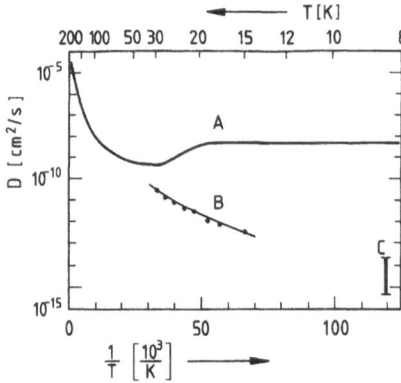

Fig. 3.9. The diffusion coefficient of hydrogen in Ta at low temperatures according to [3.6, 106–108]. See the text for more details. The figure is taken from [3.108]

turbed-angular-correlation results of *Weidinger* and *Peichl* [3.107], and *C* is an estimate for the diffusion coefficient obtained from the pinning rates of dislocations in the presence of hydrogen [3.108]. The disagreement between the data in Fig. 3.9 is obvious. From an experimental point of view, the perturbed-angular-correlation technique of *Weidinger* and *Peichl* [3.107] provides the most direct information on a diffusive process so that its results should be considered to be the most reliable ones. However, it was suggested [3.6] that the perturbed-angular-correlation measurements above were influenced by trapping effects so that they may have probed the decay of hydrogen clusters rather than the free diffusion of individual hydrogen atoms. For this reason, it seems presently not possible to definitely answer the question which ones of the results in Fig. 3.9 is most likely to represent the actual diffusion coefficient of the hydrogen in the absence of traps.

The influence of an elastic tensile stress on hydrogen diffusion was investigated by *Suzuki* et al. [3.109, 110] in resistance measurements. The authors report an increase of the diffusivities in V and Ta by an order of magnitude, an effect which they called superdiffusion. However, this unexpected effect could not be confirmed in a series of experiments carried out by other groups [3.13, 111–113]. It is therefore extremely unlikely that superdiffusion really exists, and this conclusion holds in particular since the data evaluation procedure of Suzuki et al. [3.109, 110] was shown not to be entirely correct [3.13].

(C) Molybdenum and Tungsten

Figure 3.10 presents Arrhenius plots of results for the diffusion coefficient of hydrogen in the two bcc metals Mo and W which are of technological importance because of their high melting points.

Fig. 3.10. Arrhenius plots of results for the diffusion coefficient of hydrogen in Mo (**a**) and W (**b**). The figure also shows the references for the data

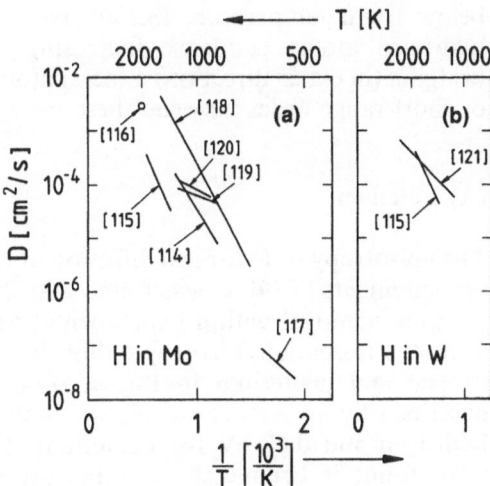

3.2.3 Diffusion in Hexagonal Metals

For hydrogen concentrations in the range of solid solution (i.e. in the respective α phases), hydrogen was found to occupy nearly exclusively tetrahedral interstitial sites in hexagonal metals (for instance in Ti [3.122, 123], Zr [3.123], Sc [3.124, 125], Y [3.126, 127], Lu [3.128, 129]). Figure 3.11 shows the primitive unit cell of a hcp metal together with tetrahedral (full circles) and octahedral (open squares) interstitial sites. A consequence of the hexagonal symmetry is that hydrogen diffusion must be described by a diffusion coefficient tensor consisting of two independent components D_c and D_a that stand for diffusion in c-axis direction or in the ab basal plane, respectively [3.16, 39].

An important characteristic of hexagonal rare earth metals such as Sc, Y and Lu is that the concentration range of the solid solution is (i) very high (~ 20 at-% for Y and Lu at room temperature) and (ii) essentially constant

Fig. 3.11. Primitive unit cell of a hcp metal (*open circles*) with tetrahedral (*full circles*) and octahedral (*open squares*) interstitial sites. The arrows indicate four different types of jump paths for the hydrogen, where the jumps of type 2, 3, and 4 cross a common octahedral site. The two different lattice parameters a and c are also indicated. The figure is taken from [3.39]

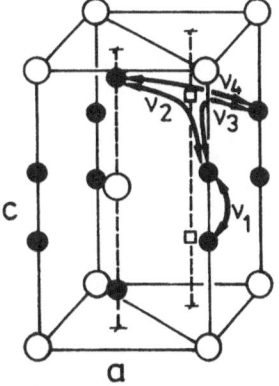

below room temperature [3.130]. However, with falling temperature, the hydrogen atoms condense increasingly into next-nearest-neighbor pairs bridging (in c-axis direction) a metal atom, and there is additionally a degree of short range order between these pairs [3.125, 127, 129].

(A) Lutetium

The anisotropy of hydrogen diffusion in Lu was investigated in Gorsky effect measurements [3.39]. It was found that the two diffusion coefficients D_c and D_a were identical within experimental accuracy ($\sim 20\%$). Figure 3.12 presents the (essentially) isotropic diffusion coefficient of the two isotopes hydrogen and deuterium in Lu, as observed in [3.39]. The results can be described by an Arrhenius relation with activation energies of 0.57 eV for hydrogen and 0.63 eV for deuterium. These values are much larger than those found in bcc metals, and they are also larger than activation energies that are typical for the fcc metals.

At a first glance, the practically isotropic diffusion behavior seems to be a surprising result, considering the fact that anisotropy in diffusion can be quite substantial in noncubic systems [3.7, 131]. However, it can be understood from the diffusive jumps that exist between the tetrahedral interstitial

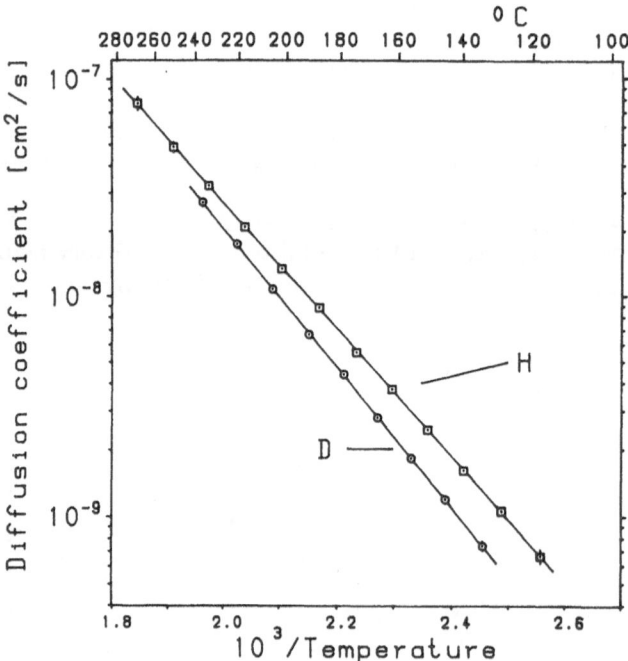

Fig. 3.12. Arrhenius plots of the diffusion coefficients of hydrogen and deuterium in polycrystalline Lu [3.39]. The figure is taken from [3.39]

sites in a hexagonal lattice [3.39]. Figure 3.11 shows four different types of jumps, characterized by their respective jump paths and jump frequencies v_i. A diffusive jump of type 1 (jump rate v_1) connects two nearest neighbor tetrahedral sites in the c-axis direction. The three other jumps represent the three (possible) types of jump paths via a nearby octahedral site, thus maintaining the largest possible distance from the host lattice atoms in the course of a jump process. Under the assumption that all the jumps crossing a given octahedral site have the same jump rate v, the two diffusion coefficients D_c and D_a are given by [3.39]

$$D_c = \frac{3v_1 v}{v_1 + 9v} \frac{3}{8} c^2 \quad \text{and} \quad D_a = 3v a^2 , \tag{3.22}$$

where c and a are the lattice parameters in the c-axis direction and in the basal plane. Equation (3.22) yields the ratio

$$\frac{D_c}{D_a} = \frac{v_1}{v_1 + 9v} \frac{3c^2}{8a^2} \tag{3.23}$$

between the two diffusion coefficients where the factor $3c^2/(8a^2)$ is 1 for the c/a ratio of an ideal hcp crystal.

According to (3.23), the essentially isotropic diffusion behavior observed in [3.39] requires v_1 being much larger than v. This condition is actually quite plausible from an inspection of the respective jump geometries. A jump of type 1 connects two nearest-neighbor tetrahedral sites with a distance of about $c/4 = 1.39$ Å (for Lu). This distance is much shorter than the corresponding ones for the three other jump types. Moreover, a type-1 jump needs to cross only once a bottleneck formed by a triangle of most closely approached host lattice atoms, whereas the other jump types require twice the passing of such a bottleneck. For these reasons, the condition $v_1 \gg v$ as required for an isotropic diffusion seems quite understandable.

Internal friction measurements [3.132], carried out between 200 and 230 K on hydrogen and deuterium doped Lu, yield for both isotopes activation energies identical to those of the diffusion coefficient measured at much higher temperatures. The absolute relaxation rates agree further with extrapolated values of the jump rate v, determined according to (3.22) under the condition $v_1 \gg v$. This is a strong indication that the jump rate v, which dominates long range diffusion, is also responsible for the internal friction peak.

(B) Yttrium

The diffusion of hydrogen in Y was studied by quasi-elastic neutron scattering between 590 and 700 K [3.133]. The measurements yielded a diffusion coefficient D_c about twice as large as D_a, both with an activation energy of roughly 0.6 eV. The fact that D_c was found larger than D_a cannot be explained by the jump diffusion model above. The authors of [3.133] assumed therefore that a small fraction ($\sim 2\%$) of the hydrogen atoms is located on

72   H. Wipf

octahedral sites, and that these atoms can rapidly jump in the c-axis direction between different octahedral sites so that the diffusion rate increases in this direction.

The activation energy found by neutron scattering agrees within experimental accuracy with that (0.57 eV) reported from internal friction measurements around 220 K [3.134], and it is somewhat higher than that (0.51 eV) derived from nuclear magnetic resonance (T_1) data in the range of 500 K [3.135].

(C) Titanium, Zirconium and Hafnium

Figures 3.13, 14 present results for the diffusion coefficient of hydrogen in Ti and Zr, respectively. Both metals show at high temperatures a phase transformation from the (low-temperature) hexagonal α phase to a bcc β phase. The full and the broken curves describe, accordingly, hydrogen diffusion in the α or in the β phase, respectively. The presence of the hydrogen stabilizes the β phase [3.67] so that the diffusion coefficients measured in this phase can extend considerably also below the transition temperature of the hydrogen-free metals.

The diffusion coefficients in Figs. 3.13, 14 were obtained from measurements on polycrystalline samples. The α phase data represent, for this reason, a type of average between the two diffusion coefficients D_c and D_a to be considered for the hexagonal lattice structure. The data for the corresponding β phases are results for the chemical diffusion coefficient which

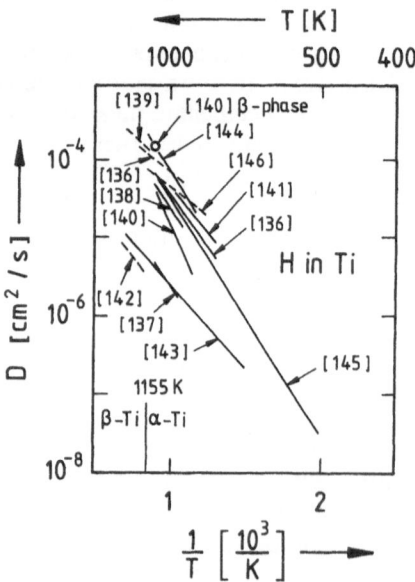

Fig. 3.13. Arrhenius plots of results for the diffusion coefficient of hydrogen in polycrystalline Ti. The full and the broken curves describe the diffusion coefficients in hexagonal α-Ti and in bcc β-Ti, respectively. The abscissa indicates the α-β phase transition at ≈ 1155 K (for hydrogen-free Ti). The figure shows also the references for the data

Fig. 3.14. Arrhenius plots of results for the diffusion coefficient of hydrogen in poly-crystalline Zr. The full and the broken curves describe the diffusion coefficients in hexagonal α-Zr and in bcc β-Zr, respectively. The abscissa indicates the α-β phase transition at ≈ 1138 K for (hydrogen-free Zr). The figure also shows the references of the data

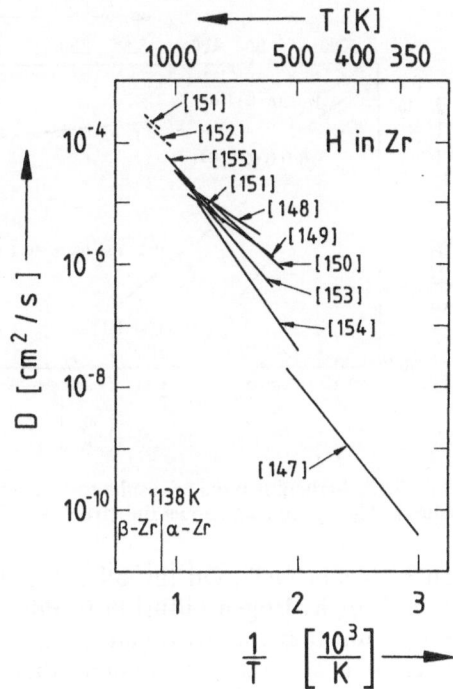

were, at least in part, obtained for higher hydrogen concentrations so that in these cases the self diffusion coefficients may exhibit different values.

The diffusion of tritium in α-Ti was investigated in [3.156]. In the case of α-Zr, the diffusion of both hydrogen and deuterium was investigated in [3.147], and the diffusion of tritium was studied in [3.156].

The diffusion coefficient of the two isotopes hydrogen and deuterium in hexagonal α-Hf was recently measured in the temperature range between 870 and 1870 K [3.157]. The diffusion of tritium in α-Hf was studied in [3.156].

3.2.4 Diffusion in Alloys

Figure 3.15 shows a compilation of results for the diffusion coefficient of hydrogen interstitials in some binary alloys. The data are valid for low hydrogen concentrations (in the solid solution range), and they apply to alloys with CsCl, A-15 and (disordered) fcc structure. The results for the three alloys with CsCl structure are remarkably different in that the activation energy for hydrogen diffusion in PdCu (≈ 0.035 eV) [3.2] is much lower than in FeTi (≈ 0.49 eV) [3.158] and in NiTi (≈ 0.48 eV) [3.159]. It is known that hydrogen (in solid solution) occupies octahedral interstitial sites in the hydrogen storage alloy FeTi (halfway between two Fe atoms) [3.161], and its comparable diffusion behavior in NiTi makes it likely that it occupies also

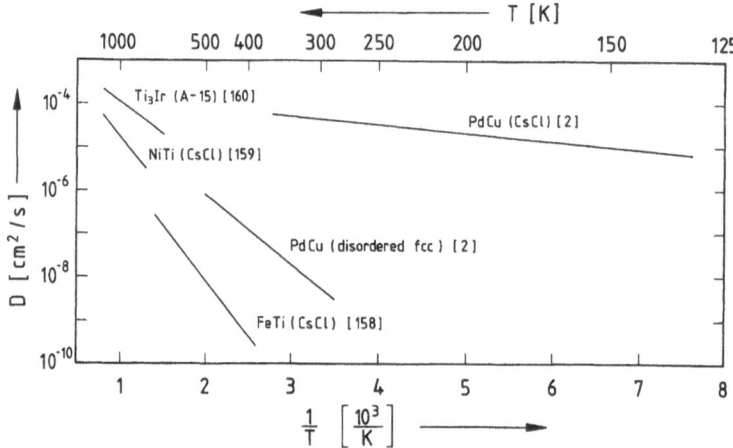

Fig. 3.15. Arrhenius plots of results for the diffusion coefficient of hydrogen in some binary alloys. The figure also shows the structure of the alloys and the references of the data

such sites in NiTi. On the other hand, the extremely low activation energy found for hydrogen diffusion in PdCu with CsCl structure [3.2] is a strong indication that the hydrogen in this alloy is located on different interstitial sites. It is suggestive to assume that these sites are tetrahedral ones, considering the fact that hydrogen diffusion in bcc metals, where the hydrogen occupies tetrahedral sites, is characterized by similarly low activation energies. The case of PdCu is particularly interesting since the high-temperature (disordered) fcc structure of this alloy can be quenched to room temperature. This made it possible to study hydrogen diffusion also in the fcc phase, and Fig. 3.15 shows that the activation energy (≈ 0.32 eV) [3.2] is much higher in this situation. In fact, the size of the activation energy is similar to that found for hydrogen diffusion in fcc metals so that the hydrogen is likely to occupy octahedral sites also in the fcc phase of PdCu.

3.2.5 Diffusion in Metallic Glasses

Both the chemical and the self diffusion of hydrogen interstitials was investigated for a great number of metallic glasses (Cu_xTi_{1-x}, Fe_xB_{1-x}, $Fe_xNi_yP_zB_{1-x-y-z}$, Fe_xZr_{1-x}, $Ni_xPd_yP_{1-x-y}$, Ni_xZr_{1-x}, $Pd_xCu_ySi_{1-x-y}$, Pd_xSi_{1-x}, Zr_xRh_{1-x}) [3.46, 162–173]. A chief effect of the configurational disorder in these glasses is that the jump rates of the hydrogen and the energies (or occupation probabilities) of the interstitial sites, which can be occupied by the hydrogen, vary drastically and randomly from site to site. This fact has characteristic consequences for hydrogen diffusion. A second consequence of the structural disorder is a suppression of the formation of hydride phases so that large hydrogen concentrations can be found, and investigated, without any substantial changes in the structure of the glass [3.174].

The chemical diffusion coefficient increases in general with rising hydrogen content. This behavior is exemplified in Fig. 3.16 which shows Gorsky effect results for the chemical diffusion coefficient in glassy $Ni_{64}Zr_{36}$, measured for three different hydrogen concentrations [3.167]. The concentration dependence can qualitatively be understood by the fact that the hydrogen tends to occupy interstitial sites with energies as low as possible. At low concentrations, the hydrogen is therefore, on the average, located on interstitial sites with lower energies than at high concentrations where it has to reside also on energetically less favorable sites. Also, at the same time, it can be expected that the potential barriers from a given site to its neighbors are, again on the average, lower when the energy of the considered site is higher. This means that with rising hydrogen concentration, the hydrogen atoms need to pass increasingly lower potential barriers in a diffusion process so that an increase of the diffusion coefficient can be expected. The qualitative considerations above were quantitatively confirmed by both analytic and Monte-Carlo model calculations carried out under consideration of a wide distribution in the energies of the interstitial sites [3.46, 163, 164].

The essentially statistical spatial distribution of the jump rates of the hydrogen in a metallic glass leads to an anomalous self diffusion behavior as observed in neutron scattering experiments on a hydrogen doped $Ni_{24}Zr_{76}$ glass [3.173]. The anomalous behavior means that it takes considerably longer than an average mean residence time until a linear relationship is found between t and the mean square distance $\langle [r(t) - r(0)]^2 \rangle$ a diffusing hydrogen walks within time t, as stated in (3.4). This means at the same time that the limits in the definition of the self diffusion coefficient according to (3.5) requires actually times t that can be much larger than the average mean residence time.

Fig. 3.16. Arrhenius plots of the chemical diffusion coefficient of hydrogen in glassy $Ni_{64}Zr_{36}$ for hydrogen-metal atom ratios $x = 0.04$ (A), $x = 0.09$ (B) and $x = 0.2$ (C). The figure is taken from [3.167]

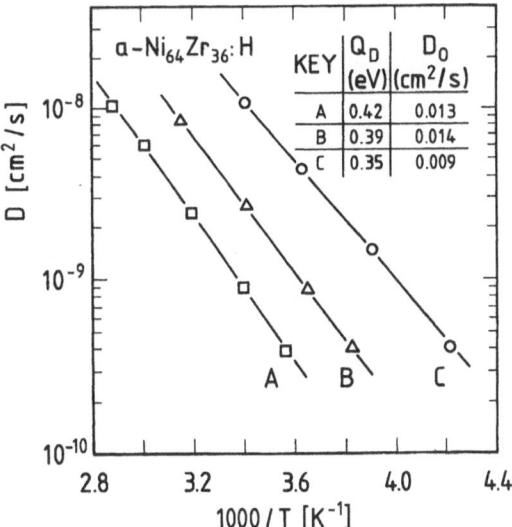

3.3 Local Dynamical Processes
of Trapped Hydrogen Interstitials

In the last decade, a great number of experimental studies were carried out on trapped hydrogen interstitials that performed a spatially restricted diffusion or tunneling process. A leading example for such a situation, to be discussed in detail in Sect. 3.3.1, is hydrogen in Nb which is trapped by O, N, or C impurities below 160 K [3.175–178]. In this case, O and N, and presumably also C, are located on octahedral interstitial sites [3.179, 180], and the trapped hydrogen occupies tetrahedral interstitial sites [3.177]. Figure 3.17 shows two bcc unit cells of Nb with an O, N, or C atom on an octahedral site and tetrahedral sites indicated by the letters a–h. According to a detailed discussion in [3.177, 178, 181], it can be concluded that the trapped hydrogen is likely to occupy tetrahedral sites such like those marked by e (note that there are in total 16 possible sites of type e since Fig. 3.17 shows only part of all the tetrahedral sites around the trapping O, N or C atom). However, it is emphasized that the occupation of e sites has not been definitely proven until now, although it represents the most likely situation which further is in complete agreement with the present experimental evidence.

Figure 3.17 exemplifies well the situation of a local diffusion process where the trapped hydrogen can jump between a limited number of sites (of type e) without diffusing away from the trapping atom. It is of particular importance that the local dynamics of trapped hydrogen interstitials can be investigated at low temperatures where, in the absence of traps, long-range hydrogen diffusion cannot be observed because of the formation of ordered hydride phases. This fact is the chief reason why local diffusion or tunneling processes were intensively investigated in recent years. It is finally worth mentioning that these local processes are subjected to the same theoretical concepts as ordinary long-range hydrogen diffusion (Chap. 2).

3.3.1 Oxygen, Nitrogen and Carbon Traps in Niobium

Hydrogen interstitials in Nb which were trapped by O, N, or C represent the system whose local dynamics was most intensively studied [3.181–208]. The experiments were carried out with specific heat, thermal conductivity, ane-

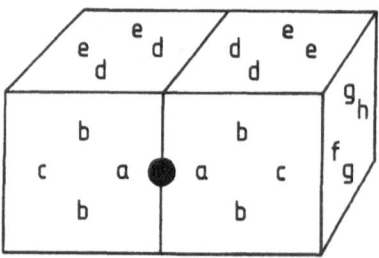

Fig. 3.17. Two bcc Nb unit cells with an O, N or C atom (full circle) on an octahedral interstitial site. The letters a–h indicate the positions of tetrahedral interstitial sites located on the visible surfaces of the unit cell. The trapped hydrogen is likely to occupy the sites labeled by e

lasticity, neutron spectroscopic, piezoresistance and nuclear magnetic resonance measurements. As stated above, the trapped hydrogen is in this case likely to occupy 16 tetrahedral sites labeled by e in Fig. 3.17. The occupation of these sites has the important consequence that always two out of the 16 sites are nearest neighbor sites with a distance of ≈ 1.17 Å that is almost three times smaller than the distance between each one of these two sites and any other one of the remaining sites of type e located around the trapping atom. This fact makes it understandable that the complete diffusion (or tunneling) dynamics of the trapped hydrogen is experimentally found to be characterized by two time scales differing by many orders of magnitude. The shorter one of these time scales can be ascribed to a movement between two nearest-neighbor sites (e.g., the two e sites on either the right-hand or the left-hand unit cell in Fig. 3.17), whereas the longer one can be attributed to jumps of the hydrogen between nonnearest trap sites with a much larger distance (e.g., jumps from an e site on the left-hand unit cell to an e site on the right-hand unit cell). The fact that the two time scales differ that widely allows a separate consideration of the two dynamical processes within our subsequent discussion. It finally seems worth pointing out that the existence of the two different time scales for the hydrogen movement, as experimentally observed, does not necessarily require the occupation of trap sites of type e. For instance, such two different time scales would also be expected if the hydrogen would be located on d sites since this would lead to a similar situation like the one discussed above where a hydrogen movement over all the respective sites involves at least two widely different jump lengths.

(A)   Hydrogen Motion Between Two Nearest-Neighbor Sites

The shorter one of the two time scales above characterizes a hydrogen motion in a double well potential consisting of two nearest neighbor tetrahedral sites. The experiments clearly identify tunneling as the dominating transport mechanism, demonstrating in fact the complete tunneling scenario for such a situation (as discussed in Chap. 2), a coherent tunneling with two tunnel-split eigenstates at low temperatures (below ≈ 10 K), and diffusive jumps (incoherent tunneling) between the two tetrahedral sites at high temperatures (above ≈ 10 K). The experiments further show that the tunneling dynamics is strongly affected by a nonadiabatic interaction of the hydrogen with the conduction electrons. This means that the tunneling dynamics cannot be explained by solely considering interatomic lattice potentials, such as defined in the conventional adiabatic or Born-Oppenheimer approximation. A detailed theoretical discussion of the tunneling behavior of the hydrogen in the presence of a nonadiabatic interaction with conduction electrons is given in Chap. 2.

The nonadiabatic influence of conduction electrons on the motion of the trapped hydrogen in Nb was first demonstrated 1984 in low-temperature ultrasonic experiments by *Wang* et al. [3.196]. In the same year, nonadiabatic electronic effects were also proposed by *Kondo* [3.209] and *Yamada* [3.210] as

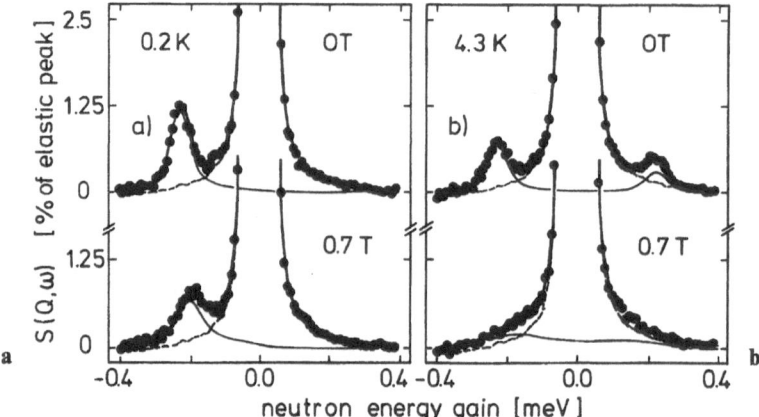

Fig. 3.18. Neutron spectra of a $NbO_{0.0002}H_{0.0002}$ sample at 0.2 K (**a**) and 4.3 K (**b**). For both temperatures, the spectra were taken in the superconducting (0 T) and normal conducting electronic state, where the latter state was achieved by a magnetic field of 0.7 T. The thick and the thin *solid lines* represent fits for the total and the inelastic scattering intensity, respectively. The *broken lines* are for the elastic intensity. The figure is taken from [3.201]

a mechanism to understand the previously unexplained temperature dependence observed for the diffusion rate of muons at low temperatures. Figure 3.18 presents more recent neutron spectroscopic results obtained for hydrogen interstitials in Nb that were trapped by O [3.201]. The figure shows four spectra taken at 0.2 K (Fig. 3.18a) and 4.3 K (Fig. 3.18b). Both temperatures are well below the superconducting transition temperature $T_c \simeq 9.2$ K of the Nb host metal. For this reason, spectra could be measured in both the superconducting and the normal conducting electronic state where the latter state was produced by application of a magnetic field. The top and the bottom spectra in Fig. 3.18 represent the results for superconductivity (0 T) and normal conductivity (0.7 T), respectively. It can be seen that both spectra taken in the superconducting state show a clearly identifiable inelastic line at ~ 0.2 meV. This demonstrates a coherent tunneling behavior with two well defined eigenstates due to a delocalization of the hydrogen between its two interstitial sites, and a value $J \simeq 0.2$ meV for the (renormalized) tunnel splitting between the two states. In the normal conducting state, the inelastic lines show a center shift and a broadening at 0.2 K, and a transition to an almost quasielastic behavior at 4.3 K. These distinct differences between superconductivity and normal conductivity demonstrate the influence of the nonadiabatic electronic interaction since any phononic effects are not expected to depend noticeably on the respective electronic state.

In the coherent tunneling regime below ≈ 10 K, the motion of the hydrogen is characterized by the (renormalized) tunnel splitting J and the relaxation rate γ between the two delocalized eigenstates. Figure 3.19 shows results for these two quantities, obtained in the neutron scattering measurements above for both superconductivity (open data points) and normal

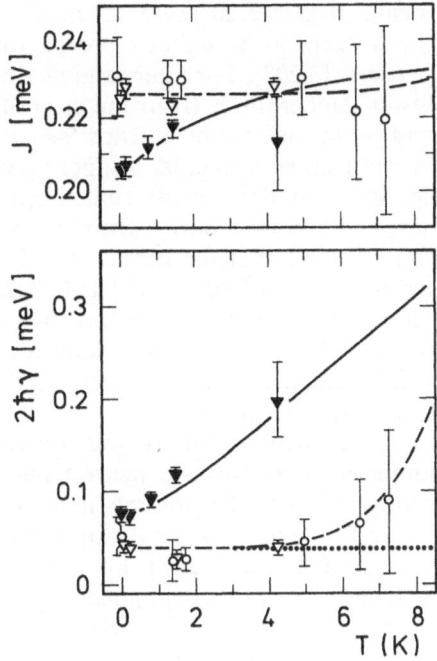

Fig. 3.19. Tunnel splitting J and relaxation rate γ (or damping $2\hbar\gamma$) in a plot versus temperature. *Triangles* and *circles* are the results for NbO$_x$H$_x$ samples with $x = 0.0002$ and 0.002, respectively. *Open data points* indicate superconductivity, and *full data points* are for normal conductivity. The *solid* and *broken curves* are explained in the text. The *dotted curve* represents an inhomogeneous contribution to the relaxation rate. The figure is taken from [3.201]

conductivity (full data points) [3.201]. In the low temperature limit $T \to 0$ K, the tunnel splitting is found to be $J(0) = (0.226 \pm 0.004)$ meV for the superconducting and $J(0) = (0.206 \pm 0.003)$ meV for the normal conducting state. The latter quantity shows an additional renormalization by 9% due to the presence of the normal conducting electrons. The relaxation rate, on the other hand, is distinctly larger in the normal than in the superconducting state.

The data in Fig. 3.19 can quantitatively be described by the theories developed for the tunneling of light particles in the presence of a non-adiabatic interaction with conduction electrons (Chap. 2 and [3.211–215]). The predictions of these theories are indicated in Fig. 3.19 by broken and full curves for the superconducting and normal conducting state, respectively. The theoretical curves were obtained with either of the two above results for the tunnel splitting in the limit $T \to 0$ K, and with the value $K = 0.055 \pm 0.005$ for the dimensionless Kondo parameter that describes the strength of the nonadiabatic coupling between the hydrogen and the conduction electrons. The fact that K is much smaller than 0.5 shows that the interaction can be described within the weak coupling limit.

Neutron scattering measurements demonstrate that the tunnel splitting is smaller in the case of N and C traps. The measurements yield $J = (0.167 \pm 0.004)$ μeV [3.208] and $J = (0.162 \pm 0.004)$μeV [3.206] for N and C, respectively, both under superconducting conditions and for $T \to 0$ K. The isotope dependence of the tunnel splitting was investigated in specific heat (O and N traps) [3.181, 202] and ultrasonic (N traps) [3.207] measurements. Figure 3.20 presents specific heat results for the two hydrogen isotopes, trapped by either O or N, in a double-logarithmic plot versus temperature [3.202]. The figure shows a large excess specific heat that is shifted to lower temperatures (i) in the case of deuterium (as compared to hydrogen) and (ii) in the case of N traps (as compared to O traps), thus demonstrating in both cases a smaller tunnel splitting. The quantitative analysis of the specific heat data yields tunnel splittings for the isotope hydrogen which agree with those reported from neutron spectroscopy [3.181, 202]. The results show further that the tunnel splittings are about ten times smaller for deuterium than for hydrogen [3.181, 202]. The ultrasonic measurements confirm the isotope dependence found in the specific heat measurements; their absolute value for the tunnel splitting of the trapped hydrogen is, however, $\approx 20\%$ smaller than those derived from the specific heat and neutron spectroscopic data [3.207].

In the temperature regime above 10 K, the hydrogen performs diffusive jumps between the two nearest neighbor tetrahedral sites (incoherent tunneling). Figure 3.21 presents neutron spectroscopic results for the jump rate where the hydrogen was trapped by either O (open data points) or N (full data points) [3.203, 208]. For both types of traps, the jump rates exhibit a distinct temperature dependence that is characterized by a minimum at about ≈ 60 K. In the case of a sole nonadiabatic coupling between the hydrogen and the conduction electrons, the above mentioned theories predict a jump rate v given by (2.118)

$$v = \frac{1}{2} \frac{\Gamma(K)}{\Gamma(1-K)} \frac{J}{\hbar} \left(\frac{2\pi k_B T}{J} \right)^{2K-1} , \tag{3.24}$$

where $\Gamma(x)$ denotes the gamma function and J is the tunnel splitting in the normal conducting state for $T \to 0$ K. The jump rate can be calculated (without any adjustable parameter) from K and J, and the broken and the solid lines in Fig. 3.21 indicate the theoretical predictions for v, calculated with values for K and J as determined from neutron scattering measurements carried out below 10 K. The theoretical curves decrease with rising temperatures, showing the T^{2K-1} power law first suggested by *Kondo* [3.209] and *Yamada* [3.210]. The data in Fig. 3.21 also show that the jump rates are smaller in the case of the N traps because of the smaller value of the tunnel splitting.

Figure 3.21 demonstrates that the experimental jump rates are well described by a dominant nonadiabatic interaction in the temperature regime up to ≈ 60 K. Above this temperature, the experimental data exceed the theoretical prediction, and they are found to increase with rising temperature. This means that the influence of an interaction with phonons, which was neglected so far, becomes significant at elevated temperatures (Chap. 2).

In the temperature range above 120 K, the jump rates of the trapped hydrogen can be compared with those derived according to (3.19) from results for the diffusion coefficient of hydrogen in Nb, as shown in Fig. 3.8. For such a comparison, the appropriate jump rate deduced from the diffusion

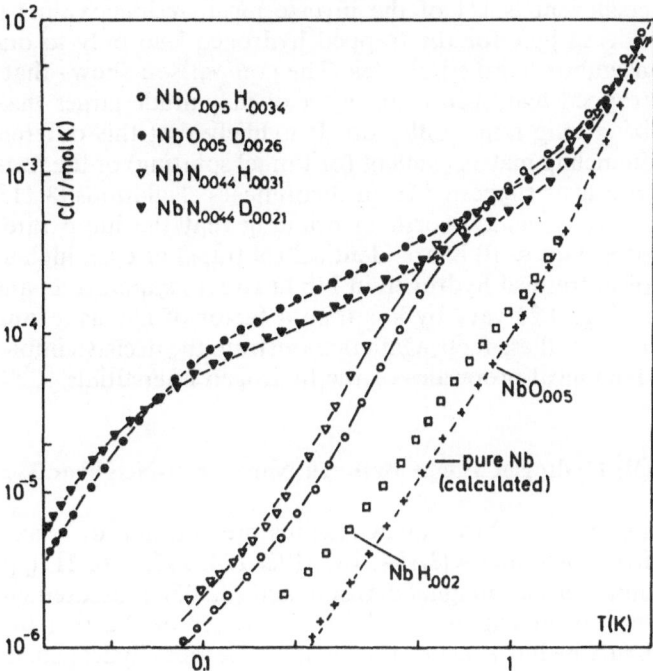

Fig. 3.20. Specific heat of (superconducting) Nb samples doped with O or N and hydrogen or deuterium. The *full lines* are fit curves which allow the determination of the tunnel splitting. The figure is taken from [3.202]

Fig. 3.21. Jump rates v of trapped hydrogen interstitials in a double-logarithmic plot versus temperature. *Open* and *full data points* apply to hydrogen trapped by O and N, respectively. The *broken* (O traps) and the *full* (N traps) *lines* are the theoretical predictions for a dominant nonadiabatic interaction with conduction electrons according to (3.24). The figure is taken from [3.208]

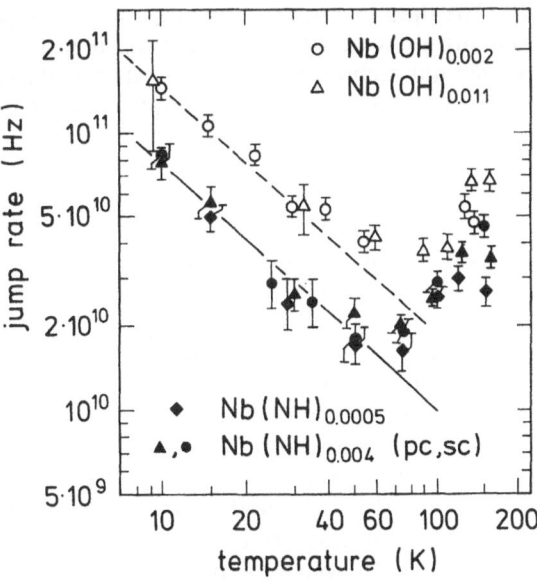

coefficient is 1/4 of the inverse mean residence time since the jumps considered here for the trapped hydrogen lead only to one out of four nearest neighbor tetrahedral sites. The comparison shows that the jump rate of the trapped hydrogen is an order of magnitude larger than that one characterizing long range diffusion. It is likely that this difference is due to a larger tunneling matrix element (or tunnel splitting) of the trapped hydrogen, which also is as suggested from theoretical calculations [3.213].

It is finally worth mentioning that the jump rates of the trapped hydrogen at ≈ 10 K are identical (N traps) or even higher (O traps) than those of untrapped hydrogen in Nb at room temperature, and that the jump rates in Fig. 3.21 vary by less than a factor of five as a function of temperature. Both of these facts again demonstrate the decisive impact of tunneling on the dynamical properties of the hydrogen interstitials.

(B) Hydrogen Jumps Between Nonnearest-Neighbor Trap Sites

Figure 3.22 shows an Arrhenius presentation of relaxation rates v reported from anelasticity [3.184, 187, 192, 194, 197, 216, 217], piezoresistance [3.204] and nuclear magnetic resonance [3.205] measurements for the hydrogen movement characterized by the larger of the two time scales above. The figure includes results for the three isotopes hydrogen, deuterium, and tritium in the presence of O traps. The relaxation rates can be expected to represent jump rates of the hydrogen between non-nearest-neighbor trap sites up to an unknown factor of about 1. It is readily seen that the greater jump lengths involved in the present case lead to relaxation (or jump) rates which are up to a factor of 10^{13} (at 40 K) smaller than those characterizing jumps between nearest neighbor sites (Fig. 3.21).

The data in Fig. 3.22, which cover a frequency range of eleven orders of magnitude, were collected with the help of a large number of different experimental techniques. The various results seem to agree surprisingly well, except for the two conflicting data sets for hydrogen below 50 K. The data demonstrate a relaxation rate that decreases with rising isotopic mass. The isotopic ratio between the relaxation rates of hydrogen and deuterium considerably exceeds the classical value of $\sqrt{2}$, at least in the temperature range below ≈ 110 K. The slope of the respective data points in the Arrhenius presentation of Fig. 3.22 decreases further with falling temperature below ≈ 75 K. This behavior can be considered as a decrease of an (apparent) activation energy similarly as it is already observed at higher temperatures for ordinary long-range hydrogen diffusion in Nb in the absence of traps (Fig. 3.8).

The situation in which hydrogen isotopes were trapped by N rather than by O was investigated in anelasticity measurements [3.187, 190, 192, 194]. The measurements demonstrate, for all three isotopes, a reduction of the relaxation rate in the case of N traps. This reduction is similar to the situation found for the hydrogen motion between two nearest neighbor sites [Sect. 3.3.1(A)] where both the tunnel splitting (below ≈ 10 K) and the jump rate of the hydrogen was also smaller for the N traps.

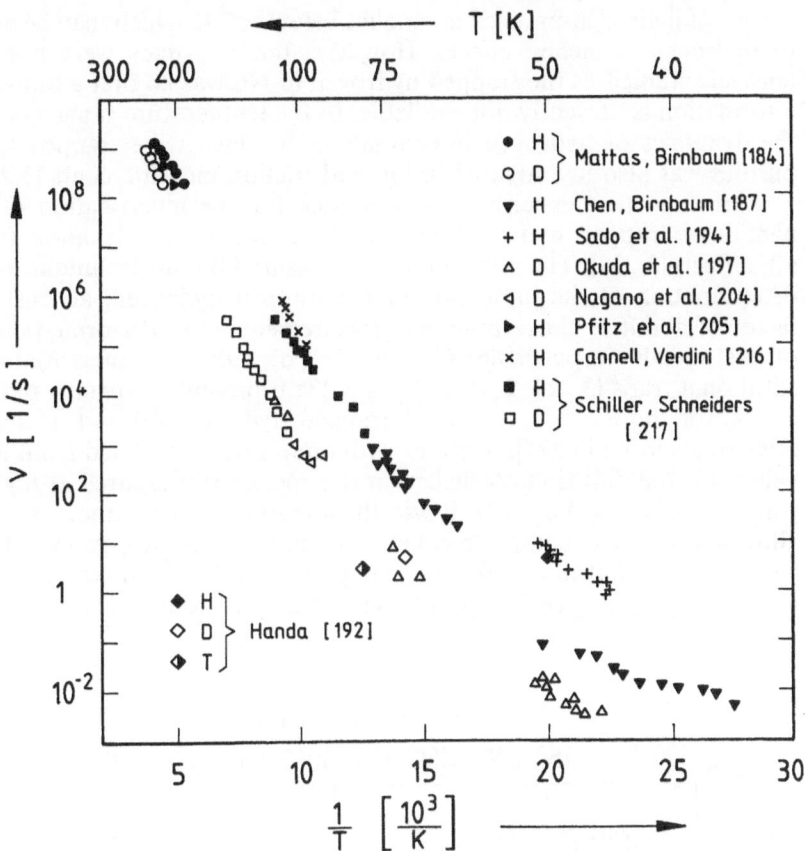

Fig. 3.22. Arrhenius plots of results for the relaxation rate of hydrogen, deuterium and tritium in Nb in the presence of O traps [3.184, 187, 192, 194, 197, 204, 205, 216, 217]. The results characterize the jump rate between nonnearest-neighbor trap sites

3.3.2 Other Trap Systems

In the range below room temperature, the dynamical properties of hydrogen and deuterium interstitials in Nb which were trapped by substitutional atoms such as Ti or Zr were investigated in internal friction measurements [3.218–220]. In the case of Ti traps, the measurements demonstrated the appearance of a relaxation peak even below 1K [3.219]. Specific heat measurements carried out in the range between 0.05 and 2 K on a $NbTi_{0.05}$ sample yielded further a large hydrogen- or deuterium-induced excess heat capacity [3.221]. Both of these results prove the existence of a low-temperature dynamical process of the trapped hydrogen. However, the presently available experimental information does not allow a clear understanding of the actual hydrogen motion which exists in this process.

In the case of the metals V and Ta, measurements of both the specific heat [3.182, 222] and the thermal conductivity [3.183] demonstrated the existence of hydrogen-induced anomalies below ≈ 2 K which may be attributed to hydrogen tunneling effects. However, these systems were not that intensively studied as the trapped hydrogen in Nb was so that a more detailed information is presently not available. In the temperature range above 50 K, the dynamics of hydrogen interstitials in V which were trapped by Ti impurities was also investigated in internal friction measurements [3.223].

An important experimental technique for the investigation of the dynamics of trapped hydrogen interstitials is magnetic relaxation (magnetic aftereffect) [3.224]. The relaxation rate measured by this technique allows the determination of the jump rate of the trapped hydrogen, at least up to a factor of about 1. Corresponding measurements were performed on a series of host metals (in particular Co, Ni, Fe, Pd) with both interstitial and substitutional traps [3.224–231]. Figures 3.23a,b present Arrhenius plots of the results for the local jump rate of trapped hydrogen (a) and deuterium (b) interstitials in Ni [3.227], together with jump rates calculated from literature values for the diffusion coefficient in this metal, as measured at higher temperatures (see also Fig. 3.4). It was the intention of the authors of [3.227] to allow a meaningful comparison between the different jump rates in Fig. 3.23, which implies that the local jump rate presented there should not be affected very much by the variations of the lattice potential due to the trapping

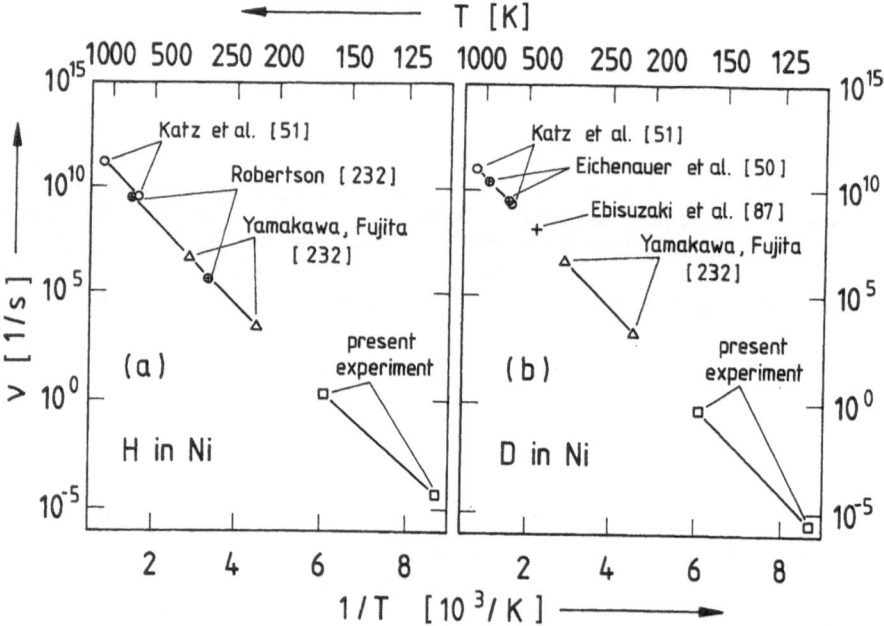

Fig. 3.23. Arrhenius plots of the jump rate ν of trapped hydrogen (**a**) and deuterium (**b**) interstitials in Ni as measured by *Hohler and Kronmüller* (indicated as present experiment) [3.227]. The figure shows also jump rates derived from literature data for the respective diffusion coefficients measured above 200 K. The figure is taken from [3.227]

impurity. For this reason, the measurements were performed with a large number of different substitutional trapping impurities which, respectively, caused (i) a different (macroscopic) lattice expansion and (ii) a (slightly) differing local jump rate of the trapped hydrogen. The reasoning leading to Fig. 3.23 was that the impurity-induced variations of the lattice potential should be expected to be smaller when the lattice expansion is smaller. The local jump rates in this figure were, accordingly, obtained from an extrapolation of the respectively measured jump rates to a zero lattice expansion.

The results in Fig. 3.23 demonstrate in fact that the local jump rates and those derived from long range diffusion represent a consistent data set. In the investigated temperature range down to ≈ 120 K, the respective data points

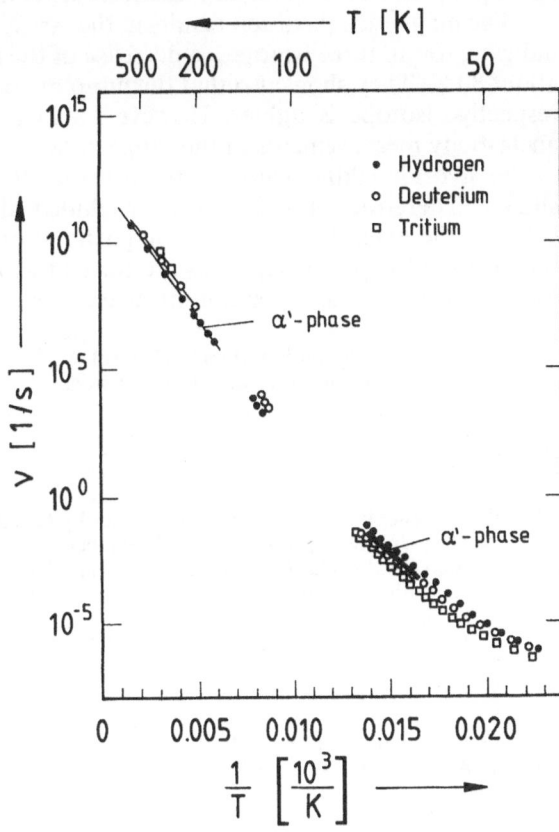

Fig. 3.24. Arrhenius plots for the jump rates of hydrogen isotopes in Pd. Full circles, open circles and open squares represent hydrogen, deuterium and tritium, respectively. The data below 80 K and around 130 K are the results of magnetic relaxation measurements where the hydrogen (deuterium, tritium) was trapped by Fe impurities (*Higelin* et al. [3.230]). The full line (for hydrogen) between 60 and 70 K, resulting from anelasticity measurements, and the NMR data around 180 K were obtained for Pd hydride [3.233]. The data above 200 K were obtained for a low hydrogen concentration (in the α phase) with the help of Gorsky effect [3.2, 88] and permeation measurements [3.90]. The figure is taken from [3.230]

do not indicate any clear evidence for a change of the slope in the Arrhenius presentation. However, they do show that the isotope hydrogen has a higher jump rate than that of deuterium.

A more recent example, investigating also a much lower temperature range, is presented in Fig. 3.24. The figure shows jump rates for all three hydrogen isotopes in Pd where the data below 80 K and around 130 K are the results of magnetic relaxation measurements in which the respective hydrogen isotopes were trapped by Fe impurities [3.230]. The full line (anelasticity results for hydrogen) between 60 and 70 K and the nuclear magnetic resonance (NMR) data around 180 K were measured from Pd hydride (α' or β phase) samples [3.233]. The data above 200 K were obtained at low hydrogen (deuterium, tritium) concentrations with the help of Gorsky effect [3.2, 88] and permeation measurements [3.90].

The magnetic relaxation results in the Arrhenius presentation of Fig. 3.24 indicate, for all three isotopes, a decrease of the slope of the jump rates below about 80 K. They show also that the jump rate is higher when the mass of the respective isotope is lighter. However, a fact to be considered is that the anelasticity measurements in the range between 60 and 70 K were performed on Pd hydride (PdH$_x$ with $x \simeq 0.76$) rather than on dilute hydrogen interstitials. Therefore, it can not be excluded that the excellent agreement between the anelasticity and the magnetic relaxation data is due to the fact that α' (or β) phase precipitates are formed around the Fe impurities in the case of the magnetic relaxation measurements.

Acknowledgement. The author thanks H. Grabert, K. Kehr, H.R. Schober, H. Teichler, U. Weiss and H. Züchner for valuable discussions.

References

3.1 K.W. Kehr: In *Hydrogen in Metals I,* ed. by G. Alefeld, J. Völkl, Topics in Applied Physics, Vol. 28 (Springer, Berlin Heidelberg 1978) p. 197
3.2 J. Völkl, G. Alefeld: In *Hydrogen in Metals I,* ed. by G. Alefeld, J. Völkl, Topics in Applied Physics, Vol. 28 (Springer, Berlin Heidelberg 1978) p. 321
3.3 M.I. Klinger: Phys. Rep. **94**, 183 (1983)
3.4 R. Messer, A. Blessing, S. Dais, D. Höpfel, G. Majer, C. Schmidt, A. Seeger, W. Zag, R. Lässer: Z. Phys. Chem. N. F., Suppl. **2**, 61 (1986)
3.5 Y. Fukai, H. Sugimoto: Adv. Phys. **34**, 263 (1985)
3.6 Y. Fukai: In Springer Ser. Mater. Sci, Vol. 21, *The Metal-Hydrogen System: Basic Bulk Properties* (Springer, Berlin Heidelberg 1993)
3.7 Y. Adda, J. Philibert: *La Diffusion dans les Solides* (Presses Universitaires de France, Paris 1966) Vols. I and II
3.8 S.R. de Groot, P. Mazur: *Non-Equilibrium Thermodynamics* (North-Holland, Amsterdam 1969)
3.9 J.L. Bocquet, G. Brébec, Y. Limoge: In *Physical Metallurgy*, ed. by R.W. Cahn, P. Haasen (North-Holland, Amsterdam 1983) Pt I, p. 385
3.10 N. Boes, H. Züchner: J. Less-Common Met. **49**, 223 (1976)
3.11 R. Schmidt, M. Schlereth, H. Wipf, W. Assmus, M. Müllner: J. Phys.: Condens. Matter **1**, 2473 (1989)
3.12 H. Wipf, G. Alefeld: Phys. Status Solidi (a) **23**, 175 (1974)
3.13 R.C. Brouwer, H. Douwes, R. Griessen, E. Walker: Phys. Rev. Lett. **58**, 2551 (1987)

3.14 J. Völkl: Ber. Bunsenges. Phys. Chem. **76**, 797 (1972)
3.15 H. Wipf, J. Völkl, G. Alefeld: Z. Physik B **76**, 353 (1989)
3.16 J.F. Nye: *Physical Properties of Crystals* (Clarendon, Oxford 1976)
3.17 G.E. Murch: Phil. Mag. A **45**, 685 (1982)
3.18 U. Potzel, J. Völkl, H. Wipf, A. Magerl: Phys. Status Solidi (b), **123**, 85 (1984)
3.19 U. Potzel, R. Raab, J. Völkl, H. Wipf: J. Less-Common Met. **101**, 343 (1984)
3.20 D.A. Faux, D.K. Ross: J. Phys. C **20**, 1441 (1987)
3.21 R. Hempelmann, D. Richter, D.A. Faux, D.K. Ross: Z. Phys. Chem. N. F. **159**, 175 (1988)
3.22 E. Wicke, H. Brodowsky, H. Züchner: In *Hydrogen in Metals II*, eds. G. Alefeld, J. Völkl, Topics in Applied Physics, Vol. 29 (Springer, Berlin Heidelberg 1978) p. 73
3.23 K. Nakazato, K. Kitahara: Prog. Theor. Phys. **64**, 2261 (1980)
3.24 G.E. Murch: Phil. Mag. A **46**, 151 (1982)
3.25 K.W. Kehr, K. Binder: In *Applications of the Monte Carlo Method in Statistical Physics*, ed. by K. Binder, Topics in Current Physics, Vol. 36 (Springer, Berlin Heidelberg 1987) p. 181
3.26 R. Kubo: J. Phys. Soc. Jpn. **12**, 570 (1957)
3.27 R. Zwanzig: Ann. Rev. Phys. Chem. **16**, 67 (1965)
3.28 R. Gomer: Rep. Prog. Phys. **53**, 917 (1990)
3.29 H. van Beijeren, K.W. Kehr: J. Phys. C **19**, 1319 (1986)
3.30 A.D. LeClaire: In *Physical Chemistry* 10, 261 (Academic, New York 1970)
3.31 H.C. Bauer, J. Völkl, J. Tretkowski, G. Alefeld: Z. Physik B **29**, 17 (1978)
3.32 H. Wipf: J. Less-Common Met. **49**, 291 (1976)
3.33 H. Wipf: In *Hydrogen in Metals II*, ed. by G. Alefeld, J. Völkl, Topics in Applied Physics, Vol. 29 (Springer, Berlin Heidelberg 1978) p. 273
3.34 P.S. Ho, T. Kwok: Rep. Prog. Phys. **52**, 301 (1989)
3.35 H. Peisl: In *Hydrogen in Metals I*, ed. by G. Alefeld, J. Völkl, Topics in Applied Physics, Vol. 28 (Springer, Berlin, Heidelberg 1978) p. 53
3.36 J.M. Rowe, K. Sköld, H.E. Flotow, J.J. Rush: J. Phys. Chem. Sol. **32**, 41 (1971)
3.37 C.P. Flynn: *Point Defects and Diffusion* (Clarendon, Oxford 1972)
3.38 S. Ishioka, M. Koiwa: Phil. Mag. A **52**, 267 (1985)
3.39 J. Völkl, H. Wipf, B.J. Beaudry, K.A. Gschneidner Jr.: Phys. Status Solidi (b) **144**, 315 (1987)
3.40 T. Schober, H. Wenzl: In *Hydrogen in Metals II*, ed. by G. Alefeld, J. Völkl, Topics in Applied Physics, Vol. 29 (Springer, Berlin Heidelberg 1978) p. 11
3.41 R.A. Oriani: Acta Met. **18**, 147 (1970)
3.42 A. Wert: In *Hydrogen in Metals II*, ed. by G. Alefeld, J. Völkl, Topics in Applied Physics, Vol. 29 (Springer, Berlin Heidelberg 1978) p. 305
3.43 Zh. Qi, J. Völkl, H. Wipf: Scripta metall. **16**, 859 (1982)
3.44 K. Kiuchi, R.B. McLellan: Acta metall. **31**, 961 (1983)
3.45 R. Kirchheim: Acta metall. **35**, 281 (1987)
3.46 R. Kirchheim, U. Stolz, Acta metall. **35**, 281 (1987)
3.47 H.H. Johnson: Metall. Trans. A **19**, 2371 (1988)
3.48 T. Mütschele, R. Kirchheim: Scripta metall. **21**, 135 (1987)
3.49 C.E. Ransley, D.E.J. Talbot: Z. Metallkde. **46**, 328 (1955)
3.50 W. Eichenauer, W. Löser, H. Witte: Z. Metallkde. **56**, 287 (1965)
3.51 L. Katz, M. Guinan, R.J. Borg: Phys. Rev. B **4**, 330 (1971)
3.52 G. Perkins, D.R. Begeal: Ber. Bunsenges. Phys. Chem. **76**, 863 (1972)
3.53 V.A. Kurakin, A.A. Kurdyumov, V.N. Lyasnikov, M.I. Popatov: Sov. Phys.- Solid State **21**, 616 (1979)
3.54 Y. Sakamoto, K. Takao: J. Jpn. Inst. Metals **46**, 285 (1982)
3.55 T. Ishikawa, R.B. McLellan: J. Phys. Chem. Solids **46**, 445 (1985)
3.56 H. Hagi: Trans. Jpn. Inst. Metals **27**, 233 (1986)
3.57 W. Eichenauer, H. Künzig, A. Pebler: Z. Metallkde. **49**, 220 (1958)
3.58 H. Katsuta, R.B. McLellan: Scripta metall. **13**, 65 (1979)
3.59 T. Ishikawa, R.B. McLellan: Acta Met. **33**, 1979 (1985)

3.60 I.Ye. Gabis, A.A. Kurdyumov, S.N. Mazayev, T.A. Ovsyannikova, N.I. Timifeyev: Phys. Met. Metall. **69**, 90 (1990)
3.61 W. Eichenauer, D. Liebscher: Z. Naturforsch. **17a**, 355 (1962)
3.62 E. Gileadi, M.A. Fullenwider, J.O'M. Bockris: J. Electrochem. Soc. **113**, 926 (1966)
3.63 Y. Ebisuzaki, W.J. Kass, M. O'Keeffe: J. Chem. Phys. **49**, 3329 (1968)
3.64 H. Katsuta, R.B. McLellan: J. Phys. Chem. Solids **40**, 697 (1979)
3.65 J. Čermák, A. Kufudakis, G. Gardavská: J. Less-Common Met. **63**, P1 (1979)
3.66 Y. Sakamoto, H. Kamohara: J. Jpn. Inst. Metals **45**, 797 (1981)
3.67 W.M. Mueller, J.P. Blackledge, G.G. Libowitz (eds): *Metal Hydrides*, (Academic, New York 1968)
3.68 W. Eichenauer, K. Hattenbach, A. Pebler: Z. Metallkde. **52**, 682 (1961)
3.69 S. Matsuo, T. Hiraba: J. Jpn. Inst. Metals **31**, 590 (1967)
3.70 M. Ichimura, M. Imabayashi, M. Hayakawa: J. Jpn. Inst. Metals **44**, 1053 (1980)
3.71 K. Papp, E. Kovacs-Csetényi: Scripta metall. **15**, 161 (1981)
3.72 R.A. Outlaw, D.D. Peterson, F.A. Schmidt: Scripta metall. **16**, 287 (1982)
3.73 E. Hashimoto, T. Kino: J. Phys. F. **13**, 1157 (1983)
3.74 T. Ishikawa, R.B. McLellan: Acta metall. **34**, 1091 (1986)
3.75 Á. Csanády, K. Papp, E. Pásztor: Mat. Sci. Eng. **48**, 35 (1981)
3.76 K. Yamakawa: J. Phys. Soc. Jpn. **47**, 114 (1979)
3.77 G. Gardavská, P. Lejček: Kristall und Technik **14**, 285 (1979)
3.78 A. Atrens, D. Mezzanotte, N.F. Fiore, M.A. Genshaw: Corrosion Sci. **20**, 673 (1980)
3.79 G. Meunier, J.-P. Manaud, M. Devalette: J. Less-Common Met. **77**, P47 (1981)
3.80 T. Tsuru, R.M. Latanision: Scripta metall. **16**, 575 (1982)
3.81 Y. Yamanishi, T. Tanabe, S. Imoto: Trans. Jpn. Inst. Metals **24**, 49 (1983)
3.82 K.A. Lee, R.B. McLellan: Scripta metall. **18**, 859 (1984)
3.83 Y. Furuya, E. Hashimoto, T. Kino: Jpn. J. Appl. Phys. **23**, 1190 (1984)
3.84 G. Matusiewicz, D.J. Duquette: Acta metall. **33**, 1637 (1985)
3.85 Y. Hayashi, A. Tahara: Z. Phys. Chem. N. F. **145**, 261 (1985)
3.86 D.L. Cummings, D.A. Blackburne: J. Nucl. Mater. **144**, 81 (1987)
3.87 Y. Ebisuzaki, W.J. Kass, M. O'Keeffe: J. Chem. Phys. **46**, 1373 (1967)
3.88 J. Völkl, G. Wollenweber, K.-H. Klatt, G. Alefeld: Z. Naturforsch. **26 a**, 922 (1971)
3.89 H. Katsuta, R.J. Farraro, R.B. McLellan: Acta metall. **27**, 1111 (1979)
3.90 G. Sicking, M. Glugla, B. Huber: Ber. Bunsenges. Phys. Chem. **87**, 418 (1983)
3.91 R.B. McLellan, M. Yoshihara: Acta metall. **35**, 197 (1987)
3.92 M. Nagano, Y. Hayashi, N. Ohtani, M. Isshiki, K. Igaki: Scripta metall. **16**, 973 (1982)
3.93 E. Riecke, K. Bohnenkamp: Z. Metallkde. **75**, 76 (1984)
3.94 H. Hagi, Y. Hayashi: Trans. Jpn. Inst. Metals **29**, 373 (1988)
3.95 C.W. Su, J.K. Wu: J. Mater. Sci. Lett. **7**, 849 (1988)
3.96 Y. Hayashi, H. Hagi, A. Tahara: Z. Phys. Chem. N. F. **164**, 815 (1989)
3.97 Th. Heumann, D. Primas: Z. Naturforsch. **21a**, 260 (1966)
3.98 A.J. Kumnick, H.H. Johnson: Acta metall. **25**, 891 (1977)
3.99 N.R. Quick, H.H. Johnson: Acta metall. **26**, 903 (1978)
3.100 M. Nagano, Y. Hayashi, N. Ohtani, M. Isshiki, K. Igaki: Trans. Jpn. Inst. Met. **22**, 423 (1981)
3.101 M. Nagano: Thesis, Tohoku University (1983)
3.102 Zh. Qi, J. Völkl, R. Lässer, H. Wenzl: J. Phys. F. **13**, 2053 (1983)
3.103 A. Klamt, H. Teichler: Phys. Status Solidi (b) **134**, 503 (1986)
3.104 H. Schober, A.M. Stoneham: Phys. Rev. Lett. **60**, 2307 (1988)
3.105 T. Eguchi, S. Morozumi: J. Jpn. Inst. Met. **41**, 795 (1977)
3.106 Y. Fukai: Jap. J. Appl. Phys. **23**, L596 (1984)
3.107 A. Weidinger, R. Peichl: Phys. Rev. Lett. **54**, 1683 (1985)
3.108 J. Lauzier, J. Hillairet, A. Vieux-Champagne, H. Schultz: J. Phys. F. **18**, 2529 (1988)
3.109 T. Suzuki, H. Namazue, S. Koike, H. Hayakawa: Phys. Rev. Lett. **51**, 798 (1983)
3.110 T. Suzuki: Trans. Jpn. Inst. Met. **26**, 601 (1985)
3.111 T.H. Metzger: Phys. Rev. Lett. **52**, 476 (1984)
3.112 T. Schober, J. Golczewski: Phys. Rev. Lett. **52**, 478 (1984)

3.113 D. Steinbinder, H. Wipf, G. Kearley, A. Magerl: J. Phys. C **20**, L321 (1987)
3.114 M.L. Hill: J. Metals **12**, 725 (1960)
3.115 L.N. Ryabchikov: Ukrains'kyi Fiz. Zhn. **9**, 293 (1964)
3.116 G.E. Moore, F.C. Unterwald, J. Chem. Phys. **40**, 2639 (1964)
3.117 P.M.S. Jones, R. Gibson, J.A. Evans: AWRE Rep. O-16/66, Aldermaston England (1966)
3.118 A.P. Zakharov, V.M. Sharapov: Fiz.-Khim. Mekh. Mat. **7**, 54 (1971)
3.119 H. Katsuta, R.B. McLellan, K. Furukawa: J. Phys. Chem. Solids **43**, 533 (1982)
3.120 H. Katsuta, T. Iwai, H. Ohno: J. Nucl. Mater. **115**, 206 (1983)
3.121 R. Frauenfelder: J. Vac. Sci. Techn. **6**, 388 (1969)
3.122 H. Pinto, C. Korn, S. Goren, H. Shaked: Solid State Commun. **32**, 397 (1979)
3.123 R. Khoda-Bakhsh, D.K. Ross: J. Phys. F. **12**, 15 (1982)
3.124 C.K. Saw, B.J. Beaudry, C. Stassis: Phys. Rev. B **27**, 7013 (1983)
3.125 O. Blaschko, J. Pleschiutschnig, P. Vajda, J.P. Burger, J.N. Daou: Phys. Rev. B **40**, 5344 (1989)
3.126 D. Khatamian, C. Stassis, B.J. Beaudry: Phys. Rev. B **23**, 624 (1981)
3.127 M.W. McKergow, D.K. Ross, J.E. Bonnet, I.S. Anderson, O. Schaerpf: J. Phys. C **20**, 1909 (1987)
3.128 R. Danielou, J.N. Daou, E. Ligeon, P. Vajda: Phys. Status Solidi (a) **67**, 453 (1981)
3.129 O. Blaschko, G. Krexner, J. Pleschiutschnig, G. Ernst, J.N. Daou, P. Vajda: Phys. Rev. B **39**, 5605 (1989)
3.130 J.P. Burger, J.N. Daou, A. Lucasson, P. Lucasson, P. Vajda: Z. Phys. Chem. N. F. **143**, 111 (1985)
3.131 D.C. Yeh, H.B. Huntington: Phys. Rev. Lett. **53**, 1469 (1984)
3.132 P. Vajda, J.N. Daou, P. Moser: J. Physique **44**, 543 (1983)
3.133 I.S. Anderson, D.K. Ross, J.E. Bonnet: Z. Phys. Chem. N. F. **164**, 923 (1989)
3.134 P. Vajda, J.N. Daou, P. Moser, P. Rémy: Solid State Commun. **79**, 383 (1991)
3.135 L. Lichty, R.J. Schoenberger, D.R. Torgeson, R.G. Barnes: J. Less-Common Met. **129**, 31 (1987)
3.136 R.J. Wasilewski, G.L. Kehl: Metallurgia **50**, 225 (1954)
3.137 H. Kusamichi, Y. Yagi, T. Yukawa, T. Noda: J. Jpn. Inst. Metals **20**, 39 (1956)
3.138 W.M. Albrecht, M.W. Mallett: Trans. Met. Soc. AIME **212**, 204 (1958)
3.139 M. Someno, K. Nagasaki, S. Kagaku: Vacuum Chemistry (Tokyo) **8**, 145 (1960)
3.140 T.P. Papazoglou, M.T. Hepworth: Trans. Met. Soc. AIME **242**, 682 (1968)
3.141 B.A. Kolachev, O.P. Nazimov, L.N. Zhuravlev, Izv. Vyssh. Ucheb. Zaved: Tsvet. Met. **12**, 104 (1969)
3.142 D.N. Kazakov, V.M. Khokhrin, L.L. Kunin, P.I. Ozhegov, Yu.A. Priselkov: Zavodskaya Lab. **4**, 441 (1970)
3.143 D.L. Johnson, H.G. Nelson: Metall. Trans. **4**, 569 (1973)
3.144 A.M. Sukhotin, É.I. Antonovskaya, E.V. Sgibnev, I.I. Kornilov, T.T. Nartova, T.V. Mogutova, A.K. Shul'man: Zashchita Metallov **11**, 430 (1975)
3.145 O.P. Nazimov, L.N. Zhuravlev, Izv. Vyssh. Ucheb. Zaved: Tsvet. Met. **19**, 160 (1976)
3.146 T. Lin, H. Zhang, W. Wong, N. Pan: Scripta Metall. **23**, 891 (1989)
3.147 E.A. Gulbransen, K.F. Andrew: J. Electrochem. Soc. **101**, 560 (1954)
3.148 C.M. Schwartz, M.W. Mallet: Trans. Am. Soc. Metals **46**, 640 (1954)
3.149 M.W. Mallet, W.M. Albrecht: J. Electrochem. Soc. **104**, 142 (1957)
3.150 A. Sawatzki: J. Nucl. Mat. **2**, 62 (1960)
3.151 M. Someno: J. Jpn. Inst. Metals **24**, 249 (1960)
3.152 V.L. Gelezunas, P.K. Conn, R.H. Price: J. Electrochem. Soc. **110**, 799 (1963)
3.153 J.J. Kearns: J. Nucl. Mat. **43**, 330 (1972)
3.154 F.M. Mazzolai, J. Ryll-Nardzewski: J. Less-Common Met. **49**, 323 (1976)
3.155 S. Naito: J. Chem. Phys. **79**, 3113 (1983)
3.156 W. Kunz, H. Münzel, U. Helfrich, H. Horneff: Z. Metallkde. **74**, 289 (1983)
3.157 S. Naito, M. Yamamoto, T. Hashino: J. Phys.: Condens. Matter **2**, 1963 (1990)
3.158 G. Arnold, J.-M. Welter: Metall. Trans. A **14**, 1573 (1983)
3.159 R. Schmidt, M. Schlereth, H. Wipf, W. Assmus, M. Müllner: J. Phys.: Condens. Matter **1**, 2473 (1989)

3.160 D. Beisenherz, D. Guthardt, H. Wipf: J. Less-Common Met. **172–174**, 693 (1991)
3.161 P. Thompson, F. Reidinger, J.J. Reilly, L.M. Corliss, J.M. Hastings, J. Phys. F. **10**, L57 (1980)
3.162 B.S. Berry, W.C. Pritchet: Phys. Rev. B **24**, 2299 (1981)
3.163 R. Kirchheim, F. Sommer, G. Schluckebier: Acta metall. **30**, 1059 (1982)
3.164 R. Kirchheim: Acta metall. **30**, 1069 (1982)
3.165 R.-W. Lin, H.H. Johnson: J. Non-Cryst. Solids **51**, 45 (1982)
3.166 C.-H. Hwang, K. Cho, Y. Takagi, K. Kawamura: J. Less-Common Met. **89**, 215 (1983)
3.167 B.S. Berry, W.C. Pritchet: J. Physique **46**, C10-457 (1985)
3.168 A.H. Verbruggen, R.C. van den Heuvel, R. Griessen, H.U. Künzi: Scripta metall. **19**, 323 (1985)
3.169 D. Richter, G. Driesen, R. Hempelmann, I.S. Anderson: Phys. Rev. Lett. **57**, 731 (1986)
3.170 A.J. Dianoux, B. Rodmacq, A. Chamberod: J. Less-Common Met. **131**, 145 (1987)
3.171 A. Hofmann, H. Kronmüller: Phys. Status Solidi (a) **104**, 619 (1987)
3.172 J.T. Markert, E.J. Cotts, R.M. Cotts: Phys. Rev. B **37**, 6446 (1988)
3.173 W. Schirmacher, M. Prem, J.-B. Suck, A. Heidemann: Europhys. Lett. **13**, 523 (1990)
3.174 R. Griessen: Phys. Rev. B **27**, 7575 (1983)
3.175 C.C. Baker, H.K. Birnbaum: Acta metall. **21**, 865 (1973)
3.176 G. Pfeiffer, H. Wipf: J. Phys. F **6**, 167 (1976)
3.177 A. Magerl, J.J. Rush, J.M. Rowe, D. Richter, H. Wipf: Phys. Rev. B **27**, 927 (1983)
3.178 T.H. Metzger, U. Schubert, H. Peisl: J. Phys. F **15**, 779 (1985)
3.179 P.P. Matyash, N.A. Skakun, N.P. Dikii: JETP Lett. **19**, 18 (1974)
3.180 N.A. Skakun, P.A. Svetashov, V.E. Storizhko, A.G. Strashinskiï: Sov. Phys.- Solid State **26**, 1919 (1984)
3.181 H. Wipf, K. Neumaier: Phys. Rev. Lett. **52**, 1308 (1984)
3.182 G.J. Sellers, A.C. Anderson, H.K. Birnbaum: Phys. Rev. B **10**, 2771 (1974)
3.183 S.G. O'Hara, G.J. Sellers, A.C. Anderson: Phys. Rev. B **10**, 2777 (1974)
3.184 R.F. Mattas, H.K. Birnbaum: Acta metallurg. **23**, 973 (1975)
3.185 P. Schiller, A. Schneiders: Phys. Status Solidi (a) **29**, 375 (1975)
3.186 P. Schiller, H. Nijman: Phys. Status Solidi (a) **31**, K77 (1975)
3.187 C.G. Chen, H.K. Birnbaum: Phys. Status Solidi (a) **36**, 687 (1976)
3.188 M. Locatelli, K. Neumaier, H. Wipf: J. Physique **39**, C6-995 (1978)
3.189 D.B. Poker, G.G. Setser, A.V. Granato, H.K. Birnbaum: Z. Phys. Chem. N. F. **116**, 39 (1979)
3.190 P.E. Zapp, H.K. Birnbaum: Acta Metallurg. **28**, 1523 (1980)
3.191 H. Wipf, A. Magerl, S.M. Shapiro, S.K. Satija, W. Thomlinson: Phys. Rev. Lett. **46**, 947 (1981)
3.192 R. Hanada: Scripta Metall. **15**, 1121 (1981)
3.193 G. Cannelli, R. Cantelli: Solid State Commun. **43**, 567 (1982)
3.194 Y. Sado, M. Shinohara, R. Hanada, H. Kimura: Proc. 3rd Int'l Congress on Hydrogen and Materials (Paris 1982) ed. by P. Azou, p. 1
3.195 G. Bellessa: J. Physique-Lett. **44**, L-387 (1983)
3.196 J.L. Wang, G. Weiss, H. Wipf, A. Magerl: In *Phonon Scattering in Condensed Matter,* ed. by W. Eisenmenger, K. Laßmann and S. Döttinger Springer Ser. Solid-State Sci., Vol. 51, (Springer, Berlin Heidelberg 1984) p. 401
3.197 S. Okuda, H. Mizubayashi, N. Matsumoto, N. Kuramochi, C. Mochizuki, R. Hanada: Acta metallurg. **32**, 2125 (1984)
3.198 E. Drescher-Krasicka, A.V. Granato: J. Physique **12**, C10-73 (1985)
3.199 K.F. Huang, A.V. Granato, H.K. Birnbaum: Phys. Rev. B **32**, 2178 (1985)
3.200 G. Cannelli, R. Cantelli, F. Cordero: Phys. Rev. B **34**, 7721 (1986)
3.201 H. Wipf, D. Steinbinder, K. Neumaier, P. Gutsmiedl, A. Magerl, A.-J. Dianoux: Europhys. Lett. **4**, 1379 (1987)
3.202 P. Gutsmiedl, M. Schiekhofer, K. Neumaier, H. Wipf: Springer Proc. Phys **17**, 158 (Springer, Berlin Heildelberg 1987)

3.203 D. Steinbinder, H. Wipf, A. Magerl, D. Richter, A.-J. Dianoux, K. Neumaier: Europhys. Lett. **6**, 535 (1988)
3.204 M. Nagano, E.A. Clark, H.K. Birnbaum: J. Phys. F **18**, 863 (1988)
3.205 T. Pfitz, R. Messer, A. Seeger: Z. Phys. Chem. N. F. **164**, 969 (1989)
3.206 K. Neumaier, D. Steinbinder, H. Wipf, H. Blank, G. Kearley: Z. Phys. B **76**, 359 (1989)
3.207 W. Morr, A. Müller, G. Weiss, H. Wipf, B. Golding: Phys. Rev. Lett. **63**, 2084 (1989)
3.208 D. Steinbinder, H. Wipf, A.-J. Dianoux, A. Magerl, K. Neumaier, D. Richter, R. Hempelmann: Europhys. Lett. **16**, 211 (1991)
3.209 J. Kondo: Physica **125** B, 279 (1984)
3.210 K. Yamada: Progr. Theor. Phys. **72**, 195 (1984)
3.211 H. Grabert, S. Linkwitz, S. Dattagupta, U. Weiss: Europhys. Lett. **2**, 631 (1986)
3.212 Yu. Kagan, N.V. Prokof'ev: Sov. Phys.–JETP **63**, 1276 (1986)
3.213 H. Teichler: *Quantum Aspects of Molecular Motions in Solids,* ed. by A. Heidemann, A. Magerl, M. Prager, D. Richts, T. Springer, Springer Proc. Phys., Vol. 17 (Springer, Berlin Heidelberg 1987) p. 167 and priv. communication
3.214 S. Dattagupta, H. Grabert, R. Jung: J. Phys.: Condens. Matter **1**, 1405 (1989)
3.215 U. Weiss, M. Wollensak: Phys. Rev. Lett. **62**, 1663 (1989)
3.216 G. Cannelli, L. Verdini: Ric. Sc. **36**, 98 (1966)
3.217 P. Schiller, A. Schneiders: Proc. Int'l Conf. on Vacancies and Interstitials in Metals, (Jülich 1968) ed. by J. Diehl et al., Jül-Conf-2(II), p. 871
3.218 G. Cannelli, R. Cantelli, M. Koiwa: Phil. Mag. A **46**, 483 (1982)
3.219 G. Cannelli, R. Cantelli, F. Cordero, K. Neumaier, H. Wipf: J. Phys. F **14**, 2507 (1984)
3.220 O. Yoshinari, N. Yoshikawa, H. Matsui, M. Koiwa: J. Physique **46**, C10–95 (1985)
3.221 K. Neumaier, H. Wipf, G. Cannelli, R. Cantelli: Phys. Rev. Lett. **49**, 1423 (1982)
3.222 G.J. Sellers, M. Paalanen, A.C. Anderson: Phys. Rev. B **10**, 1912 (1974)
3.223 S. Tanaka, M. Koiwa: Scripta Metall. **15**, 403 (1981)
3.224 H. Kronmüller: In *Hydrogen in Metals I,* ed. by G. Alefeld, J. Völkl, Topics Appl. Phys. Vol. 28, (Springer, Berlin, Heidelberg 1978) p. 289
3.225 H. Kronmüller, N. König, F. Walz: Phys. Status Solidi (b) **79**, 237 (1977)
3.226 J.J. Au, H.K. Birnbaum: Acta Metall. **26**, 1105 (1978)
3.227 B. Hohler, H. Kronmüller: Phil. Mag. A **43**, 1189 (1981)
3.228 B. Herbst, H. Kronmüller: Phys. Status Solidi (a) **66**, 255 (1981)
3.229 B. Hohler, H. Schreyer: J. Phys. F. **12**, 857 (1982)
3.230 G. Higelin, H. Kronmüller, R. Lässer: Phys. Rev. Lett. **53**, 2117 (1984)
3.231 H. Kronmüller, G. Higelin, P. Vargas, R. Lässer: Z. Phys. Chem. N.F. **143**, 161 (1985)
3.232 W.M. Robertson: Z. Metallkd. **64**, 436 (1973)
3.232 K. Yamakawa, F.E. Fujita: Jpn. J. Appl. Phys. **16**, 1747 (1977)
3.233 R.R. Arons, H.G. Bohn, H. Lütgemeier: Solid State Commun. **14**, 1203 (1974)

4. Nuclear Magnetic Resonance in Metal Hydrogen Systems

With 16 Figures and 1 Table

Nuclear Magnetic Resonance (NMR) methods have continued to make strong contributions to understanding atom locations and electronic structure in Metal-Hydrogen (M-H) systems, as well as the changes that occur in these properties at structural and electronic phase transitions. NMR methods also afford a powerful approach to the study of hydrogen motion (diffusion). Atomic nuclei interact with the static and dynamic microscopic local electromagnetic fields *via* the nuclear magnetic dipole and electric quadrupole moments, and since these moment properties are invariant for a given nuclear species, the interactions probe the magnitudes and symmetries of the magnetic and electric fields at the nuclear sites. Both steady-state (wide-line) and transient (pulsed) measurement techniques may be applied to study the NMR of all three hydrogen isotopes (^{1}H,^{2}D,^{3}T) as well as a considerable number of transition metal nuclides.

4.1 Background

The past decade has seen a surge of applications of NMR methods to the study of M-H systems. The range of systems studied has been greatly expanded to include transition metal alloy-hydrogen systems, intermetallic and amorphous hydrides, as well as the classical binary systems. Experimental capabilities have been extended and improved, in particular in the area of high-field (superconducting) magnet systems and correspondingly high resonance frequencies. The temperature range over which NMR parameters are studied has been extended in both low- and high-temperature regimes, resulting in the discovery of new phenomena at both extremes. Developments in underlying theory, especially that of spin relaxation, have been significant and have contributed to improved interpretation of experimental results.

In a broad outline, the resonance second moment, which depends on the internuclear magnetic dipolar interaction, is sensitive to hydrogen locations in the lattice, particularly in the case of the proton (^{1}H) itself. The electric quadrupole interaction of large spin nuclides (e.g., ^{93}Nb,^{139}La), and also of the deuteron (^{2}D), reflects the strength and symmetry of the second-order Crystalline Electric Field (CEF) at the nuclear site. Measurements of the quadrupole interaction parameters are especially useful in studies of struc-

Topics in Applied Physics, Vol. 73
Wipf (Ed.)
© Springer-Verlag Berlin Heidelberg 1997

tural phase transitions and hydrogen superlattice formation. Both the shift of the resonance position (Knight shift) and the Spin-Lattice Relaxation (SLR) rate R_1 depend on the electronic density-of-states at the Fermi level, $N(E_F)$, in metallic materials. The dependence of these parameters on hydrogen concentration and temperature shows how hydrogen and hydriding alters $N(E_F)$ and can reveal electronic structure transitions. With respect to hydrogen motion and diffusion, the relaxation rates R_1 and R_2 (spin-spin relaxation rate) as well as the linewidth are strongly influenced by the modulation of the magnetic dipolar interaction caused by hydrogen motion. The electric quadrupole interaction is also modulated, so that hydrogen motion may be monitored in the NMR of both the diffusing hydrogen isotope and the stationary metal nuclear species. Finally, the diffusion itself may be measured by so-called Pulsed Field Gradient (PFG) techniques.

Table 4.1 summarizes the nuclear properties of the hydrogen isotopes and of those metal nuclides of greatest significance in studies of M-H systems. Here, the quantity $\gamma/2\pi$ is the nuclear magnetogyric ratio in practical (SI) units, Q is the nuclear quadrupole moment, and $(1 - \gamma_\infty)$ is the Sternheimer antishielding factor which is important for quadrupole interactions. The remaining entries are self-explanatory. To date, the NMR of all three hydrogen isotopes has been utilized in studying M-H systems, and, in addition, the nuclei of the metals Sc, Ti, V, Y, Zr, Nb, and La have been employed to greater or lesser degree.

In this survey, developments in theory and experiment, in types of systems studied, and in information gained that have emerged in the period since the publication of the review by *Cotts* [4.1] will be outlined. The con-

Table 4.1. Nuclear Moment properties of hydrogen isotopes and metal nuclei of significance for NMR studies of metal-hydrogen systems

Metal-Hydrogen system	Nucleus	Natural abundance [%]	Spin	$\gamma/2\pi$ [MHz/T]	Q [barns]	$(1-\gamma_\infty)$
Sc–H	^{45}Sc	100	7/2	10.3	0.22	8
Y–H	^{89}Y	100	1/2	2.1	0	–
La–H	^{139}La	≈ 100	7/2	5.6	0.21	≈ 80
Lu–H	^{175}Lu	97.4	7/2	4.9		≈ 80
Ti–H	^{47,49}Ti	8, 5	5/2, 7/2	2.4	0.29, 0.24	10
Zr–H	^{91}Zr	11	5/2	4.0	–0.21	30
Hf–H	^{177,179}Hf	18, 14	7/2, 9/2	1.3, 0.8		≈ 80
V–H	^{51}V	≈ 100	7/2	11.2	0.04	≈ 10
Nb–H	^{93}Nb	100	9/2	10.4	–0.36	≈ 25
Ta–H	^{181}Ta	100	7/2	5.1	~ 3	≈ 60
Cr–H	^{53}Cr	9.5	3/2	2.4	0.03	≈ 10
Pd–H	^{105}Pd	22	5/2	1.7		
	^{1}H	≈ 100	1/2	42.6	0	
	^{2}D		1	6.5	0.0028	≈ 1
	^{3}T		1/2	45.4	0	

cepts and theoretical bases for the analysis of experimental results presented in that contribution remain for the most part valid today. Accordingly, in this chapter only further advances and modifications will be discussed, insofar as is practicable, together with reviews of experimental findings.

4.2 Structural Information

A problem fundamental to most M-H system studies is that of determing the locations of hydrogen in the metal lattice. Whether concerned with hydrogen solubility in the α-phase, with a simple binary hydride MH_x, or with more complex intermetallic hydrides, accurate knowledge of hydrogen locations is usually essential to the interpretation of other measurements (e.g., NMR and optical) and to constructing correct theoretical descriptions (e.g., band structures). Hydrogen is essentially undetected by X-rays, so that the principal experimental methods available for determing the lattice sites which it occupies are neutron scattering and NMR. NMR can yield information on hydrogen locations and structure in two ways: (1) through the magnetic dipole interaction between nuclei, and (2) through the interaction of quadrupolar nuclei with the gradient of the CEF at the nuclear site. Both interactions are sensitive to structural phase transitions and order-disorder transitions.

4.2.1 Magnetic Dipolar Second Moments

As discussed in [4.1], the resonance linewidth and shape are frequently determined by the magnetic dipole interaction between nuclei. In practise, the actual shape is almost never calculated, instead the dipolar second moment is calculated for a known or assumed structure and compared with measurement. The general theoretical expression for the proton second moment in a powder sample, expressed in (angular) frequency units, is [4.1],

$$\langle \Delta \omega_d^2 \rangle = (3/5)\, C_I \sum_n r_{mn}^{-6} + (4/15)\, C_S \sum_{n'} r_{mn'}^{-6}, \tag{4.1}$$

where $C_I = (\gamma_I^2 \hbar)^2 I(I+1)$, $C_S = (\gamma_I \gamma_S \hbar)^2 S(S+1)$, I and γ_I are the spin and magnetogyric ratio of ^1H, S and γ_S are those of the host lattice metal nuclei, and $\hbar$ is Planck's constant divided by 2π. The sum $\sum_n r_{mn}^{-6}$ is over all hydrogen sites, n, from a generic site m, and $\sum_{n'} r_{mn'}^{-6}$ is over all metal sites n' having nuclear spins S, from a generic proton site m. The two terms on the right-hand side of (4.1) are often written as $\langle \Delta \omega_{II}^2 \rangle$ and $\langle \Delta \omega_{IS}^2 \rangle$ to identify them as the homonuclear and heteronuclear contributions to $\langle \Delta \omega_d^2 \rangle$.

The second moment can be measured by direct evaluation from the resonance line shape or from the initial curvature of the free-induction decay signal $G(t)$ following a radiofrequency (rf) pulse at the resonance frequency [4.2]. Both methods entail some difficulties, especially when $\langle \Delta \omega_d^2 \rangle$ is large. In addition to the wide-line and pulse methods described in [4.1] for measuring $\langle \Delta \omega_d^2 \rangle$, several pulse techniques based on the solid echo method [4.3] have been employed to circumvent these difficulties. These are the "magic" dipolar echo [4.4] and a simplified version thereof [4.5].

Conventional proton second moment measurements are most discriminating as to hydrogen locations in the lattice when the metal spin contribution $\langle \Delta \omega_{IS}^2 \rangle$ is small. This occurs when the metal nuclei have small magnetic moments (e.g.,^{89}Y) and/or are of low abundance as is the case with the NMR isotopes of Ti and Zr, for example (Table 4.1). However, as detailed in [4.1], when $\langle \Delta \omega_{IS}^2 \rangle \gg \langle \Delta \omega_{II}^2 \rangle$ the solid echo method yields information on $\langle \Delta \omega_{II}^2 \rangle$ directly.

When isolated close pairs of protons occur, the strong nearest-neighbor dipolar interaction results in a characteristic two-peak lineshape, the spacing of the peaks being independent of applied field, B_o, and temperature, T. This "Pake doublet" structure of the ^{1}H resonance was first observed for water molecule protons in $CaSO_4 \cdot 2H_2O$ [4.6]. For polycrystalline (powder) samples, the splitting (in field units) is given by: $\Delta B = 3\gamma_I^2 \hbar^2 / 2r^3$, where r is the ^{1}H-^{1}H spacing. Hence, r can be inferred from measurements of the splitting ΔB. This dipolar splitting has only rarely been reported for protons in metal-hydrogen systems, because the short ^{1}H-^{1}H separation needed to yield a value of ΔB greater than the dipolar broadening due to all other, more distant spins, is less than the typical distance of closest approach of hydrogen atoms in a metallic solid, i.e., the "Switendick criterion", ~ 2.1 Å [4.7]. Reported examples of Pake doublet structure are given in 4.5.2d and 4.53B. Other proton configurations which also yield a simple doublet dipolar structure include a planar arrangement in which each spin has three equally spaced nearest neighbors [4.8] and a linear arrangement in which each spin has two equally spaced nearest neighbors. In both cases, the ^{1}H-^{1}H spacing r required to yield a particular splitting ΔB is greater than that required in the single close-pair case.

4.2.2 Electric Quadrupole Interactions

The nuclear quadrupole interaction is treated in detail in [4.1]. As discussed there, nuclei with spin $I > 1$ usually possess an electric quadrupole moment, Q, which measures the departure of the nuclear charge distribution from spherical symmetry. Table 4.1 reveals that virtually all host metal nuclei in M-H systems, as well as the deuteron (^{2}D), possess quadrupole moments. Q interacts with the Electric Field Gradient (EFG) tensor produced by the lattice of point charges and the conduction electrons at the nuclear site. In its principal axis system (x, y, z), the EFG tensor is conventionally characterized by the "field gradient," $eq = \partial^2 V / \partial z^2$, and the asymmetry parameter, $\eta = (\partial^2 V / \partial x^2 - \partial^2 V / \partial y^2)/eq$, where $V(x, y, z)$ is the (total) electric potential at the nuclear site, and the principal axes are chosen such that $|\partial^2 V / \partial z^2| > |\partial^2 V / \partial y^2| > |\partial^2 V / \partial x^2|$. This, combined with the fact that $\partial^2 V / \partial x^2 + \partial^2 V / \partial y^2 + \partial^2 V / \partial z^2 = 0$ for charges external to the nucleus, ensures that $0 \leq \eta \leq 1$. The EFG vanishes at sites having cubic symmetry, but interstitial sites occupied by hydrogen (deuterium) in cubic host lattices frequently have less-than-cubic symmetry, e.g., tetrahedral sites in the bcc lattice.

In most cases of interest, the quadrupole coupling, e^2qQ, is weak enough in comparison to the nuclear Zeeman energy, $\gamma \hbar B_0$, that it can be treated as a perturbation of the latter, resulting in a splitting of the resonance line into a central transition and "satellite" lines. With powder samples, the resultant "powder" spectrum yields values of eq and η, but the orientation of the EFG tensor with respect to crystal axes can only be determined from measurements on single crystals. Nonetheless, in many cases in which the crystal structure and site occupancies are known, the orientation of the EFG tensor is evident.

Because of its integer spin $(I = 1)$ and quadrupole moment, deuteron magnetic resonance is especially sensitive to structural phase transitions. The integer spin means that when an appreciable quadrupole interaction occurs, the ^2D spectrum consists of two satellite lines lacking a central transition, whereas when $eq = 0$ the two transitions coincide, yielding a single unsplit resonance [4.9]. Deuteron spectra become more complex when $\eta \neq 0$ and/or in strong magnetic fields (e.g., 8 T) where the magnetic (Knight) shift causes comparable changes in the positions of the resonance features [4.10]. The values of eq and η reflect primarily short-range rather than long-range order in consequence of the fact that eq results from charges on neighboring ions, including other deuterons, which are relatively strongly screened in metallic solids. The conduction electron contribution to eq is less important since the conduction electron states in the vicinity of hydrogen sites are predominantly s-like.

4.3 Motion and Diffusion of Hydrogen

The motion and diffusion of hydrogen in metallic hosts has been the most emphasized objective of NMR investigations. This topic alone has been the subject of reviews by *Cotts* [4.11], *Fukai* and *Sugimoto* [4.12], and *Seymour* [4.13]. The temperature dependence of the spin relaxation rates R_1 and R_2 (or the times, $T_1 = R_1^{-1}, T_2 = R_2^{-1}$), and of R_1 in the rotating frame, $R_{1\rho}$ or $T_{1\rho}$, as well as the resonance linewidth, Δv or ΔB, are strongly affected by modulation of the internuclear magnetic dipolar interaction caused by hydrogen motion. Moreover, electric quadrupole interactions are also modulated in this way and are responsible for the effect of diffusion on R_1 and R_2 of the deuteron in deuterides and on R_1 of stationary metal nuclei. Diffusive motion also results in an "averaging" of quadrupole-split spectra of both the deuteron and metal nuclei [4.1]. Consequently it has been possible to monitor hydrogen motion in some systems from the standpoint of both the diffusing species and the stationary host lattice nuclei [4.14, 15].

4.3.1 Mean Residence Times and Relaxation Rates

The relaxation rates that reflect atomic motion, R_{1d}, R_{2d}, and $R_{1\rho}$, are described by sums of one or more terms, each of the general form

$$R = \langle \Gamma^2 \rangle J(\omega_I, \omega_S, \tau_c) , \tag{4.2}$$

where $\langle \Gamma^2 \rangle$ is that part of the interaction of the nuclear spin with its electromagnetic environment that is caused to fluctuate in time by the motion, and $J(\omega_I, \omega_S, \tau_c)$ is a spectral density function (power spectrum) that describes the dependence of the fluctuations in Γ on the resonance frequencies ω_I and ω_S of the resonant and nonresonant nuclei, respectively, and on the correlation time τ_c for the fluctuations. In turn, τ_c is approximately equal to the mean residence time, τ_d, of the diffusing hydrogen which is (almost always) assumed to follow Arrhenius behavior,

$$\tau_d = \tau_0 \, \exp(E_a/k_B T) , \tag{4.3}$$

where $\tau_0^{-1} = v_0$ is the jump attempt frequency, E_a the activation energy for diffusive hopping, k_B Boltzmann's constant, and T the absolute temperature. Hence, J is also a function of temperature.

(A) Magnetic Dipolar Relaxation

For dipolar interactions between protons and between protons and metal nuclei, the rates are expressed in terms of the spectral densities of the randomly varying dipolar fields [4.1, 2]:

$$R_{1d} = \frac{3}{2} C_I \left[J^{(1)}(\omega_I) + J^{(2)}(2\omega_I) \right]$$
$$+ C_S \left[\frac{1}{12} J^{(0)}(\omega_I - \omega_S) + \frac{3}{2} J^{(1)}(\omega_I) + \frac{3}{4} J^{(2)}(\omega_I + \omega_S) \right], \tag{4.4}$$

$$R_{1\rho} \simeq \frac{3}{8} C_I J^{(0)}(2\omega_1) + \frac{1}{6} C_S J^{(0)}(\omega_1) , \tag{4.5}$$

$$R_{2d} = \frac{3}{8} C_I \left[J^{(0)}(0) + 10 J^{(1)}(\omega_I) + J^{(2)}(2\omega_I) \right]$$
$$+ \frac{1}{2} C_S \left[\frac{1}{3} J^{(0)}(0) + \frac{1}{12} J^{(0)}(\omega_I - \omega_S) + \frac{3}{2} J^{(1)}(\omega_I) \right.$$
$$\left. + 3 J^{(1)}(\omega_S) + \frac{3}{4} J^{(2)}(\omega_I + \omega_S) \right] . \tag{4.6}$$

The spectral densities $J^{(q)}(\omega)$ are the Fourier transforms of the spatial correlation functions, $G^{(q)}(t)$, of the dipole-dipole interaction,

$$F_{ij}^{(q)}(t) = d_q Y_{2,q}(\Omega_{ij})/r_{ij}^3 , \tag{4.7}$$

with $d_0^2 = 16\pi/5$, $d_1^2 = 8\pi/15$, and $d_2^2 = 32\pi/15$, and the $Y_{2,q}$ are the normalized spherical harmonics with coordinates $(r, \theta, \phi)_{ij}$ made time-dependent by hydrogen motion. Accordingly,

$$G^{(q)}(t) = \sum_j F_{ij}^{(q)}(t') F_{ij}^{(q)*}(t' + t) \tag{4.8}$$

and

$$J^{(q)}(\omega) = \int_{-\infty}^{+\infty} G^{(q)}(t)\,e^{-i\omega t}\,dt. \tag{4.9}$$

In (4.8) the brackets $\langle\ \rangle$ indicate the ensemble average over all spin pairs (i, j). The factors C_I and C_S are the same as those in the second moment expression (4.1), and in (4.5) $\omega_1 = \gamma B_1$ is the Larmor (precession) frequency in the rf field B_1. The terms that include C_S as a factor arise from the dipolar interaction with nonresonant spins. For proton relaxation rates these terms are important at low hydrogen concentrations, as in solid solution phases, and when the metal nuclei possess substantial magnetic moments, e.g., ^{45}Sc, ^{93}Nb. Equations (4.4–6) are valid in the motionally narrowed regime where $\langle\Delta\omega_d^2\rangle\tau_d \ll 1$.

Despite very significant advances in the development of lattice-specific correlation and spectral density functions, most relaxation rate measurements have been analyzed using the simple exponential correlation function and associated Lorentzian spectral densities introduced by *Bloembergen* et al. [4.16], the so-called BPP functions. These have the merit of simple analytic form, and in view of other factors that often complicate experimental data (distributions of parameters, paramagnetic impurities, etc.), are often capable of yielding sufficiently reliable analyses.

The basic assumption of the BPP formulation is that the $G^{(q)}(t)$ are isotropic and decay exponentially with decay constant τ_c. For powder samples, $J^{(0)}(\omega) : J^{(1)}(\omega) : J^{(2)}(\omega) = 6:1:4$, with

$$J^{(1)}(\omega) = \frac{2}{15}\frac{2\tau_c}{1 + \omega^2\tau_c^2}\sum_k r_k^{-6}. \tag{4.10}$$

For the interaction between stationary host lattice nuclei (S) and ^{1}H, τ_c is just τ_d, whereas that of ^{1}H-^{1}H interactions is $\tau_d/2$. Full analytic expressions for R_{1d}, $R_{1\rho}$, and R_{2d} in terms of these spectral densities have been given in many papers and reviews [4.1, 2, 17]. The sum $\sum_k r_k^{-6}$ in (4.10) extends over all occupied hydrogen sites in those terms containing C_I and over all metal nuclear sites in those containing C_S, paralleling the sums in the dipolar second moment, (4.1). This can be useful in assessing whether the observed maximum rates of R_{1d} and $R_{1\rho}$ are reasonable.

The BPP form predicts a maximum R_{1d} when $\omega_0\tau_c \approx 1$, with maximum value proportional to ω_0^{-1}. At high temperatures, i.e., when $\omega_0\tau_c \ll 1$, R_{1d} is independent of resonance frequency and proportional to τ_c. At low temperatures ($\omega_0\tau_c \gg 1$) R_{1d} is proportional to $\omega_0^{-2}\,\tau_c^{-1}$. As already noted, R_{1d} depends on the rigid lattice second moment. If τ_d conforms to an Arrhenius relation, the conventional graph of $\ln(R_{1d})$ vs T^{-1} should be symmetrical and exhibit equal and opposite asymptotic slopes far from the maximum rate.

These features are illustrated in Fig. 4.1, which shows R_{1d} and $R_{1\rho}$ for protons as functions of reciprocal temperature in a metal-hydrogen system, using parameters typical of **fcc** dihydrides. The measured rate R_1 is the sum of the diffusion controlled dipolar rate R_{1d} and the electronic contribution, R_{1e} (Sect. 4.4), which is insensitive to hydrogen motion and is expected to dominate R_1 at low and very high temperatures outside the region of the R_{1d} maximum (e.g., see Sects. 4.3.2A,B). R_{1e} usually needs to be determined in order to reliably obtain the temperature dependence of R_{1d}. The rate $R_{1\rho}$ is so much faster than R_{1d} that it is not normally affected by R_{1e}, and its maximum rate occurs at a lower temperature, which is sometimes advantageous.

The BPP formulation fails to yield absolute values of τ_d better than a factor of 2 or 3. This makes it difficult to distinguish between different jump processes, i.e., elementary jump lengths L, by combining τ_d and diffusion (D) measurements through the relation, $D = f_t L^2 / 6\tau_d$, where f_t is the tracer correlation factor for the process. All theories for the correlation functions yield the same asymptotic slopes in $\ln(R_{1d})$ vs $\ln\tau_d$ plots [or $\ln(R_{1d})$ vs T^{-1} plots if τ_d obeys (4.3)], and therefore the same value of E_a sufficiently far from the R_{1d} maximum. In fact, E_a values based solely on R_{1d} measurements and BPP spectral densities normally agree well with those based on direct measurements of the diffusion. Nonetheless, what is needed, as first shown by *Torrey* [4.18], is to express the $G^{(q)}(t)$ in terms of the specific features of the

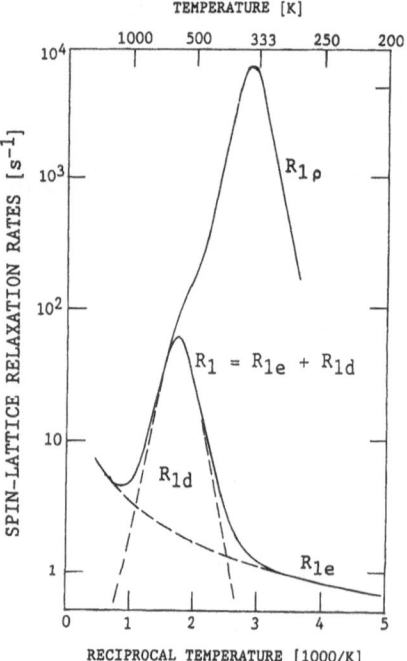

Fig. 4.1. Temperature dependence of the proton R_{1d} and $R_{1\rho}$, based on BPP spectral densities [4.16], and of R_{1e} and $R_1 = R_{1e} + R_{1d}$, using parameters typical of transition metal dihydrides at a resonance frequency of 40 MHz

random jumps among the interstitial sites of the hydrogen sublattice. The problem is more complex than finding the tracer correlation function, in part because it deals with relative displacements of a pair of particles.

Three general approaches have been used to calculate the $G^{(q)}(t)$ or $J^{(q)}(\omega)$. The calculations usually, but not always, assume jumps between nearest-neighbor sites and neglect repulsions between neighboring hydrogens, except to forbid double occupancy of any site.

First, exact results have been obtained in the limits of high and low frequency, i.e., low and high temperatures, and spin concentrations [4.19–24]. For example, using multiple-scattering theory, *Sankey* and *Fedders* [4.25, 26] tabulated the concentration dependence of $J^{(q)}(\omega)$ in the high- and low-temperature limits for the fcc lattice, e.g., PdH_x [4.25] and for the tetrahedral (T) sites in bcc lattices, e.g., α-NbH_x [4.26]. In the fcc case only ^1H-^1H dipolar interactions are considered, but in the latter both ^1H-^1H and ^1H-metal spin interactions are included. For the bcc lattice these calculations were extended to include octahedral (0) site occupancy and jumps between both T and O sites [4.27]. *Sholl* [4.21] has shown that for diffusive motion in 3 dimensions the rate R_{1d} depends on resonance frequency ω_0 and diffusivity D as $D^{-3/2}\omega_0^{-1/2}$, provided that $(\omega_0 \tau_d) \ll 1$. Measurement of the frequency dependence of R_{1d} at temperatures above that of the R_{1d} maximum furnishes a means of determining D. This dependence was verified experimentally by *Salibi* and *Cotts* [4.28] with measurements on $TiH_{1.63}$. For diffusive hopping in 2 dimensions, i.e., planar diffusion, $R_{1d} \propto D^{-1}\ln \omega_0$ [4.21], and this dependence on ω_0 has been reported [4.29] for the proton R_{1d} in $ZrBe_2H_2$ in which the hydrogen sites are located within the Zr planes. The Be atoms block direct H jumps from one Zr plane to the next.

The second approach is calculations made for approximate analytic theories for general values of frequency and spin concentrations [4.19, 30].

For the third approach, Monte Carlo (MC) calculations have been employed, beginning with the work of *Bustard* [4.31] for the sc lattice in the fcc dihydrides. This method has been applied to nearest-neighbor site hopping in powder samples for the case of ^1H-^1H dipolar interactions, using the MC procedure to determine the correlation function for short distances and after short times, with the large-distance, long-time behavior approximated by the diffusion equation. The results are presented in the form of tables and graphs for a range of concentrations in the sc, bcc, and fcc lattices [4.32]. The tabulated results also can be used to determine $R_{1\rho}$ and R_{2d} [4.33]. These calculations were extended to include the effect of pair-wise two-body interactions (both repulsive and attractive) between diffusing particles [4.34] for the sc and fcc lattices.

Because of the close proximity of nearest-neighbor T-sites in the bcc lattice and the repulsive electronic interactions between H atoms at close range, an H atom at a given site tends to reduce the occupation of neighboring sites (site-blocking). This was already evident in the first measurements of R_{1d} and D in $NbH_{0.6}$ [4.35]. The MC procedure has been applied to this case [4.36–38], taking account of both ^1H-^1H and ^1H-metal spin dipolar interactions and blocking of all sites as far as the second or third neighbor, for several hydrogen concentrations. The MC results obtained at [H]/[M] = 0.12 are shown in Fig. 4.2 in the form $\ln(R_{1d}/A)$ vs $\ln(\omega_0 \tau_d/2)$ for the cases of no blocking and for blocking to the third neighbor, for ^{1}H-^{1}H interactions only [4.36]. Also shown in Fig. 4.2 are the predictions of the BPP [4.16] and *Torrey* [4.38] models for the two cases. Although the latter both provide reasonable agreement with the MC data for the maximum rate at all blocking distances, both give poor estimates of the peak position, i.e., the value of $\omega_0 \tau_d$ at the maximum. This is especially significant for obtaining reliable estimates of τ_d.

The numerical results of MC and other lattice-specific theories are inconvenient to use in analyzing experimental data in comparison to the simple BPP functions. *Sholl* [4.40] has introduced a new analytic form for the spectral densities which adequately represents the numerical results of the MC calculations. Writing the spherically averaged form of $J^{(q)}(\omega)$ as [4.40],

$$J_{av}^{(q)}(\omega) = \frac{d_q^2 \tau_d}{4\pi b^6} \, g(y),$$ (4.11)

where b is the lattice parameter and $y = \omega \tau_d / 2$, the rate R_{1d} for ^{1}H-^{1}H interactions can then be written as

$$R_{1d} = A y_0 \left[g(y_0) + 4g(2y_0) \right]$$ (4.12)

with

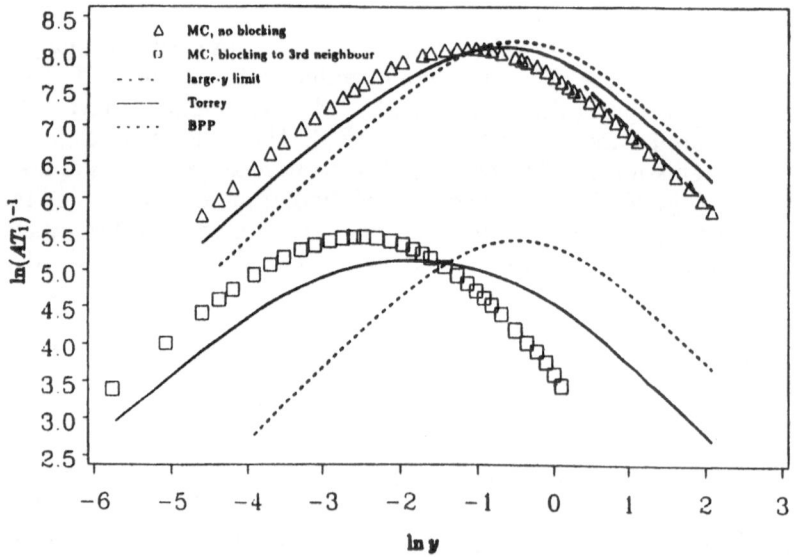

Fig. 4.2. Monte Carlo simulation results for the proton R_{1d} for hydrogen diffusing on the tetrahedral site sublattice in bcc metals, taking account only of the ^{1}H-^{1}H dipolar interaction, and assuming no site-blocking and blocking to the third neighbor [4.36]. The corresponding results for the Torrey [4.39] and BPP theories [4.16] are also shown

$$A = 2\gamma^4 \hbar^2 I(I+1)c/5b^6\omega_0 , \tag{4.13}$$

where c is the fraction of occupied sites. The BPP form of $g(y)$ is [4.40]

$$g(y) = S_1/(1+y^2) , \qquad \text{with} \qquad S_1 = b^6\sum_k r_k^{-6} ,$$

[compare with (4.10)].

The new form for $g(y)$ is

$$g(y) = S(a_1 + a_2 y^{1/2} + a_3 y^u + a_4 y^v + y^2)^{-1} , \tag{4.14}$$

with the seven parameters chosen to fit the numerical, e.g., MC, data. Tables of the parameter values are given in [4.40] for the sc, bcc, and fcc lattices for hydrogen concentrations ranging from $c \to 0$ to $c \to 1$ for ^{1}H-^{1}H dipolar interactions only. It is estimated [4.40] that the accuracy achieved using only the values of S, a_1, and a_2 (neglecting a_3, a_4, u, v) is better than 2% over the entire range of y for all three lattices. Tables of parameter values for this $g(y)$ have also been given for the case of nearest-neighbor interactions in the sc and fcc lattices [4.34].

The effects of short-ranged (electronic) interactions between hopping particles on R_{1d} have also been evaluated in the cases of low H concentrations on a sc lattice [4.41], and in a model in which strong repulsions at the saddle point make hopping to a vacant site difficult unless there is another vacancy nearby [4.42]. The latter case may be appropriate to fcc dihydrides at high hydrogen concentrations. Also, for hydrogen in solid solution in the bcc metals, the effects on R_{1d} caused by the presence of localized excited states of the interstitial H have been calculated and compared with experimental data [4.41].

(B) Distributions of Motional Parameters

It is frequently found that the slope on the low-temperature side of the proton $\ln R_{1d}$ vs T^{-1} maximum is significantly less than on the high-temperature side, and the frequency dependence on the low-temperature side is weaker than ω_0^2, as expected on the basis of theory. The more sophisticated lattice-specific theories for $J(\omega)$ yield R_{1d} maxima which are somewhat broader than BPP and the asymptotic limits are approached less rapidly, but the ω_0 and τ_d dependences of the slopes are the same as for BPP. However, for systems in which metal-proton dipolar interactions are weak, all calculations of the maximum rate are in remarkably good agreement, the lattice-specific theories yielding values only about 15% weaker than BPP for the principal hydrogen sublattices [4.36].

A contribution to R_1 due to low levels of paramagnetic impurities (Sect. 4.3.3) can result in a weaker low-temperture side slope, and even in an apparent weak secondary maximum that could be interpreted as arising from a second motional process, e.g., on the O-site sublattice. A far more typical reason for the occurrence of such asymmetric R_{1d} maxima is the presence of a distribution of the motional parameters, E_a and v_0 [4.44, 45]. This is certainly the origin of the effect found in amorphous [4.46] and disordered (e.g., alloy) [4.47] systems in which a distribution of both site and barrier energies and attempt frequencies is to be expected.

The calculation of R_{1d} follows from (4.4) with the spectral density "distributed" by introducing a distribution function, $G(\tau_d)$, which is folded with the spectral density. In most applications of this procedure, $G(\tau_d)$ has been attributed to a distribution of activation energies, $G(E_a)$, usually taken to be Gaussian for simplicity and ease of visualization of the results. Any variations in v_0 that may occur are neglected or approximated by use of the Zener relation [4.47], $v_0 = v_{00} \exp(\beta E_a)$, with $\beta = 5 \times 10^{-4} \mathrm{K}^{-1}$ as a fixed parameter. The calculations of R_{1d} then become [4.46, 47]

$$R_{1d} = \int_0^{\infty} R_{1d}\,(E_a, \tau_d)\,G(E_a)\,dE_a . \tag{4.15}$$

This procedure has the advantage of simultaneously fitting R_{1d} data at several different resonance frequencies with a single set of parameters, the average E_a, $\overline{E_a}$, the distribution width, ΔE_a, and the jump attempt frequency, v_0. The effects of this are to [4.46]: (i) reduce the maximum rate, (ii) broaden the width of the peak, (iii) cause the peak to become asymmetric, and (iv) reduce the frequency dependence on the low-temperature side. However, the peak position changes only slightly in temperature at a given NMR frequency. The strongest and simplest indications of a distribution of E_a values are the reduction in maximum relaxation rate from the expected value based on the second moment for the proton spin system [4.46] and the dependence of the resulting τ_d values on the resonance frequency [4.47].

An excellent example of the applicability of this approach is given by measurements of both R_{1d} and $R_{1\rho}$ in amorphous $Zr_3RhH_{3.5}$ [4.46]. By including $R_{1\rho}$, these measurements cover frequencies from 25 kHz to 38.9 MHz, providing a significant test of these ideas. The measured rates, together with fitted curves based on (4.15), are shown in Fig. 4.3. An activation energy distribution having mean energy, $\overline{E}_a = 0.46$ eV/atom, and width 0.12 eV/atom, independent of temperature, were used in the fitting. The jump attempt frequency, $v_0 = 5 \times 10^{13}$ s^{-1}, was fixed and independent of E_a. The fit was made using BPP functions for R_{1d} and $R_{1\rho}$ multiplied by a numerical factor so that the peak rates agree with those from lattice-specific numerical calculations [4.32].

(C) Electric Quadrupole Relaxation

Measurements of the spin relaxation rates of host lattice metal nuclei can provide an excellent means of studying both particle and particle vacancy motion on the hydrogen sublattice when the dominant SLR mechanism is electric quadrupole. The quadrupolar relaxation rate, R_{1Q}, of a stationary nucleus depends on site occupancy probabilities only. The spin label which

Fig. 4.3. Temperature dependence of the proton R_1 and $R_{1\varrho}$ for the amorphous metal hydride, $Zr_3RhH_{3.5}$, together with the best simultaneous fit to all of the data points (solid curves) based on a distribution of E_a values using (4.15) with BPP spectral densities [4.16]. scaled to approximate lattice specific results [4.46]

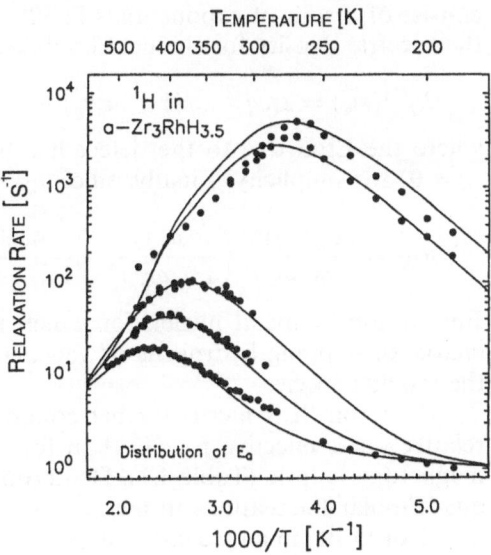

distinguishes particles in the magnetic dipolar case (e.g., the hopping proton) does not influence the EFG fluctuations which cause the relaxation. Also, contributions to the SLR of the stationary nucleus from paramagnetic impurities in the lattice are negligible at high temperatures, but may contribute at low temperatures. Measurement of the temperature dependence of R_{1Q} of a stationary metal nucleus should reflect directly the hydrogen or hydrogen vacancy hopping rate, depending on whether the hydrogen sublattice is mostly filled or mostly empty. Unfortunately, as can be seen in Table 4.1, almost all metal nuclear species in hydrides require strong magnetic fields for effective study of relaxation processes because of large spins and/or low abundance. In addition, even when the quadrupole moment is small (e.g., ^{51}V) the electronic contribution R_{1e} to the total rate may be too great, making detection of R_{1Q} difficult, if not impossible. An exception is ^{45}Sc, which has proved to be an extremely useful nucleus in both the solid solution and dihydride phases, see Sect. 4.5.1A.

The quadrupole relaxation rate is given by [4.48, 49]

$$R_{1Q} = \frac{9\,\alpha_Q c}{160} \left[J_Q^{(1)}(\omega_0) + J_Q^{(2)}(2\omega_0) \right], \tag{4.16}$$

where

$$\alpha_Q = (eQ)^2 (2I + 3)/(\hbar I)^2 (2I - 1), \tag{4.17}$$

with ω_0 the resonance frequency of the quadrupolar nucleus and c the concentration of moving (hopping) defects. This formulation assumes that a common spin temperature is maintained among the spin sublevels of the quadrupolar nucleus by, for example, sufficiently strong dipolar interactions. Otherwise, the magnetization recoveries monitored in the R_{1Q} measurement consist of a sum of exponentials [4.50]. Again, within the BPP framework, the spectral density for the field gradient fluctuations is given by [4.48],

$$J_Q^{(1)}(\omega_0) = 2(eq)^2 \tau_c / (1 + \omega_0^2 \tau_c^2) \,, \tag{4.18}$$

where the EFG due to the defect has been taken to be axially symmetric ($\eta = 0$) for simplicity. Finally, since $J_Q^{(2)}(\omega) = 4\,J_Q^{(1)}(\omega)$ for powders,

$$R_{1Q} = \frac{9\,\alpha_Q(eq)^2 c}{80\,\omega_0} \left(\frac{\omega_0 \tau_c}{1 + \omega_0^2\,\tau_c^2} + \frac{4\omega_0 \tau_c}{1 + 4\,\omega_0^2\,\tau_c^2} \right). \tag{4.19}$$

For stationary metal nuclei, for which the EFG results from the effective charge of hopping hydrogens or vacancies, $\tau_c = \tau_d$, the mean dwell time of the moving species.

Deuteron SLR merits further comment. Magnetic dipolar relaxation is relatively less effective for ^{2}D than for ^{1}H because of the deuteron's small magnetogyric ratio (Table 4.1). Deuteron relaxation is usually dominated by quadrupolar fluctuations in metal-deuterium systems. Although a deuteron at a T or O site in an fcc lattice experiences no EFG due to the metal lattice, fluctuating EFG's arise from transient occupation of some neighboring interstitial sites by other deuterons. On the other hand, the T sites in the bcc lattice are less-than-cubic and possess an EFG, but three different orientations of the EFG z axis occur, so that a deuteron hopping among these sites experiences an EFG whose orientation constantly changes. At low deuterium concentrations in the bcc lattice, $\tau_c = \tau_d$, but at high concentrations and in the fcc dideuterides, $\tau_c = \tau_d/2$. In this case, (4.19) has the same form as R_{1d} for ^{1}H-^{1}H dipolar interactions alone. Consequently, the temperature dependence of R_{1Q} for ^{2}D and of R_{1d} for ^{1}H should maintain a constant proportionality. This is not found to be the case.

An important difference between magnetic dipole and electric quadrupole relaxation is that in the former the nuclear dipole moment interacts separately with each of the surrounding dipoles, whereas the quadrupole moment interacts with the total EFG due to all surrounding charges taken together. Hence, 3-particle as well as 2-particle correlation functions are needed to adequately describe quadrupole relaxation, whereas only 2-particle functions are necessary to give a good approximation for dipolar relaxation. It has been shown that the 3-particle term tends to cancel the 2-particle term in R_{1Q}, removing the proportionality between quadrupolar and dipolar rates and shifting the R_{1Q} maximum to larger values of $\omega_0\tau_d/2$ with respect to the R_{1d} maximum [4.51].

In a number of fcc dideuterides, e.g., YD$_{1.88}$ [4.52] and HfD$_{1.7}$ [4.53], the R_{1Q} maxima occur at lower temperatures than in the corresponding hydrides at the same resonance frequency, opposite to the effect of the classical isotope dependence of diffusive hopping. Using mean field theory for these deuterides [4.54], it has been shown that the 3-body term increases the value of $\omega_0\,\tau_d$ at the R_{1Q} maximum by a factor of 2–2.5, depending on deuterium concentration and whether 0 or T sites are occupied, and in the correct sense to explain most observations [4.13]. A thorough discussion of the relations between relaxation in the laboratory and rotating frames for dipolar and quadrupolar relaxation has been given by *Kelly* and *Sholl* [4.55].

4.3.2 Other Relaxation Mechanisms

In addition to the "classical" SLR mechanisms, three additional processes have been shown to contribute significantly in certain systems and temperature ranges, thus interfering with determinations of both R_{1e} and R_{1d}. These are (i) SLR by paramagnetic impurity ions, R_{1p}, (ii) cross relaxation between protons and quadrupolar nuclei of the host lattice, R_{1c}, and (iii) anomalous SLR at high temperatures above the R_{1d} maximum, R_{1x}. Of these, R_{1p} can be effective at all temperatures of interest, interfering with determinations of both R_{1e} and R_{1d}; R_{1c} is most effective in the rigid-lattice regime, i.e., in the absence of hydrogen diffusive motion and hence interferes primarily with the determination of R_{1e}; and R_{1x} interferes with the determination of R_{1e} at high temperatures. In addition, R_{1x} appears to result from yet an unknown mechanism. Each of these relaxation mechanisms is reviewed in the following sections.

(A) Relaxation by Paramagnetic Impurity Ions, R_{1p}

The impurity-induced contribution R_{1p} has been shown [4.56] to markedly affect the temperature dependence of R_1. Neglecting R_{1p} can lead to serious misinterpretation of data. The magnetic rare earth ions are the most troublesome in this respect, although Mn^{2+} and Cr^{3+} have also been shown to influence R_{1d} determinations [4.57, 58]. Figure 4.4 shows the temperature dependence of the proton T_1 in a series of samples of $YH_{\sim 2}$ prepared from high-purity yttrium metal containing progressively increasing controlled levels of Gd impurity [4.56]. R_{1p} contributes to the measured rate over the entire temperature range investigated, not just at low temperatures, so that the measured T_1 is reduced to half its value in the purest sample by as little as 20 parts-per-million (ppm) of Gd. The other characteristic feature of these data is the appearance of a subsidiary (or secondary) minimum on the low-temperature side of the T_{1d} minimum associated with hydrogen diffusion. At high impurity levels this secondary minimum "wipes out" the normal T_{1d}, but at low impurity levels this extra minimum may be mistakenly interpreted to originate from different stages of hydrogen motion having different activation energies. Entirely similar effects have also been found for Nd, Dy, and Er impurities [4.59]. However, the cases of Ce and Pr are somewhat different (see further below).

For rare earth ions in the Sc, Y, La, and Lu hydrides, R_{1p} appears to result solely from fluctuations of the magnetic dipolar field of the impurity moments. Protons close to the impurities are directly relaxed by this mechanism. At low temperatures, ^1H magnetization distant from an impurity ion diffuses via the mutual spin-exchange process (spin diffusion) towards the impurity where relaxation occurs. At intermediate temperatures, hydrogen

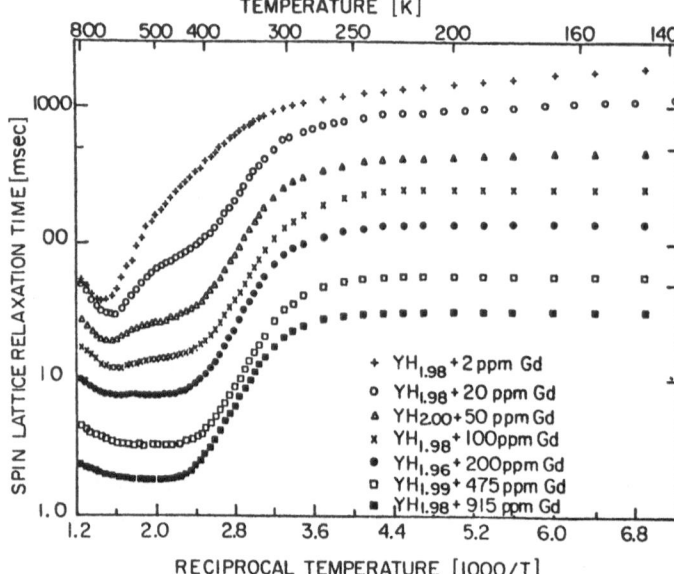

Fig. 4.4. Temperature dependence of the measured proton relaxation rate, T_1, in a series of yttrium dihydride samples containing progressively increasing controlled levels of Gd impurity, at a resonance frequency of 40 MHz [4.56]

atom diffusion becomes faster than spin diffusion and becomes the means of transporting distant ^{1}H spin magnetization to impurity ions. This latter process is atom diffusion limited and, since atom diffusion increases exponentially with temperature, relaxation in the intermediate region can become 25–30 times greater than at low temperatures in the spin diffusion regime. At still higher temperatures, hydrogen diffusion becomes so rapid that the duration of individual encounters with impurity moments becomes too short for efficient relaxation, and in the case of the heavy rare earth ions R_{1p} passes through a broad maximum and then decreases slowly with further increasing temperature, as seen in Fig. 4.4.

The strength of the proton-impurity moment interaction is defined by $\tau^{-1}(r) = Cr^{-6}$, where $\tau^{-1}(r)$ is the relaxation rate that a single proton would have at a distance r from the ion. For powder samples, C is given by [4.56]

$$C = (2/5)\, \gamma_p^2 \gamma_I^2 \hbar^2 J\,(J+1)^2 \left(\frac{\tau_c^*}{1 + \omega_0^2\,\tau_c^{*2}} + \frac{7\tau_c^*}{3(1 + \omega_e^2\,\tau_c^{*2})} \right), \qquad (4.20)$$

where J, γ_p, and ω_e are the angular momentum, magnetogyric ratio, and Larmor frequency of the ion, respectively, and γ_I and ω_0 are the corresponding quantities for the proton. The correlation time τ_c^* is determined by both the fluctuation rate, τ_i^{-1}, of the ion moments and the hopping rate of the hydrogens, τ_d^{-1}, according to

$$\tau_c^{*-1} = \tau_i^{-1} + \tau_d^{-1}. \qquad (4.21)$$

The term in brackets is just the ^{1}H-metal spin spectral density function of (4.4), the ion moment being the nonresonant one, and since $\omega_e \gg \omega_0$ the factors $(\omega_0 \pm \omega_e)^2$ reduce to ω_e^2, all in the context of BPP spectral densities. The relative importance of direct relaxation characterized by C and that due to diffusion, D, whether spin or atom diffusion, is described by the pseudo-potential radius, $\beta = (C/D)^{1/4}$ [4.60], which is the distance from the impurity at which the rate of diffusion to the impurity equals the relaxation rate by the impurity. In the slow diffusion (strong collision) limit the relaxation time is the average time for a diffusing spin to first encounter an impurity. It is assumed that relaxation effectively occurs once an impurity is encountered. The rate R_{1p} is then given by

$$R_{1p} = (8\pi/3)\,NC^{1/4}D^{3/4}, \tag{4.22}$$

where N is the impurity ion concentration per unit volume. When the condition, $\beta < a_1$, is reached at some temperature, where a_1 is the nearest neighbor proton-impurity spacing, many encounters with the impurity are required to produce relaxation (weak collision limit), and the rate becomes

$$R_{1p} = (4\pi/3)\,N\,Ca_1^{-3} \tag{4.23}$$

which is independent of D. This limit is appropriate to temperatures high enough that $\tau_d^{-1} \gg \tau(a_1)$ and is the well-known result for the relaxation of nuclei in liquids due to dissolved paramagnetic ions.

These results are applicable whether D represents spin diffusion, D_s, or atom diffusion, D_a. The transition from spin to atom diffusion occurs when $D_a > D_s$ which corresponds to $\tau_d^{-1} > R_2^{-1}$, the latter being effectively the "dwell" time for spin diffusion. The exponential temperature dependence of D_a is responsible for the rapid decrease in T_{1p} seen in Fig. 4.4 when $D_a > D_s$.

To illustrate these effects, R_{1p} data are shown in Fig. 4.5 for YH$_{1.98}$ doped with 500 ppm Ce, Pr, and Gd, respectively [4.60]. These three impurities exemplify three distinct types of ionic behavior in the metallic YH$_{\sim 2}$ host. Gd^{3+} and the other rare earth Kramers ions (except Ce^{3+}) are "slow relaxing" ions, i.e., the ion spin fluctuation rate τ_i^{-1} is relatively slow. Because Gd^{3+} is an S-state ion, τ_i^{-1} is determined essentially entirely by the conduction electron-ion moment interaction, i.e., $\tau_i^{-1} \propto T$. The weak maximum in R_{1p} at 12 K reflects the condition $\omega_0 \tau_i = 1$ which maximizes the first term in brackets in (4.20) (τ_d^{-1} is negligible, so $\tau_c^* = \tau_i$, and the second term is negligible because $\omega_e \tau_i \gg 1$). Thus, at 12 K, $\tau_i^{-1} = \omega_0 = 2.5 \times 10^8$ s^{-1} for Gd^{3+} in YH$_{\sim 2}$. From 12 K to ≈ 100 K, $R_{1p} \propto T^{-0.21}$, essentially in keeping with the $T^{-0.25}$ dependence expected for slow spin-diffusion, (4.22). The sharp increase in R_{1p} starting at ≈ 250 K occurs when hydrogen atom diffusion becomes faster than proton spin diffusion and reflects the exponential increase of D_a with temperature in contrast to the temperature independent D_s. In this region, $R_{1p} \propto D_a^{3/4}$. Finally, for $T > \approx 500$ K, the fast atom diffusion regime is reached, and R_{1p} decreases with temperature, as predicted by (4.23).

In contrast to Gd^{3+}, Ce^{3+} is a "fast-relaxing" ion with $\tau_i^{-1} \simeq 6 \times 10^{11}$ s^{-1} at 77 K (100 times faster than for Gd^{3+}). Consequently, R_{1p} is already

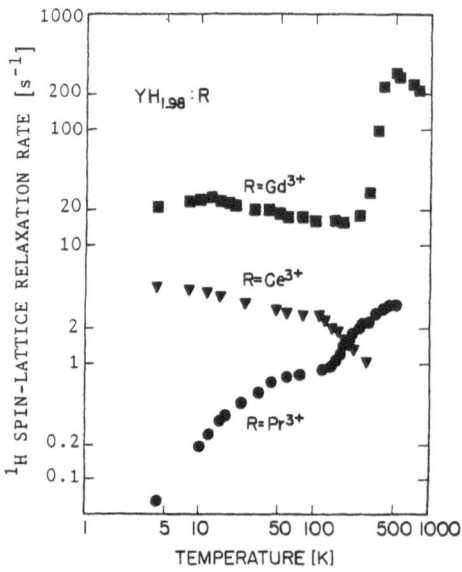

Fig. 4.5. Temperature dependence of the paramagnetic-impurity-induced proton relaxation rate, R_{1p}, in $YH_{1.98}$ containing 500 ppm of Gd, Ce, and Pr [4.60]

proportional to τ_i^{-1} and independent of spin or atom diffusion at 250 K when $D_a > D_s$, and the sharp increase in R_{1p} seen in the Gd^{3+} case is completely absent for Ce^{3+}. Instead, R_{1p} decreases with temperature, first approximately as $T^{-0.2}$ up to ≈ 100 K, consistent with slow spin-diffusion behavior expressed in (4.22). The weak R_{1p} maximum that appears at 12 K for Gd^{3+} would not appear until ≈ 0.03 K for Ce^{3+} at the 40 MHz ^1H resonance frequency used in these measurements. For $T > \approx 100$ K, R_{1p} decreases more rapidly, approximately as T^{-1}, as expected for fast diffusion, (4.23). These temperature dependences indicate conduction electron relaxation of the Ce^{3+} ion moment.

The behavior of R_{1p} for Pr^{3+} illustrates the third type, reflecting the non-Kramers nature of Pr^{3+}. The fact that $R_{1p}(Pr) = 0.063$ s^{-1} at 4.2 K, whereas $R_{1p}(Ce) = 4.5$ s^{-1} and $R_{1p}(Gd) = 21$ s^{-1} at 4.2 K, shows at once that the Pr^{3+} CEF ground state is a nonmagnetic singlet level. In sharp contrast to Ce^{3+}, $R_{1p}(Pr)$ increases over the entire temperature range investigated. Up to ≈ 100 K the rapid increase of R_{1p} in the slow spin diffusion regime is consistent with a CEF level scheme having the first excited level 62 K above the ground state. The further increase of R_{1p} for $T > \approx 100$ K results from the onset of atom diffusion as in the Gd^{3+} case, as well as from the excitation of higher CEF levels.

(B) Cross-Relaxation to Quadrupolar Metal Nuclei, R_{1c}

Proton spins may cross-relax with the spins of quadrupolar metal nuclei due to the dipolar interaction between the two spin systems, R_{cr}, when the Zeeman energy splitting of a proton spin happens to equal some combined

Zeeman-quadrupole splitting of a nearby metal nuclear spin [4.61]. The metal spins in turn transfer this energy to the lattice via their own fast relaxation rates – usually predominantly electronic in origin in M-H systems – thereby effectively short-circuiting the much slower proton R_{1e} rate. A crucial feature for this mechanism is the presence of substantial structural disorder, even in ordered hydrides. This disorder leads to a distribution of EFG's at metal nuclei, as well as of orientations of the EFG axes with respect to the applied field B_0. The lattice plays no part in R_{cr}, apart from providing the EFG, since the energy of the total spin system is conserved. When hydrogen diffusion occurs, R_{cr} is progressively weakened in the same way that spin-spin relaxation is averaged by such motion. In the absence of diffusion, R_{cr} is essentially temperature independent, and if relaxation of the metal spins to the lattice is sufficiently fast not to present a bottleneck, the overall effect on the ^{1}H spins is also temperature independent. Under these circumstances, R_{1c} has been found to dominate the total R_1 at low temperatures in polycrystalline (powder) samples, resulting in relaxation rates up to 400 times greater than those expected from the ^{1}H R_{1e} alone [4.61].

Calculation of R_{1c} is complicated since it is determined both by the cross-relaxation rate R_{cr} and the SLR rate R_{1S} of the metal nuclei. In most materials studied thus far, R_{1S} has been fast enough to maintain a common spin temperature between the lattice and the S spin system, so that R_{1S} may be ignored. Still, the calculation of R_{1c} involves essentially three steps: (i) determination of the energy levels and transition frequencies of the combined Zeeman-quadrupolar Hamiltonian of the metal nuclei, (ii) the smearing of these by EFG inhomogeneities, and (iii) calculation of the cross-relaxation rate R_{cr}. Complete details are given in [4.61].

An example of cross-relaxation behavior is shown in Fig. 4.6 [4.61]. The measured R_1 for ^{1}H at temperatures below ~ 100 K in NbH$_{0.21}$ is strongly dependent on the resonance frequency, increasing by a factor of ~ 400 as the frequency decreases from 90 to 4.5 MHz. In this temperature range the NMR signal arises from protons in the orthorhombic hydride, Nb$_3$H$_4$. The relatively small quadrupole interaction strength of ^{93}Nb in Nb$_3$H$_4$ is characterized by $v_0 = 1.2$ MHz, with $\eta = 0.6$, determined by wide-line ^{93}Nb measurements [4.62], enabling the frequency dependence of R_{1c} to be estimated. Perhaps most surprising is the degree to which cross-relaxation is extended above the highest pure quadrupole frequency (≈ 4.8 MHz) by the Zeeman effect of the applied field and consequent mixing of metal spin levels. If the metal nuclear quadrupole interaction is much stronger, as is the case for ^{181}Ta [4.61] and ^{175}Lu [4.63], then the entire cross-relaxation "spectrum" is shifted to higher resonance frequencies.

If proton R_1 measurements can be made at only a single resonance resonance frequency, the presence of cross-relaxation can be determined by low frequency (10^2–10^4 Hz) modulation of the applied field B_0 or by slow sample spinning around an axis perpendicular to B_0 [4.64]. Both techniques enhance R_{1c}. The former scans the ^{1}H resonance over a range of frequencies and causes overlap with more Zeeman-quadrupole frequencies of the metal nucleus, and the latter changes the Zeeman-quadrupole frequencies themselves,

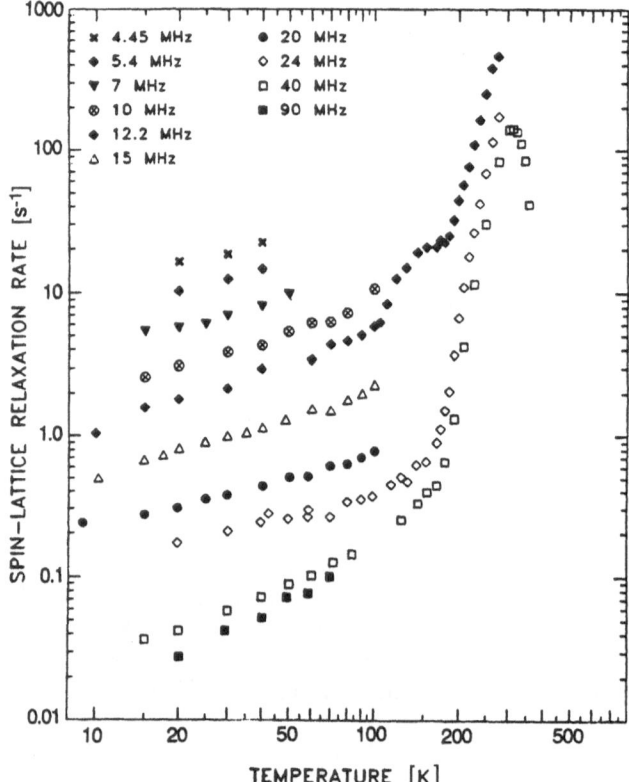

Fig. 4.6. Temperature and frequency dependence of the measured proton spinlattice relaxation rate, R_1, in niobium hydride, $NbH_{0.21}$. The strong frequency dependence at temperatures < 100 K is a consequence of $^1H - {}^{93}Nb$ cross-relaxation, R_{1c} [4.61]

causing increased overlap. The utility of these methods has been demonstrated in the cases of $TaH_{0.45}$ and $Zr_2PdH_{1.9}$ [4.64]. No change in R_1 is observed when these methods are applied in sufficiently high magnetic fields, thereby placing the ^1H resonance well-above any transition in the spectrum of the metal nucleus.

Cross-relaxation has also been observed [4.65] between the dipolar and rotating frame Zeeman nuclear-spin subsystems, ^{89}Y and ^1H, respectively, using pulsed spin-locking measurements in single-resonance experiments on the ^{89}Y spins in YH_2 and YD_2. The measured cross-relaxation spectra are described in terms of proton dipolar fluctuations.

(C) Anomalous High-Temperature Relaxation, R_{1x}

Anomalous behavior of the proton R_1 at high temperatures has been found in the dihydride phases of Sc, Ti, Y, Zr, and La [4.52, 66, 67] and in the solid solution phases of V, Nb, and Ta [4.68, 69]. It has also been found for the deuteron R_1 in the Y and Sc dideuterides [4.52]. The anomaly is that, in addition to the normal R_1 maximum due to H(D) diffusion, R_1 passes through a minimum and increases sharply again at higher temperatures instead of returning to the value R_{1e} or after returning briefly to R_{1e}. This behavior appears to imply that $\tau_c(T)$, and hence $\tau_d(T)$, passes through a minimum and then increases at high temperatures, in marked contrast to the diffusion D, which continues to increase throughout the same temperature region [4.70–73], showing that τ_d continues to decrease normally with increasing temperature. Similar behavior is found for the metal nucleus ^{45}Sc both in the solid solution phases α-ScH(D)$_x$ [4.74] and in the dihydrides (dideuterides) ScH$_2$ and ScD$_2$ [4.75]. Two significant features of the proton results are that R_{1x} (1) is nearly independent of resonance frequency at low frequencies but shows a significant dependence at high frequencies [4.76], and (2) appears to result from an activated process, so that $R_{1x} = A \exp(-U/k_B T)$ provides a phenomenological fit to the data [4.67, 74]. The latter feature also applies to the ^{45}Sc results.

Such anomalous relaxation is illustrated in Fig. 4.7 which shows the temperature dependence of R_1 for ^{1}H, ^{2}D, and ^{45}Sc at approximately the same ScH$_2$ (ScD$_2$) composition [4.77]. The onset temperature of the anomalous rate R_{1x}, taken as that of the minimum in R_1, increases in the sequence ^{45}Sc (710 K), ^{2}D (950 K), ^{1}H (100 K), suggesting that the anomaly is determined by the H(D) diffusion rate, since in these systems H(D) vacancy diffusion is much faster than H(D) atom diffusion so that the ^{45}Sc anomaly occurs at a lower temperature than the others. This assumes that $R_{1x}(^{45}$Sc$)$ is due to quadrupolar fluctuations arising from vacancy diffusion. However,

Fig. 4.7. Temperature dependence of R_1 and ^{45}Sc in ScH$_{1.83}$ and for ^{2}D in ScD$_{1.82}$. Data for ^{1}H, ^{45}SC, and ^{2}D are represented by ●, +, and ▲ symbols, respectively. The temperatures for the onset of the anomaly are $T_c = 1100$, 950 and 710°K for ^{1}H,^{2}D and ^{45}Sc. For all the data the NMR Larmor frequency was 12.2 MHz [4.77]

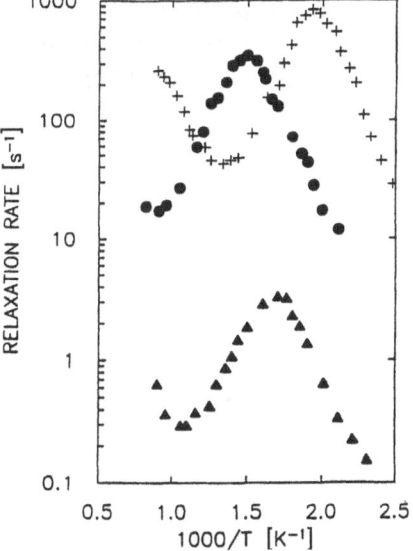

the assumption that $R_{1x}(^1H)$ and $R_{1x}(^{45}Sc)$ are both directly related to atom (vacancy) diffusion is invalidated by the failure to detect $R_{1x}(^1H)$ in α-ScH$_x$, even at 1400 K, whereas $R_{1x}(^{45}Sc)$ is clearly evident [4.74].

It has been shown for one system, Nb$_{0.5}$V$_{0.5}$H$_x$ [4.78], that $R_{1x}(^1H)$ results from fast relaxation in the H$_2$ gas, where $R_1 \simeq 5 \times 10^3$ s^{-1} at 860 K, in equilibrium with the solid phase, facilitated by the very fast exchange of hydrogen between the solid and gas phases. This mechanism is expected to apply whenever the equilibrium H$_2$ pressure approaches ~ 1 atm (10^5 Pa), hence it may account for R_{1x} in the metal dihydrides in which such pressures are reached at temperatures ~ 1000 K. The failure to detect $R_{1x}(^1H)$ in α-ScH$_x$, for example, is also consistent with this mechanism, since the equilibrium H$_2$ pressure is so low ($\sim 10^{-1}$ Pa at 875 K [4.74]) that the ratio of protons in the gas phase to those in the solid is $\sim 10^{-10}$. Additionally, the same mechanism may account for $R_{1x}(^2D)$ since the gas-phase ^2D relaxation rate is also very fast at high temperatures [4.79].

For ^{45}Sc, the anomalous rate remains unexplained. It is possible that the fast difffusion of paramagnetic iron impurity ions, usually present in Sc metal, could be responsible [4.74]. An exactly similar anomalous increase in relaxation has been reported for ^{93}Nb in Nb metal, presumably containing negligible hydrogen [4.80], which suggests that the R_{1x} of metal nuclei may result from a mechanism intrinsic to the metallic state.

4.3.3 Measurement of Hydrogen Diffusion

NMR methods of measuring hydrogen tracer diffusion, D (or D_t), directly are based on the fact that if the nuclear spins change their locations in the lattice during the time between the 90° and 180° rf pulses of an R_2 measurement and experience different values of magnetic field during this time interval, then the transverse magnetization is no longer completely refocussed, and the echo signal intensity is reduced. Experimentally, this is accomplished by applying two field-gradient pulses of strength g_a and duration δ along the direction of the applied field B_0, one before and one after the 180° rf pulse. The rate at which the echo signal decreases is measured as a function of g_a and δ, thereby measuring the distance spins diffuse along the direction of B_0 during the time between the gradient pulses. All such pulsed field gradient (PFG) experiments have in common that the attenuation of a spin-echo amplitude is related to D by an equation of the form [4.81]

$$\ln m(g_a, t_c) - \ln m(0, t_c) = -C_n \gamma^2 D (g_a \delta)^2 t_D, \qquad (4.24)$$

where C_n is a constant depending upon the particular sequence of rf and field-gradient pulses and t_D is approximately equal to the time between application of two (or two pairs of) gradient pulses. For good measuring accuracy, the magnitude of the right-hand side of (4.24) should usually exceed unity. When D is small, large gradient pulse areas ($g_a \delta$) and/or long diffusion time t_D are needed. Large pulse areas are limited by the power available from the pulse circuitry and by the relaxation times that limit δ. In practice, the technique is usually limited to diffusion rates, $D > 10^{-6} \text{cm}^2\text{s}^{-1}$.

The fundamental aspects of the PFG technique are described and discussed in [4.1]. A basic difficulty that has to be overcome in PFG experiments arises from the presence of a distribution of magnetic field gradients, the so-called background gradient, g_0, associated with nonuniform sample magnetization. This is particularly characteristic of powder samples typically used in metal hydride studies. The use of foil samples can sometimes (but not always) reduce g_0. The background gradient gives rise to an additional term proportional to the product $D\delta^2 g_0 g_a$ in the expression for the echo attenuation, thereby limiting the accuracy of the measurements see [Ref. 4.1, Eq. (9.30)].

The difficulties posed by the $g_a g_0$ term can be reduced or circumvented altogether in several ways. The first is the multiple-pulse spin-echo technique [4.1] which can significantly reduce the effects of g_0. A second method [4.82, 83] utilizes gradient pulses of opposite polarity (bipolar pulses) in a modified two-pulse experiment to eliminate the $g_a g_0$ term. This method was applied to measuring D in LaNi_5H_x [4.82]. More recently, the stimulated spin-echo (SSE) sequence [4.84] has been combined with the PFG experiment in three new sequences that reduce substantially the importance of both the $g_a g_0$ term and a further term in $g_0{}^2$ (described in detail in [4.81]). The sequence having the largest coefficient C_n in (4.24) is a 17-interval sequence utilizing bipolar field-gradient pairs [4.81]. This so-called stimulated-echo alternating pulsed field-gradient (SEAPFG) method has been used to measure hydrogen diffusion in $\text{Nb}_{1-y}\text{V}_y\text{H}_x$ alloys [4.85].

Further refinements in the SSE technique [4.86] have yielded field gradients up to 25 T/m (2500 Oe/cm), enabling accurate measurements of D values on the order of $10^{-8} \text{cm}^2\text{s}^{-1}$ in ZrH_x.

4.4 Electronic Structure

The magnetic resonance of nuclei in metallic solids is affected in two ways by the electronic structure of the solid through the hyperfine fields produced at the nuclear sites and the electronic density-of-states (DOS) at the Fermi level, $N(E_F)$. The static (time-average) component of the hyperfine field parallel to B_0 causes a shift in the resonance frequency in the metal, ν_m, compared to that in a reference salt, ν_s. This is the Knight shift [4.87], defined as $\kappa = (\nu_m - \nu_s)/\nu_s = \Delta\nu_m/\nu_s$ for measurements made at constant field B_0, or in the case of small shifts, as $\kappa = (B_s - B_m)/B_s = \Delta B_m/B_s$ at constant resonance frequency ν_0. The dynamic (fluctuating) components of the hyperfine fields contribute a conduction electron component, R_{1e}, to the spin-lattice relaxation rate that is proportional to the temperature, the so-called Korringa rate [4.88].

Since κ and R_{1e} have their origins in the hyperfine fields, their magnitudes are smallest for ^1H and ^2D and greatest for heavy metal nuclei, ranging from values as small as 0.001% for protons to $\sim 1\%$ for some metal nuclei. The

Knight shift may be positive (paramagnetic), meaning that the field seen by the resonant nucleus is enhanced over that of the external (applied) field B_0, or negative (diamagnetic), meaning that the applied field is shielded. The Korringa parameter, $C_e = R_{1e}/T = (T_{1e}T)^{-1}$, which characterizes the electronic contribution to R_1 [4.88], ranges from values as weak as 10^{-3} $(s\,K)^{-1}$ for protons to as strong as 1 $(s\,K)^{-1}$ for metal nuclei.

Within the free-electron approximation, and assuming that only s and d orbitals contribute to the hyperfine interactions, κ and C_e may be separated into three components [4.89]

$$\kappa = \kappa_s + \kappa_d + \kappa_{orb} \tag{4.25}$$

$$\kappa = 2\mu_B[\rho H_s + (1-\rho)H_{cp}]N(E_F) + (N_A\mu_B)^{-1}H'_{orb}\,\chi_{orb}\,, \tag{4.26}$$

$$C_e = 4\pi k_B\,\gamma^2\{[\rho H_s]^2 + q[(1-\rho)H_{cp}]^2 + p[(1-\rho)H_{orb}]^2\}N(E_F)^2\,, \tag{4.27}$$

with $\rho = N_s(E_F)/N(E_F)$, i.e., the fractional s character of the total DOS at the Fermi level. Here, N_A is Avogadro's number, χ_{orb} is the van Vleck paramagnetic susceptibility, and p and q are reduction factors that result from d-electron degeneracy at E_F [4.90]. The hyperfine fields at the resonant nucleus have their origins as follows: H_s results from the Fermi contact interaction with unpaired s electrons at E_F, H_{cp} from core polarization of spin-paired s orbitals at energies below E_F by unpaired d electrons at E_F, and H_{orb} and H'_{orb} from orbital motion of d electrons. H_{orb} represents only d electrons at E_F whereas H'_{orb} is an average over all contributing states in the d band, both above and below E_F. For simplicity, the approximation $H'_{orb} \simeq H_{orb}$ is usually employed. Although $H_s \gg H_{cp}$ or H_{orb}, the large $N_d(E_F)$ in most transition metal systems means that the H_{cp} and H_{orb} terms will usually dominate κ and C_e. Also, for transition metals, H_{cp} is negative whereas H_{orb} is always positive. Consequently, significant cancellations can occur in κ, giving rise to positive, negative, or even zero resultant shifts. On the other hand, C_e is determined by the squares of the hyperfine fields, which are always additive. As is evident from (4.27), in most cases $C_e^{1/2}$ is directly proportional to $N(E_F)$.

In simple metals in which the Fermi contact terms dominate both κ and C_e, the quantity

$$q_{expt} = (\hbar\gamma_e^2/4\pi k_B\,\gamma_I^2)(C_e/\kappa^2)\,, \tag{4.28}$$

where γ_e is the electron magnetogyric ratio, has the value 1 if electron-electron interactions are excluded [4.90]. For the core-polarization terms, $q_{expt} = q$, the reduction factor for R_{1e} due to d electrons, which for cubic lattices is given by [4.90]

$$q = (1/3)f^2(t_{2g}) + (1/2)[1 - f(t_{2g})]^2\,, \tag{4.29}$$

where $f(t_{2g})$ is the fractional character of t_{2g} d orbitals at the Fermi surface.

The magnetic susceptibility can provide additional useful information for analyses of κ and R_{1e},

$$\chi = \chi_e + \chi_d + \chi_{orb} \tag{4.30}$$

(after correcting for the diamagnetic contribution, χ_{dia}). The Knight shift can also be written as

$$\kappa = (\mu_B N_A)^{-1} \left(H_s \chi_s + H_{cp} \chi_d + H'_{orb} \chi_{orb} \right),\tag{4.31}$$

making use of the fact that $\chi_s = 2\mu_B^2 N_A N_s(E_F)$, etc. Finally, it must be emphasized that the hyperfine fields and densities of states have different values at metal and hydrogen sites.

For transition metal nuclei, measurement of κ combined with C_e and magnetic susceptibility data frequently permits experimental resolution of $N(E_F)$ into components $N_s(E_F)$ and $N_d(E_F)$. An excellent example is furnished by the analysis [4.91] of such measurements for nearly all of the metal nuclei that occur in the fcc dihydrides, i.e., ^{45}Sc, ^{89}Y, ^{139}La, ^{47}Ti, ^{91}Zr, ^{51}V, and ^{93}Nb. The resulting DOS agree well with those derived from specific heat and band structure calculations. The analysis of C_e supports the picture of nearly exclusive d-character of $N(E_F)$ in these systems.

Similar measurements for protons can provide this information at the hydrogen site. Since present-day band-structure calculations readily yield partial wave decompositions of $N(E_F)$ [4.92], there is strong motivation to compare experiment with theory. Unfortunately, $\kappa(^1\text{H})$ is intrinsically small in comparison to those of the metal nuclei, both because of the smaller hyperfine field and because of the greatly reduced $N(E_F)$ at hydrogen sites. Moreover, the shifts are small in comparison to typical ^{1}H resonance linewidths ($\Delta B \approx 10$ 0e). Typical $\kappa(^1\text{H})$ values correspond to actual field shifts, $\Delta B_m \simeq 0.15$–1.5 0e, at a resonance frequency of 60 MHz ($B_0 \simeq 15$ K0e). Accordingly, more sophisticated techniques have been employed to improve the reliability of such measurements. These techniques utilize resonance line-narrowing so that the shift is enhanced relative to ΔB. Since ΔB arises principally from the ^{1}H-^{1}H dipolar interaction, which is greatly reduced by fast hydrogen diffusion, the simplest method is to measure κ at sufficiently high temperature that the resonance is motionally narrowed. The two other principal methods employ magic-angle sample spinning (MASS) [4.93] and multipulse (transient) sequences [4.94] to create the same averaging effect without actual hydrogen diffusion. These latter two methods are applicable at lower temperatures than the former; however, they are more complex. In addition, measuring κ in as strong an applied field as possible enhances ΔB_m over the dipolar width and is advantageous.

For hydrides of the group V metals V, Nb and Ta which have high magnetic susceptibilities, an additional source of line broadening arises from the demagnetizing fields due to the distribution of particle shapes in the usual powder samples required for adequate penetration of the rf field. In addition, the demagnetizing field due to the overall sample shape may introduce a correction to the measured shift. These difficulties become more severe at higher B_0 values, but may be circumvented by employing a single metal foil as sample. In this case, measurement of κ as a function of the angle between the field and the normal to the foil surface yields both the true shift and the susceptibility χ.

With a single foil, the actual shift, $S(\theta)$, measured with respect to protons, in water, taken as a standard reference, is given by [4.95]

$$S(\theta) = S(0) - 4\pi \chi_v \sin^2 \theta, \tag{4.32}$$

where θ is the angle between B_0 and the surface of the foil and χ_v is the volume susceptibility in cgs units. The term $4\pi\chi_v \sin^2 \theta$ is the demagnetizing field induced oppositely to B_0 in an infinite plate. Plotting $S(0) - S(\theta)$ as a function of $4\pi \sin^2 \theta$ determines χ_v, and the proton κ with respect to the bare proton is obtained by correcting for the Lorentz cavity field and molecular shielding constant of protons in water, taken to be 25.6×10^{-6} [4.95]:

$$\kappa(^1\mathrm{H}) = S(0) - (4/3)\pi \chi_v - 25.6 \tag{4.33}$$

in units of 10^{-6}. An example [4.95] of such a series of measurements is shown in Fig. 4.8. In this case, the sample is $\mathrm{TaH}_{0.61}$ powder that has been compressed into a single foil at a pressure of 22.5 kbar, and the measurements were made at a temperature of $70\,^\circ\mathrm{C}$.

By comparison, measurement of the proton C_e is relatively simple, since it can be determined, in principle, from measurments of R_1 at temperatures well below or well above the temperature region of the R_{1d} maximum, i.e., when $R_{1e} \gg R_{1d}$ (Fig. 1.1). However, as discussed in Sect. 4.3.2A, low levels of paramagnetic impurities can easily dominate R_1 at low temperatures. In addition to using samples of demonstrated high purity, the impurity contribution R_{1p} can be suppressed by replacing most of the ^{1}H with ^{2}D, thereby inhibiting ^{1}H spin-diffusion and consequently R_{1p} [4.56]. Another complicating factor in proton C_e measurements can be dipolar cross-relaxation to the metal nuclear spins (Sect. 4.3.2B). This effect can be re-

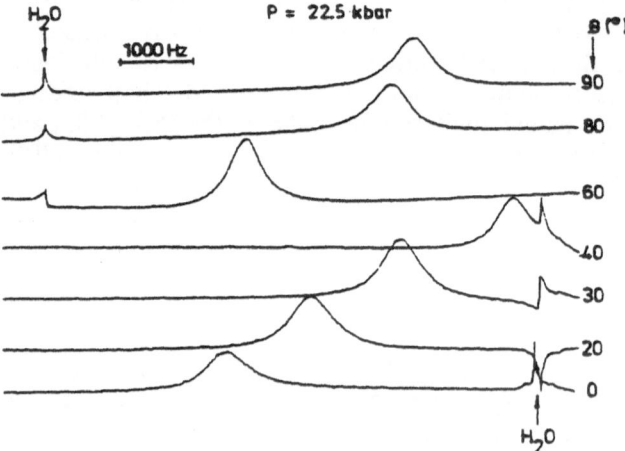

Fig. 4.8. Fourier transform ^{1}H NMR spectra of $\mathrm{TaH}_{0.61}$ powder compressed into a single foil specimen under 22.5 kbar pressure. The dependence on the orientation of the foil in the applied field B_0 is shown [(4.32) in text] at a resonance frequency of 79.54 MHz at $T = 70\,^\circ\mathrm{C}$. The resonance position moves to lower frequencies as θ increases from 0 to 90°, passing the H_2O reference signal (frequency increases to the left in the figure) [4.95]

cognized by the frequency dependence of R_{1e}, which is not otherwise frequency dependent. Finally, the anomalous rate R_{1x} which appears at temperatures above the R_{1d} maximum in most systems studied interferes with unambiguous determination of C_e at such temperatures (Sect. 4.3.2C).

κ and C_e of both metal and hydrogen nuclides can show anisotropy when the site symmetry is less than cubic. Such behavior is most clearly revealed by measurements on single crystals [4.96]. For κ, it can also be seen in powder samples [4.10]. However, in the case of C_e, the measured rate R_{1e} for a powder is an average over all orientations, so is C_e. Work on noncubic single crystals has been directed toward determining such anisotropy [4.97, 98].

4.5 Applications

The following sections comprise brief reviews of NMR investigations of the major classes of metal-hydrogen systems.

4.5.1 Binary Metal-Hydrogen Systems

(A) Group III Systems

Solid Solution (α) Phase
NMR determination of hydrogen site occupancies based on application of the ^{1}H second-moment formula (4.1) is best achieved when the metal nuclear moments contribute negligibly. The Y-H system is excellent in this respect because of the extremely small ^{89}Y moment [Table 4.1]. In α-YH$_{0.20}$, $\langle\Delta\omega_d^2\rangle$ measurements could not distinguish between predominantly T- or O-site occupancy. For predominantly T-site occupancy, subsequently shown by neutron scattering to be the case [4.99], the measurements found 86% of the hydrogen in T sites and 14% in O sites [4.100].

Numerous measurements of electronic and H diffusion parameters in the α-phases of these systems have been made based primarily on R_1 measurements [4.15, 101–103, 63]. The case of α-ScH$_x$ is especially interesting, since the H diffusion (hopping) parameters determined from both ^{1}H and ^{45}Sc R_1 measurements are the same within experimental uncertainty [4.15]. This is quite different from the situation in ScH$_{\sim 2}$ in which ^{45}Sc relaxation results from H vacancy motion whose hopping rate is much higher than that of the hydrogen. In α-ScH$_x$, ^{45}Sc relaxation results from H hopping, as does ^{1}H relaxation.

The temperature dependence of the proton R_1 has been studied in α-ScH$_x$ [4.104, 105], α-YH$_x$ [4.104–106], and α-LuH$_x$ [4.63, 105] over the temperature range (10–300 K) in which hydrogen pairing occurs [4.107]. At low temperatures (10–120 K), localized motion of H between closely spaced (~ 1 Å) T-sites gives rise to a peak in the ^{1}H R_1 in α-ScH$_x$ and α-LuH$_x$. Results for α-ScH$_x$ [4.104] are shown in Fig. 4.9. The maximum rate, on the order of 1 s^{-1}, should be compared with the R_{1d} peak due to long-range diffusion at high

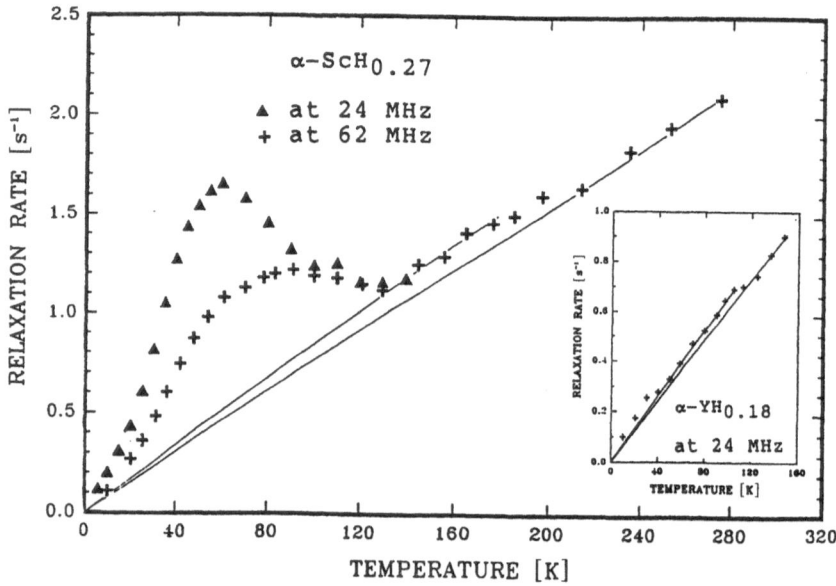

Fig. 4.9. Temperature dependence of the $^1H R_1$ in α-ScH$_{0.27}$ measured at 24 and 62 MHz. The two solid lines represent the parameter $C_e = R_{1e}/T$ for temperatures above and below ~ 170 K. Inset: ^{1}H R_1 in α-YH$_{0.18}$ at 24 MHz [4.105]

temperatures, $\sim 65\,s^{-1}$. The extreme weakness of the peak results partly from the motional averaging of only a small portion of the total dipolar interaction and partly because the localized motion occurs between hydrogen potential wells of unequal depth. A change in the Korringa rate R_{1e} is also evident in Fig. 4.9, showing that an electronic structure transition accompanies or is caused by completion of the pairing. The R_1 peak results primarily from modulation of the dipole field of nearest-neighbor metal nuclei (^{45}Sc or ^{175}Lu) by the localized H motion, and it is not found in α-YH$_x$ because of the extremely small ^{89}Y moment (inset in Fig. 4.9). The case of α-LuH$_x$ is complicated by unusually strong ^{1}H-^{175}Lu cross-relaxation, R_{1c}, which prevents determination of the localized contribution to R_{1d} [4.63, 105].

The low-temperature peak has been interpreted in terms of classical over-barrier hopping [4.104, 105] or tunnelling through the barrier [4.108] between the close sites. The dependence of the peak in α-ScH$_x$ on temperature and resonance frequency shows characteristics typically found in amorphous and disordered systems, which have been interpreted in terms of two level systems (TLS), and leads to the suggestion that pair formation results effectively in a "proton glass" [4.105]. The measurements are consistent with motion governed by a distribution of activation energies having an average value of 50 meV/atom and half-width 30 meV/atom.

Dihydride Phase

The fcc dihydrides MH_x exhibit considerable H concentration ranges, and much effort has been expended to determine site occupancies. The pioneering NMR study of LaH_x by *Schreiber* and *Cotts* [4.109] found O-site occupancy starting at $x = 1.95$, based on second-moment measurements. In YH_x, such measurements revealed substantial O-site occupancy, up to 15% at $x = 1.98$, increasing with x within the dihydride phase [4.110]. Such determinations are not feasible in the case of ScH_x because of the strong ^{45}Sc contribution to $\langle \Delta\omega_d{}^2 \rangle$ by conventional measurements, but would be if solid-echo techniques were used [4.1].

The La-H system (as well as Ce-H and Pr-H) differs from those of Sc, Y, and Lu in two important aspects: (i) H solubility in La is very low at normal temperatures, thus there has been no NMR work on the solid solution phase, and (ii) the fcc dihydride structure is retained as the composition ranges continuously from ~ 1.95 to the trihydride LaH_3 as H fills the interstitial O-sites in the lattice. However, ordered H superlattices occur in several composition ranges [4.111], accompanied (not always) by small tetragonal distortions of the metal lattice [4.111, 112]. These orderings have been detected in both ^2D and ^{139}La spectra which show quadrupole interaction splittings resulting from the noncubic site symmetries [4.113, 114]. These splittings become evident at moderately low temperatures when deuteron motion has slowed sufficiently that the residence time in a site exceeds the interaction frequency.

Values of the ^1H electronic structure parameter C_e depend critically on sample purity in these dihydrides, being strongly influenced by low levels (< 10 ppm) of magnetic rare-earths [4.56]. Measurements made on samples of proven high purity found $C_e = (3.00 \pm 0.25) \times 10^{-3}(s\,K)^{-1}$ in the Sc, Y, La, and Lu dihydrides [4.115], essentially independent of H concentration, with the exception of LaH_x where C_e decreases rapidly for $x > 2$ [4.109, 116]. In the group III dihydrides it is expected that the orbital hyperfine field, H_{orb}, in (4.27) can be neglected in comparison to the H_s and H_{cp} terms, so that $C_e{}^{1/2}$ should reflect directly the x-dependence of $N(E_F)$, provided the partitioning into s and d densities is not itself x-dependent. If electrons are uniformly withdrawn from the Fermi level as x increases from 2 to 3, the electron density is $n \propto (3 - x)$. In the simple free-electron approximation, $N(E_F) \propto n^{1/3}$, so that $N(E_F) \propto (3 - x)^{1/3}$. Therefore, $C_e{}^{1/2}$ may reflect this dependence, and this was found to be the case [4.116]. The ^{139}La Knight shift results [4.109, 116] follow closely free-electron behavior as well. This system furnishes an unusual example in which the dependence of $N(E_F)$ on the H concentration has been followed by measurements on both metal and hydrogen nuclei.

The unusual metal-nonmetal transition that occurs in LaH_x for $x \gtrsim 2.80$ [4.117] has also received attention [4.116, 118–120]. Proton R_1 measurements are complicated by paramagnetic impurities in the La (Sect. 4.3.2A). However, by using highest-purity La containing controlled low levels of Gd, the temperature dependence of the Gd^{3+} ion spin fluctuations, characterized by the ion spin-lattice relaxation time, τ_i, was determined from measurements of the impurity-induced rate, R_{1p} [4.116]. For $x < 2.8$, $\tau_i \propto T^{-1}$ at all temperatures, indicating metallic character, but for $x \gtrsim 2.8$ and $T > 250$ K, $\tau_i \propto T^{-5}$, consistent with the two-phonon relaxation process typical of insulating solids.

Activation energies and jump attempt frequencies have been derived from ^{1}H R_{1d}, $R_{1\rho}$, and linewidth measurements at various compositions in the dihydride phases of Sc [4.121, 122], Y [4.56, 110], and La [4.109, 123]. For ScH_x and LaH_x, ^{45}Sc [4.14, 75] and ^{139}La [4.109] measurements have also been made. The ScH_x, YH_x, and LaH_x dihydride (and trihydride) systems all show E_a decreasing with increasing x. In YH_x and LaH_x this decrease is pronounced and sudden, dropping from $E_a \simeq 1$ eV/atom at the lower limiting composition of the dihydride phase to ≈ 0.4 eV/atom at $x \simeq 2$ in YH_x and to ~ 0.15 eV/atom at $x \simeq 3$ in LaH_x. This behavior contrasts sharply with that in the Ti and Zr dihydrides where E_a remains essentially constant with x in TiH_x and increases markedly with x in ZrH_x (Sect. 4.5.1B), presumably reflecting increasing O-site occupancy in the group III hydrides [4.109, 110] but not in those of group IV. Neutron scattering results for H diffusion in YH_x and TiH_x [4.71] show behavior similar to the NMR results and have been interpreted on the basis of the model assumptions of a concentrated lattice gas [4.71].

Anomalous R_1 behavior at high temperatures has been reported in all of these systems [4.52, 66, 75, 77]. Figure. 4.5 shows examples for the ScH_x and ScD_x dihydride and dideuteride.

Magnetic Rare-Earths: Rare-Earth Nuclei

The rare-earth hydrogen systems form an important subclass of the group III systems. In magnetically dense materials, magnetic resonance of the nuclei of the magnetic atoms (ions) is generally not feasible because of the extreme broadening of the resonance caused by the internal hyperfine field. This is certainly the case for Kramers ions, since these always possess at least 2-fold degenerate magnetic CEF levels which give rise to strong hyperfine fields at the nucleus and consequently fast nuclear spin relaxation and broad resonance lines. The situation differs, however, for the non-Kramers ions for which nonmagnetic singlet CEF levels can occur.

The best opportunity to observe NMR occurs when the CEF ground-state level is a singlet (van Vleck paramagnets), since then the hyperfine interaction with the induced moment of the singlet state is relatively weak and does not result in efficient spin relaxation. Among the rare-earth hydrides, those of Pr and Tm are most suitable for this. The nuclides ^{141}Pr and ^{169}Tm are 100% abundant, and of these two ^{169}Tm is most attractive because it has spin 1/2 so that quadrupole interaction complications cannot arise (^{141}Pr has spin 5/2 and a small quadrupole moment).

The NMR of ^{169}Tm has been investigated at low temperatures in TmH_2 [4.124], for which magnetic susceptibility measurements showed the CEF ground state to be a singlet Γ_2 with the first excited magnetic triplet state $\Gamma_5^{(2)}$ at about 170 K. The field for resonance is strongly enhanced (shifted) relative to the applied field B_0:

$$H_{res} = B_0 + H_{hf} = (1 + \kappa_{VV})B_0 \qquad (4.34)$$

with

$$\kappa_{VV} = -A\chi_{VV}/g_I\mu_I N_A g_J\mu_J, \qquad (4.35)$$

being in effect a giant Knight shift. Here g_I and g_J are the nuclear and electronic g-factors, respectively, and μ_I and μ_J are the nuclear and Bohr magnetons. Experimentally, the hyperfine enhancement factor, $(1 + \kappa_{VV}) = 13.99 \pm 0.03$ at 3 K [4.124]. This agrees well with the "free-ion" estimate of 13.7 based on the hyperfine coupling, $A/h = 388.8$ MHz, with $g_I = -0.462$, and $\chi_{VV} = 8.1 \times 10^{-2}$ emu mol^{-1}.

The ^{169}Tm signal could not be observed above ≈ 40 K because of increasingly fast spin relaxation. The temperature dependence of the rate R_1 shows a low-temperature $(T \leq 12$ K$)$ range in which Korringa behavior holds, with $R_{1e}/T = (1.18 \pm 0.03) \times 10^2$ (s K)$^{-1}$ (after allowing for a small temperature-independent contribution), and a strongly temperature-dependent regime above ~ 12 K [4.124].

The NMR of ^{141}Pr cannot be observed in PrH$_2$ because the Pr^{3+} ground state is a magnetic Γ_5 triplet. However, the possibility occurs in PrH$_{2.5}$ where the orthorhombic symmetry at Pr sites results in singlet levels only [4.60].

Magnetic Rare-Earths: Hydrogen Nuclei

Resonance of the nuclei of nonmagnetic atoms, e.g., ^{1}H, in magnetically dense solids is nearly always possible in the paramagnetic state. There has been extensive NMR study of the Ce-H and Ce-D systems [4.125, 126] and to a lesser degree of most other rare-earth dihydrides. The proton R_1 is dominated by the strong magnetic coupling between the rare-earth ion moments and the ^{1}H spin, in comparison to which R_{1d} is negligible. There have been no NMR studies of hydrogen diffusion in these systems, with the exception of SmH$_{2.12}$ for which motional narrowing of the linewidth yielded an estimate [4.127] of $E_a = 0.38 \pm 0.03$ eV/atom.

The temperature dependence of the ^{1}H Knight shift has been measured [4.128] in the dihydrides of Pr, Nd, Sm, Gd, and Tb. With the exception of SmH$_2$ [4.128, 129], the shifts follow behavior of the form:

$$\kappa = \kappa_0 + (g_J - 1) H_{eff} \chi_f/N_A g_J\mu_J, \qquad (4.36)$$

Where H_{eff} is the effective hyperfine field due to the 4f electron configuration and χ_f is the 4f electron susceptibility. κ_0 is the temperature-independent Knight shift, which may be inferred from measurements on LaH$_2$ [4.109]. The shift data for all four dihydrides (i.e., except SmH$_2$) are fit by (4.36) with $H_{eff} = -(1.67 \pm 0.17) \times 10^5$A/m (i.e., 0.21 T). The negative value of H_{eff} is typical of lanthanide and actinide compounds and indicates an anti-ferromagnetic spin polarization at the proton. The magnitude of H_{eff} is about a factor of 3 smaller than in uranium hydride [4.130], (Sect. 4.5.1D).

Relaxation rate measurements have been utilized to study the CeH_x and CeD_x systems [4.125, 126]. In common with LaH_x, PrH_x, and NdH_x, these retain the fcc structure as x increases from 2 to 3. In CeH_x, $(2.10 \leq x \leq 2.92)$, for $100 \leq T \leq 375$ K, the ^1H R_1 is characterized by a single relaxation rate at each x and T. To a good approximation, $R_1 = A/T + R$, where A/T is attributed to direct dipolar coupling between ^1H and Ce^{3+} moments, and R is an essentially temperature independent term. The x dependence of A is consistent with dipolar coupling. The constant R decreases steeply as x increases above ≈ 2.65, just below the range $2.75 \leq x \leq 2.85$ where the metal-nonmetal transition occurs [4.131]. For this reason, R is ascribed to the indirect (RKKY) ^1H-Ce^{3+} moment coupling, and because of the low magnetic ordering temperature in CeH_x, this mechanism is essentially temperature-independent.

In CeD_x $(2.01 \leq x \leq 2.90)$, for $77 \leq T \leq 573$ K, the duteron R_1 are in substantial agreement with those for ^1H when the latter are multiplied by the square of the ratio of magnetogyric ratios, $(\gamma_2/\gamma_1)^2$, showing that any nuclear quadrupole contribution is negligible [4.126]. The temperature dependence of R_1 again shows that direct dipolar interaction between ^2D and Ce^{3+} moments is the dominant interaction.

The temperature dependence of the ^1H R_1 in $PrH_{2.0}$ and $PrH_{2.5}$ is entirely accounted for by the ^1H interaction with fluctuating Pr^{3+} moments. Its behavior in $PrH_{2.0}$ differs markedly from that in $PrH_{2.5}$ [4.60]. Due to the different ordering of the Pr^{3+} CEF levels, R_1 increases with decreasing temperature in $PrH_{2.0}$, but increases with increasing temperature in $PrH_{2.5}$.

The trihydrides of the heavy rare-earths crystallize in the hexagonal HoD_3 structure and are probably insulators or semiconductors having low magnetic ordering temperatures compared to the dihydrides. The ^1H linewidth and shift in GdH_3 have been measured over the temperature range 78–293 K [4.132]. GdH_3 orders only below 1.8 K. Although 3 inequivalent H sites occur in this structure, only a single resonance was discernable. The resonance shift is negative and proportional to the susceptibility, as expected from (4.36). The effective hyperfine field, $H_{eff} = -0.14$ T, is smaller than that in the dihydrides [4.128]. Ytterbium trihydride, $YbH_{3-x}(x \simeq 0.45)$, has also been investigated using wide-line techniques [4.133]. The asymmetry of the ^1H resonance shape and the field dependence of the second moment suggest the presence of Yb^{3+} ions in this hydride.

(B) Group IV Systems

Solid Solution (α) Phase

On the basis of R_1 measurements [4.134] the solid-solution-dihydride transition was studied in $TiH_{0.08}$, and estimates were made of E_a and C_e in the α-phase. An $E_a \simeq 0.56$ eV/atom was obtained, similar to values for the group III hcp solid solutions.

Dihydride Phases

Several series of proton R_1 and R_{1e} measurements have been made on these Ti-H [4.135–139] and Zr-H [4.71, 140–145] phases, yielding the hydrogen concentration dependence of E_a, v_o, and C_e based on partitioning R_1 between R_{1e} and R_{1d} contributions and employing BPP spectral densities. For the Ti dihydrides, E_a is found to remain essentially constant at about 0.5 eV/atom, independent of x, whereas in ZrH_x E_a increases sharply near the stoichiometric limit, $x = 2$ with $E_a = 1.1$ eV/atom at $x = 1.98$ [4.67, 142]. The hydrogen diffusivity D has also been measured using the PFG technique (Sect. 4.3.3) in TiH_x for $x = 1.55, 1.71$ [4.146, 147] and $1.65 \leq x \leq 2.02$ [4.148], and in ZrH_x for $1.58 \leq x \leq 1.98$ [4.86]. Together with determinations of τ_d based on lattice-specific theories for spectral densities [4.31, 40], these measurements establish the elementary jump distance L as the nearest-neighbor hydrogen T-site distance in both systems.

In both Ti-H and Zr-H dihydrides, the PFG measurements of D [4.86, 149] reveal a decrease in D and an increase in the effective, i.e., Arrhenius fit, activation energy with increasing x. At intermediate x values, plots of log D vs $10^3/T$ show non-Arrhenius behavior, becoming steeper with increasing temperature. Results for ZrH_x are shown in Fig. 4.10 [4.149]. This behavior can be understood in the context of the lattice-gas approach [4.71] by introducing hydrogen occupancy of a second interstitial site (X site) in addition to the normal tetrahedral (T) site [4.149]. This X site must be energetically higher than the T site by ΔE and is presumably not the octahedral (0) site, since in those dihydrides in which 0 sites become occupied, D increases and E_a decreases as x increases [4.71]. The solid lines in Fig. 10 are calculated with

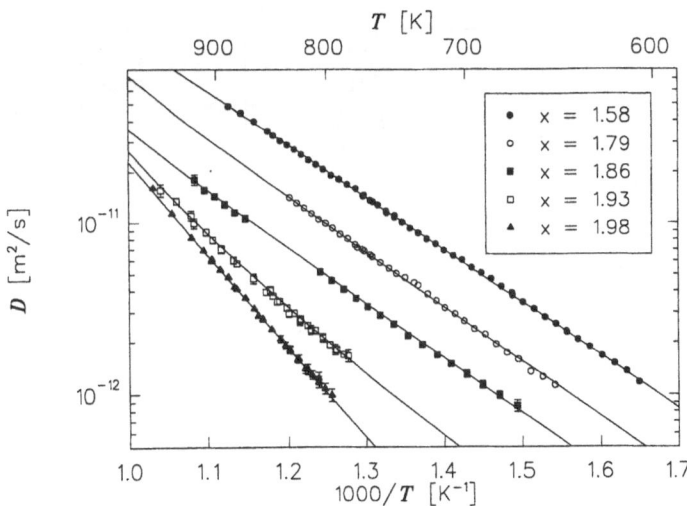

Fig. 4.10. PFG-NMR data of hydrogen diffusion in ZrH_x[257]. The solid lines are calculated taking into account a certain fraction of hydrogen atoms located on so-called X-sites, which require an excitation energy $\Delta E = 0.85$ eV/atom for occupancy. Values of the diffusion parameters are given in the text. Reprinted with permission from [4.149]

the energy parameters $\Delta E = 0.85$, $E_{TT} = 0.62$, $E_{XX} = 0.40$, and $E_{TX} = 1.10$ eV/atom, and attempt frequencies $v^0_{TT}, v^0_{XX}, v^0_{TX}$ between 10^{13} and 2×10^{13} s^{-1} for the entire concentration range. The prefactors are in good agreement with the hydrogen vibration frequency $v_N = 3.4 \times 10^{13}$ s^{-1} observed by neutron scattering [4.150], as expected for a classical diffusion mechanism. Owing to the high activation energy, $\Delta E = 0.85$ eV/atom, the fraction of H atoms occupying X sites is well below 1% even at 1000 K for $x = 1.98$. Despite this low concentration, the X sites play an important role for H diffusion in ZrH$_x$ because E_{XX} is substantially smaller than E_{TT}. Similar parameters hold for TiH$_x$ except that $E_{TT} = 0.53$ eV/atom. Identification of the X sites has not been established. However, one may speculate [4.148] that they are associated with the "superabundant" metal atom vacancies [4.151] that arise at high temperatures and H concentrations [4.152].

The effects of 3d transition metal impurities on the ^{1}H R_1 have been investigated in the dihydrides of Ti [4.57] and Zr [4.58]. The rate R_{1p} due to paramagnetic impurities in TiH$_x$ increases sharply with increasing x, showthat Mn, which is not a hydride-forming metal, is avoided by hydrogen. Increasing x forces hydrogen closer to Mn and increases the rate R_{1p}. Hence, the Mn ions act as "antitrapping" centers. Similar behavior is found for Mn and Cr impurities in ZrH$_x$.

The ^{1}H Knight shift and square root of the Korringa rate parameter C_e in these systems trace out the peak in the DOS that occurs as x is varied. This behavior is shown in Fig. 11 for the dihydride phases of Zr at several temperatures [4.140]. Similar results have been obtained for the Ti dihydrides [4.136, 139]. The ^{91}Zr Knight shift has been measured in ZrH$_x$(1.55

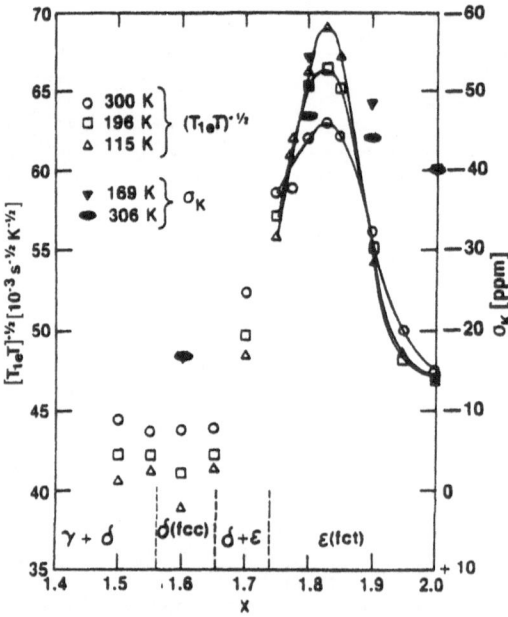

Fig. 4.11. Composition dependence of the $(T_{1e}T)^{-1/2}$ i.e., $(R_{1e}/T)^{1/2}$, and Knight shift σ_K, i.e., κ, in ZrH$_x$ at several temperatures [4.140]

$\leq x \leq 2.0$), with possible contributions to the lineshape due to shift aniso-torpy and quadrupole interactions ignored [4.153]. $\kappa(^{91}\text{Zr})$ reaches a mini-mum at $x \simeq 1.80$ where $C_e^{1/2}(^1\text{H})$ and $\kappa(^1\text{H})$ reach maxima, presumably due to the interplay between positive d-orbital and negative d-core contributions to the Knight shift [4.153]. Both the isotropic and axial κ parameters and C_e of ^{91}Zr have been measured in $\text{ZrH}_{2 \pm 0.05}$ [4.154, 155], and of $^{47,49}\text{Ti}$ in TiH_2 [4.156].

An anomalously sharp increase in the ^1H R_1 is found at all hydrogen concentrations in $\text{ZrH}_x(1.58 \leq x \leq 1.98)$ at temperatures above $\approx 900 \text{ K}$ [4.67]. The anomalous behavior depends only weakly on resonance frequency and shows no correlation with the peak in the electronic DOS at $x \simeq 1.80$. The anomalous rate R_{1x} is described by $R_{1x} = A'\exp(-U/k_\text{B}T)$, with $U = 0.4$–1.0 eV/atom. No differences were found in the behavior of pure and impurity-doped (Cr, Mn, and Fe) samples.

The relative effectiveness of atomic motion on dipolar and quadrupolar relaxation was studied in $\text{HfH}_{1.7}$ and $\text{HfD}_{1.7}$ [4.53], in what appears to be the only NMR work on Hf hydrides.

(C) Group V Systems

Solid Solution (α and α') Phases
In the solid solution phase the group V transition metals can absorb hy-drogen continuously up to about [H]/[M] $= 0.8$ with no change in the bcc structure of the host lattice. These α (low-concentration) and α' (high-con-centration) regions of the same phase provide an unusual opportunity to study the effects of changing hydrogen concentration on physical properties of the metal. Hydrogen atoms occupy the interstitial T-sites essentially randomly, and occupancy of one site appears to block occupancy of nearby sites through the second nearest-neighbor (Sect. 4.3.1A). For these and other reasons, measurement of hydrogen diffusion in these phases, utilizing pulsed-field-gradient (PFG) methods (Sect. 4.3.3) has been an active area for NMR investigation.

For TaH_x, PFG measurements have covered both the low ($0.011 \leq x \leq 0.15$) [4.157] and high ($0.12 \leq x \leq 0.77$) [4.158] hydrogen con-centration ranges. At low x values, the measurements of *Hampele* et al. [4.157] covered the temperature range 140–900 K, as shown in Fig. 4.12. These results show that E_a values derived from Arrhenius fits depend on the temperature range of the fit, i.e., $|$ d $\ln D/\text{d}(1/k_\text{B}T)$ $|$ versus $1/T$ increases with increasing temperature for both hydrogen and deuterium diffusion (also shown in Fig. 4.12).

The upper inset in Fig. 4.12 compares the x-dependence of $D(^1\text{H})$ at 670 K with different site-blocking models: The ---- line is for blocking up to second nearest neighbor; - - - - is blocking up to third nearest-neighbor; and — is the exclusion of double occupancy combined with an x-dependent E_a such that $\text{d}E_a/\text{d}x = 100$ meV/atom $(x)^{-1}$. The ratio $D(^1\text{H})/D(^2\text{D})$ at $x = 0.1$, shown in the lower inset in Fig. 4.12, decreases with increasing temperature, and its extrapolation passes through the classical value $2^{1/2}$ at about 1000 K.

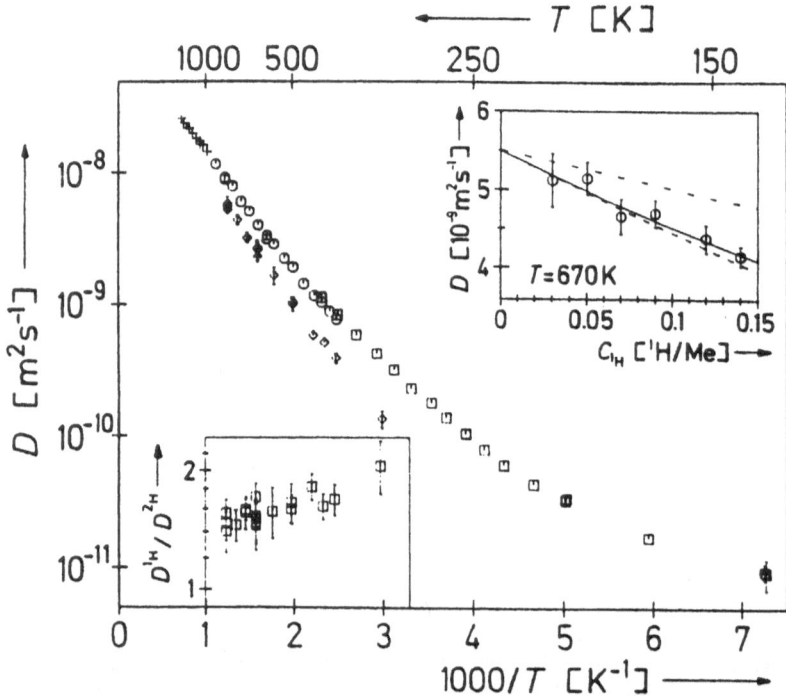

Fig. 4.12. Diffusivity of ^{1}H and ^{2}D in Ta determined by PFG measurements [4.157] $\bigcirc$, $x_H \approx 0.04$; $\square$, $x_H \approx 0.011$; $\diamondsuit$, $x_D \approx 0.1$. Also shown are the results of absorption experiments (+) [4.159]. The upper inset compares the x-dependence of $D(^1H)$ with different models of site blocking (see text), and the lower inset shows the ratio of ^{1}H and ^{2}D diffusivities [4.157]

Since for $T < 800$ K, this ratio exceeds $2^{1/2}$, tunneling processes are important at least up to this temperature. *Hampele* et al. [4.157] conclude that the data are fully compatible with the adiabatic mechanism.

Activation energies derived from Arrhenius fits to PFG measurements of $D(^1H)$ at higher x-values in VH$_x$ [4.160], NbH$_x$ [4.158], and TaH$_x$ [4.158] are summarized in Fig. 4.13. Also included in Fig. 4.13 are results for the bcc phase of TiH$_x$ [4.161]. For all bcc systems having $x \gtrsim 0.1$, experiments show that at a constant temperature addition of hydrogen causes a decrease in the tracer diffusion coefficient, D (or D_t), due principally to an increase in E_a. The relative change is greatest in α'-VH$_x$ where (Fig. 4.13) E_a for $x = 0.7$ is more than twice that at $x \approx 0$. This behavior contrasts with that of the chemical diffusion coefficient D_c [4.162] which first decreases, passes through a minimum, and may finally increase strongly with increasing hydrogen concentration. The difference in behavior results from the x-dependence of the thermodynamic factor [4.158, 162], $f_{therm} = (\rho/k_B T)(\partial\mu/\partial\rho)$, where ρ is the hydrogen atom volume density and μ the chemical potential. $(D_c/D) = f_{therm}$ approximately, and at high x-values in bcc systems, $f_{therm} > 10$, e.g., in the case of TaH$_{0.61}$ [4.163]. Such large values of f_{therm} are consistent with increased H-H atom repulsion [4.162].

Fig. 4.13. Dependence of E_a for hydrogen diffusion in bcc VH_x, TiH_x, NbH_x, and TaH_x. Pulsed field gradient data are shown by: X for VH_x, $\triangle$ for TiH_x, $\bullet$ for NbH_x, and $\circ$ for TaH_x. Quasielastic neutron scattering data are shown by solid and open squares. The dashed lines are guides to the eye [4.160]

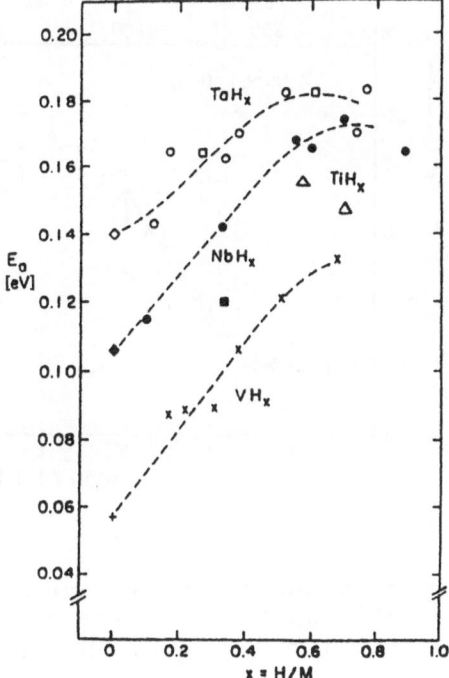

Fukai et al. [4.12, 164] measured the diffusion of hydrogen as a function of increasing deuterium concentration y in foil samples of $\alpha(\alpha')$-NbH_xD_y and $\alpha(\alpha')$-TaH_xD_y using the PFG technique. For fixed total H + D concentration $x + y$, E_a for hydrogen diffusion approaches that for deuterium (determined by Gorsky effect measurements [4.162]) as $x \to 0$, leading to the conclusion that the interaction between hydrogen atoms is strong enough to make the migration process strongly correlated [4.164].

Trapping of hydrogen by oxygen impurities in Nb has also been studied using NMR. Other experimental methods, e.g., quasielastic neutron scattering and internal friction, see [4.27] and references therein, have shown that O reduces hydrogen mobility in the temperature range 150–300 K, and that H atoms are permanently trapped below $\approx$ 90 K. *Messer* et al. [4.27, 165] measured the $^1H\,R_1$ in NbH samples doped with 0.1 or 1 at-% O over the temperature range 130–450 K at two frequencies. A relatively weak R_1 peak at about 200 K was found, as shown in Fig. 4.14, whose dependence on O concentration c_O and frequency ω_0 are well described by an additional rate contribution, R_{1t}, to the rate $R_1 = R_{le} + R_{ld}$ in pure (undoped) Nb, and given by [4.27]

$$R_{1t} = A\,c_0\,\tau_t/(1 + \omega_0^2\,\tau_t^2), \qquad (4.37)$$

where τ_t is the correlation time associated with the trapping-detrapping process. The NMR results show that O reduces the long-range diffusion

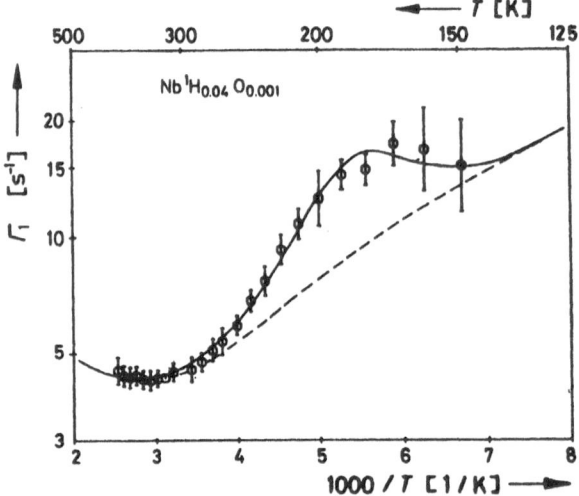

Fig. 4.14. Temperature dependence of $R_1(^1H)$ in an oxygen-doped sample, $NbH_{0.04}O_{0.001}$. Dashed line: updoped $NbH_{0.04}$ sample; the full line represents a fit to (4.37) using a pair binding energy of 90 mev/atom [4.27, 165]

process by trapping and that in the temperature and c_O range studied only one H is trapped at an O atom at any one time.

Hydrogen trapping by lattice defects in β-NbH_x $(0.71 \leq x \leq 0.86)$ has been observed in single foil specimens in high resolution ^1H spectra [4.166]. Three resonances having significantly different Knight shifts were found and attributed to (i) protons in the undisturbed lattice, (ii) protons trapped at dislocations, and (iii) protons possibly trapped at O impurities.

Determination of $C_e(^1H)$ in the $\alpha(\alpha')$ phases at arbitrary H concentrations is made difficult by two complicating factors. First, sharply reduced H solubility at temperatures below the solid solution $\Longleftrightarrow$ hydride phase transition boundary progressively extinguishes the $\alpha(\alpha')$ proton resonance, so that an $\alpha(\alpha')$ resonance signal is not obtained at temperatures below the R_{1d} maximum which occurs on the order of 100 K. (Note: this is not the case in Nb-V alloys, for example, where hydride formation can be completely suppressed [4.61].) At high temperatures in the $\alpha(\alpha')$-phase, an anomalous contribution R_{1x} appears (Sect. 4.3.2B) [4.68]. However, in α-TaH_x a substantial temperature range exists in which R_1/T = constant, indicative of electronic relaxation [4.24, 65, 143], and the same appears to hold in α-VH_x but over a rather narrow temperature range [4.167]. In the latter case, C_e decreases from $2.2 \times 10^{-2}(s\ K)^{-1}$ at $x = 0.042$ to $7.5 \times 10^{-3}(s\ K)^{-1}$ at $x = 0.736$, consistent with decreasing $N(E_F)$ as H concentration increases. Proton Knight shifts have been measured in the solid solution phases using, for example, the single foil technique (Sect. 4.4) in α-VH_x [4.168], $\alpha(\alpha')$-NbH_x [4.168, 169], and $\alpha(\alpha')$-TaH_x [4.95, 168]. In all cases, the true shifts are small and negative, reflecting the dominance of the κ_d term, i.e., the core-polarization hyperfine field, in (4.26). Both κ and C_e of ^{51}V and ^{93}Nb were measured in the solid solution phases and found to decrease with increasing H concentration [4.1, 170].

The much weaker deuteron relaxation rates compared to protons, which are a consequence of the small ^{2}D γ value and quadrupole moment (Table 4.1) causing the rate R_{1e} to scale as γ^2 [4.171], and resulting in a very weak R_{1Q}, make experimental determinations of the rates difficult indeed. *Salibi et al.* [4.171] determined E_a and C_e for deuterons in α'-TaD$_x$ ($0.43 \leq x \leq 0.76$) from R_1 measurements over the temperature range 300–600 K, i.e., on the high-temperature side of the R_{1Q} maximum. E_a was found to be essentially constant, $E_a = 0.20 \pm 0.02$ eV/atom, for this composition range. Similar measurements on α'-NbD$_x$ ($x = 0.55, 0.88$) yielded two-component R_1 behavior with fast and slow relaxing components [4.171]. In the case of α-TaH$_x$ at $x \approx 0.08$, measurements of $R_1(^2$D) over the temperature range 300–900 K were consistent with quadrupolar and electronic relaxation up to about 800 K, beyond which an anomalous increase in R_1 was observed [4.157], see Fig. 4.12 for PFG measurements of deuterium diffusion [4.157].

Triton (^{3}T) spin-lattice relaxation has been studied in α-NbT [4.27, 164]. In contrast to ^{1}H measurements, $R_1(^3$T) reaches a maximum at $\omega_0\tau_c \approx 1$ between 150 and 200 K at 86.3 MHz resonance frequency. Analysis of R_{1d}, assuming jumps between nearest-neighbor T-sites, yielded the temperature dependence of τ_d from which triton diffusivities were calculated [4.27]. These were found to be one to two orders of magnitude smaller than those of protons.

Hydride Phases

Proton R_1 measurements in hydride phases at low temperatures are complicated by cross-relaxation to ^{93}Nb and ^{181}Ta spins, causing the apparent C_e to depend strongly on resonance frequency (Sect. 4.3.2B, Fig. 4.6). This difficulty can be overcome by measuring R_1 at sufficiently high resonance frequencies, e.g, 400 MHz. In the case of Ta$_2$H, the ^{1}H cross-relaxation spectrum, i.e., the frequency dependence of R_{1c}, enabled the ^{181}Ta quadrupole coupling parameters to be determined [4.61]. The spectrum at 130 K in a sample of TaH$_{0.32}$ is shown in Fig. 4.15. At this temperature, the ^{1}H NMR originates essentially from protons in the orthorhombic Ta$_2$H phase. The fit to the experimental points, shown by the several curves in Fig. 4.15 (see caption), taking account of the spin-lattice relaxation rate R_{1S} of the ^{181}Ta, yielded $\nu_Q = 43$ MHz, $\eta = 0.35$, with $C_e(^{181}$Ta) = 0.38 (s K)$^{-1}$. The complexity and width of the spectrum are consequences of substantial EFG inhomogeneity, reflecting disorder in Ta site environment, and of the mixing of ^{181}Ta spin states caused by the applied field which produces Zeeman energies comparable to the pure quadrupole energies.

Measurements of diffusion parameters have been made for some hydride phases. In NbH, the temperature dependence of $D(^1$H) was calculated from values of τ_d derived from proton R_1 and R_2 measurements [4.27] over the temperature range 140–375 K spanning the ε-, ζ-, and β-phase regions. The activation energy increases progressively in the sequences $\beta \rightarrow \zeta \rightarrow \varepsilon$ as hydrogen becomes progressively more ordered on a subset of the interstitial sites.

Proton relaxation rates and lineshapes have been studied most intensively in the hydride phases of vanadium [4.172–175]. Site assignments in

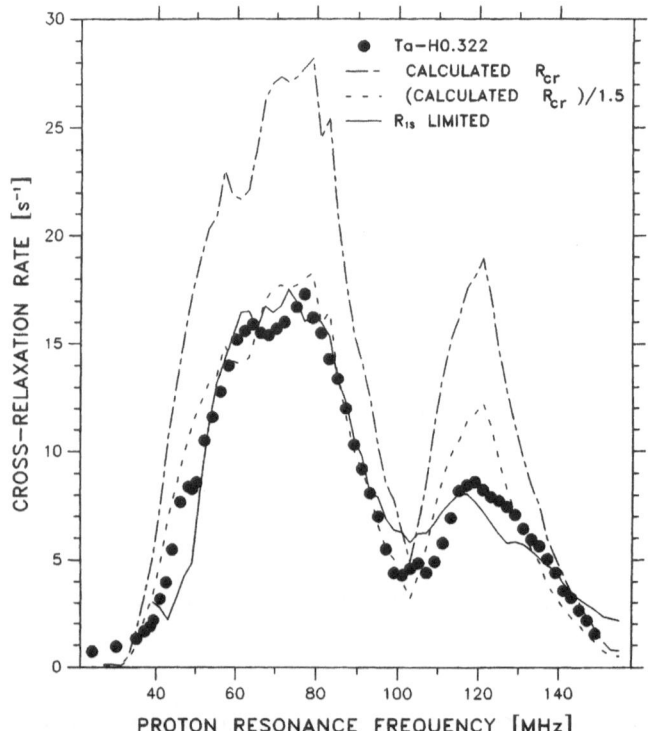

Fig. 4.15. Frequency dependence of $R_1(^1H)$ in TaH$_{0.32}$ at 130K (solid points) compared with calculated estimates of R_{1c} in the fast and slow R_{cr} limits [4.61]. Dash-dot line: R_{cr} calculated directly using $\nu_Q = 43$ MHz, $\eta = 0.35$. The calculated values are greater than the data due either to slow $R_1(^{181}Ta)$ or to approximations in the model. Dashed line: $R_{cr}/1.5$ which, using the above parameters, best fit the data. Solid line: R_{1c} in fast R_{cr} limit, using $\nu_Q = 43$ MHz, $\eta = 0.6$. In this case, the overall magnitude was adjusted to match the data [4.61]

V$_2$H, as well as in V$_2$D and V$_2$T, were discussed by *Bowman* et al. [4.174]. The complex behavior of $R_1(^1H)$ *vs* T in V$_2$H and of $R_1(^3T)$ *vs* T in V$_2$T have also been elaborated [4.172, 174, 175].

A number of investigations have focussed on ^2D quadrupole spectra in the deuteride phases: in VD$_x$ [4.171, 176, 177]; in NbD$_x$ [4.10, 171, 178–180]; and in TaD$_x$ [4.171, 181–183]. A principal objective of these studies has been determination of phase boundaries and transition regions *via* the transition from quadrupole-split spectrum to single line characteristic of the solid solution phases (Sect. 4.2.2). The ^2D quadrupole coupling parameter, ν_Q, falls in the range 30–35 kHz in NbD$_x$ and TaD$_x$, essentially independent of deuterium concentration. This shows that the EFG results principally from the nearest-neighbor metal ion environment, the deuterons themselves contributing negligibly. Somewhat greater ν_Q values are found in VD$_x$, in the range 50–100 kHz. Deuteron relaxation rates in these phases have also been measured [4.171], leading to the conclusion that while ^2D relaxation associated with hopping among T-sites in the $\alpha(\alpha')$ phases is mostly quadrupolar, in the warm regions of the hydride phase the dominant relaxation is magnetic dipolar because hopping occurs among an ordered subset of interstitial sites. At lower temperatures, quadrupolar relaxation again dominates [4.171].

The ^{1}H and metal Knight shifts and R_1 have been measured in the cubic (fcc) dihydrides, VH_2 [4.184, 185] and NbH_2 [4.186]. The linear temperature dependence of $R_1(^1H)$ and $R_1(^{51}V)$ and of $R_1(^{93}Nb)$ shows that these dihydrides are still metallic [4.185, 186]. Both $\kappa(^1H)$ and $\kappa(^{51}V)$, as well as $\kappa(^{93}Nb)$, are small and negative. Analysis of these results, along the lines of (4.25–27), shows for example that $\kappa(^{51}V)$ is the result of strong competition between a negative κ_d and positive κ_{orb} [4.184]. These results, together with those for the other transition metal dihydrides, have been reviewed by *Zogal* and *Nowak* [4.91]. The ^{51}V and ^{93}Nb spectra and Knight shifts have also been studied in the noncubic hydride and deuteride phases: in $VH(D)_x$ [4.187], and in $NbH(D)_x$, [4.62, 178, 180, 188].

(D) Group VI Systems

The sole example from this group to have been studied with NMR is chromium hydride, CrH_x. This compound forms in the hexagonal anti-NiAs structure, with hydrogen occupying the octahedral interstitial sites. CrH_x is paramagnetic, the susceptibility being both temperature and composition dependent. Ignoring possible lineshape complications due to shift anisotropy and quadrupole interactions, the ^{53}Cr Knight shift in $CrH_{0.93}$ was found to be $\kappa = (0.53 \pm 0.02)\%$, independent of temperature for $150 \leq T \leq 294$ K [4.189]. The Knight shift and susceptibility were partitioned into spin and orbital contributions in the usual manner [4.189]. Earlier measurements of $\kappa(^1H)$ were probably influenced by high level of paramagnetic impurities [4.190].

(E) Palladium-Hydrogen

Palladium hydride was probably the first hydride to be studied by NMR [4.191], and interest in this system has not subsided, in part, because it is one of very few in which hydrogen occupies an fcc interstitial sublattice, i.e., octahedral sites. The following paragraphs concern primarily studies not reported in [4.1].

Measurements of $R_{1d}(^1H)$ in β-PdH_x at several frequencies were quantitatively explained for $T > 195$ K by diffusion *via* octahedral site jumps [4.1, 192]. Below 195 K interpretation of the data was uncertain due to effects of paramagnetic impurities. For Pd-Ce-H alloys, the maxima in R_{1d} are unshifted from those in PdH_x, leading to the conclusion that the permeation increase for Pd-Ce alloys is probably associated with surface phenomena [4.192].

PFG measurements of hydrogen diffusion in β-$PdH_{0.70}$ prepared by loading Pd foil with H from the gas phase were made at temperatures in the range 396–413 K [4.1, 193]. Combined with τ_d values from R_{1d}, these PFG results affirm the O-O jump path of the diffusing hydrogen. Diffusion in the α'-phase of PdH_x and $(Pd-Ag)H_x$ was also studied through R_{1d} and PFG measurements [4.194]. The dependence of D on x results largely from the blocking factor $(1 - x)$. The quantity $D/(1 - x)$ decreases and E_a increases with increasing Ag content in the alloys. Measurements of the anisotropy in $R_1(^1H)$ in a single crystal of α'-$PdH_{0.7}$ at 238 K as the applied field was aligned along the [001], [111], and [110] directions were also entirely consistent with O-site occupancy [4.195].

Two features of the ^1H resonance in β-PdH_x at low temperatures have been attributed to hydrogen tunneling between adjacent O-sites [4.196] (i) The lineshape consists of a weak ($\approx 1\%$ of total intensity) narrow line superposed on a broad symmetrical resonance [4.197, 198]; (ii) The second-moment of the broad component is $\approx 35\%$ less than the value predicted by (4.1) which applies to a "rigid lattice" [4.199]. By employing a pulse sequence originally utilized in the study of low-frequency tunnel splittings in CH_3 and NH_4 groups in solids [4.200], spin-polarization torsional spectroscopy (SPOTS), *Avram* and *Armstrong* [4.196] measured the x-dependence of the two lowest frequency tunneling frequencies, ω_T, at 40 K, finding that ω_T ranged from ~ 15 kHz to ~ 70 kHz, decreasing with increasing x.

Trapping of deuterium by dislocations in cold-rolled Pd foil samples of $PdD_{0.6}$ has been studied by $^2D\,R_1$ and R_2 measurements [4.201, 202]. The experiments reveal clearly the presence of two spin-spin relaxation rates, $R_{2a} < R_{2b}(\approx 11$ and $170\,s^{-1}$, respectively, at 300 K), both much faster than $R_1(\leq 0.42\,s^{-1})$. R_{2a} describes relaxation of ^2D in normal interstitial sites whereas R_{2b} is attributed to deutrons trapped in dislocation core regions. The temperature dependence of R_{2b} yields $E_a \approx 0.024$ eV/atom for deuteron diffusion within the core regions. From the signal amplitudes it is estimated that trapping can cause almost all dilated deuteron sites to be filled out to a radius of 30 Å around dislocation cores for a dislocation density $\approx 10^{12}cm^{-2}$ [4.201].

Measurements of triton relaxation rates were made on tritide-free aged (~ 8 yrs) PdT_x as part of a study of helium retention in microbubbles [4.203]. Most of the tritium was removed from nominal (aged) $PdT_{0.65}$ samples, and the NMR measurements of R_1, R_{1d}, and R_2 were made on $PdT_{0.08}$. The results indicate a value of the activation energy for T diffusion much less than for hydrogen in PdH_x, suggesting that T is not distributed uniformly among O-sites in this material. The low E_a value may be associated with short-range motion of trapped tritium, perhaps at internal surfaces surrounding He microbubbles [4.203].

The NMR of ^{105}Pd was studied in α-PdH_x at 75 °C as a function of hydrogen pressure [4.204], yielding a linear relation between $\kappa(^{105}Pd)$ and x and also between κ and the susceptibility χ. $\kappa(^{105}Pd)$ becomes less negative with increasing x ($\kappa = -(1.80 \pm 0.14)\%$ in pure Pd [4.204]) due to the decreasing 4d spin susceptibility within the α-phase. The ^{105}Pd resonance could not be detected in β-PdH_x. Measurements of $\kappa(^1H)$ were also made.

(F) Actinide Hydrides

The thorium hydrides, ThH_2 and Th_4H_{15}, are especially interesting because the latter becomes superconducting at a relatively high temperature ($T_c \approx 9K$) and because all of the Th nuclides have zero nuclear moments, so that dipolar interactions are purely ^1H-^1H in these. Early work [4.205, 206] focussed on hydrogen diffusion based on proton R_1 and R_2 measurements and Torrey spectral densities. *Lau* et al. [4.207] measured $R_1, R_{1\rho}$, and R_2 in two Th_4H_{15} samples prepared under different conditions (essentially low and high temperatures and pressures) over the temperature range 200–500 K and determined C_e and E_a. E_a values were substantially greater than previously reported. The ^1H Knight shift was measured using multipulse line-narrowing methods over the temperature range 40–300 K and found to be strongly temperature dependent for $T \gtrsim 200$ K. *Peretz* et al. [4.208] found $C_e = 1.8 \times 10^{-3}(s \ K)^{-1}$ in ThH_2 and independent of temperature for $78 \leq T \leq 300$ K, but temperature dependent in Th_4H_{15}, ranging from $6.1 \times 10^{-3}(s \ K)^{-1}$ at 78 K to $4.35 \times 10^{-3}(s \ K)^{-1}$ at 300 K, using samples prepared at moderate temperatures and pressures. Measurements of the deuteron R_{1Q} in Th_4D_{15} yielded values of the pre-exponential factor τ_0 consistent with classical behavior, i.e., $\tau_0^D \approx \sqrt{2}\,\tau_0^H$, based on BPP analysis of both $R_{1d}(^1H)$ and $R_{1Q}(^2D)$ [4.208].

Uranium hydride, β-UH_3, which is a nominally stoichiometric hydride, ferromagnetic for $T < 180$ K, has been the subject of several NMR studies. Linewidth, second-moment, and κ measurements [4.209] yielded $E_a \simeq 0.36$ eV/atom and showed κ directly proportional to the molar susceptibility. The ^2D quadrupole coupling was measured in UD_3. In further work, $R_1(^1H)$ was studied at temperatures $190 < T < 700$ K [4.210] and resolved into contributions from R_{1d}, R_{1e}, and R_{1p}, the latter due to interaction with the localized $5f$ electrons, yielding $E_a = 0.84$ eV/atom, $C_e = 0.135$ (s K)$^{-1}$, and indicating that the U ion occurs as U^{3+}. A later study [4.211] was concerned with H diffusion, based on $R_2(^1H)$ measurements, and concluded that E_a is the sum of energies of vacancy formation and barrier height and that the pre-exponential τ_0 contains an entropy change factor.

The fcc plutonium dihydrides, PuH_x, $1.78 \leq x \leq 2.78$, have been studied [4.212]. These are analogous to the LaH_2-LaH_3 system, having hydrogen in T–sites for $x \lesssim 2$, with additional H occupying O-sites. Lineshapes, R_1, and R_2 were measured in the temperature range 77–300 K. The results are consistent with paramagnetic phases at high temperatures, with localized Pu $5f$ moments, and possible magnetically ordered phases at low temperatures, depending on H concentration. Hydrogen diffusion parameters were estimated from R_2 measurements and found to be similar to those in LaH_x, with E_a decreasing with increasing x.

4.5.2 Intermetallic Hydrides

The highly favorable hydrogen storage properties of intermetallic hydrides such as $FeTiH_x$ and $LaNi_5H_x$ have made these and other systems the subject of many investigations. They are especially interesting from the NMR standpoint because of the variety of behavior shown by their spectra and relaxation rates that frequently result from hydrogen occupancy of multiple interstitial site sublattices and structural phase transitions. Major results are briefly reviewed in the following subsections. It should be emphasized that for lack of lattice-specific theories (Sect. 4.3.1A) for the complex H sublattices in these systems, all analyses of ^1H relaxation rates are based on BPP spectral densities. Systems are classified according to structural type.

(A) LaNi₅-H and Related Systems

The results of NMR studies of $LaNi_5H_x$ and related hydrides have been discussed by *Spada* et al. [4.213]. Briefly, in the case of $LaNi_5H_x$, ^1H and ^2D wide-line measurements [4.214] and ^1H relaxation rate measurements [4.213, 215, 216] indicate multiple site occupancy and two (or more) stages of hydrogen motion, most probably localized motion superposed on long-range diffusion [4.213]. Characteristic features of the ^1H NMR in $LaNi_5H_x$ include the appearance of both narrow and broad resonances within certain temperature ranges [4.214] and unequal slopes on the low and high-temperature sides of the $R_{1\rho}$ and R_{1d} maxima, resulting in substantially different (factors of 2) estimates of E_a for hydrogen diffusion [4.213]. A PFG measurement of $D(^1H)$ in $LaNi_5H_{6.5}$ yielded $E_a = 0.42$ eV/atom in the temperature range 330–375 K [4.82], consistent with the high-temperature values deduced from R_{1d} measurements [4.213]. Also, the frequency dependence of $R_{1\rho}$ at low temperatures was found to be $\omega_1^{1.35}$, in contrast to ω_1^2 expected on the basis of BPP spectral densities, and typical of that found in amorphous and glassy materials [4.105].

In β-$LaNi_{5-y}Al_yH_x (0 < y < 1.5)$, Al substitution for Ni increases E_a with subsequent reduction in the apparent diffusion coefficients by more than two orders of magnitude relative to those for β-$LaNi_5H_6$ when $y \geq 1$ [4.217]. Although the temperature dependence of R_{1d} and $R_{1\rho}$ fails to show clear evidence of intermediate maxima, E_a derived from the low-temperature slopes is again much less than that of the high-temperature slope. Similar behavior was found in $LaNi_4BH_{1.5}$ [4.213].

In the case of $LaCu_5H_{3.4}$ proton R_1 and $R_{1\rho}$ measurements [4.218] showed clear evidence for the presence of two independent motional processes, both substantially less rapid than reported for $LaNi_5$ hydride. The temperature dependence of $R_{1\rho}$ exhibits an intermediate maximum consistent with the observed R_{1d} maximum. However, the R_{1d} maximum corresponding to the principal $R_{1\rho}$ maximum could not be reached because the high desorption pressures for $LaCu_5H_x$ prevented heating the sample beyond ≈ 400 K [4.218]. The activation energy for the low-temperature motion is roughly half that for the high-temperature motion.

(B) TiFe-H and Related Systems

Measurements of ^{1}H linewidths and relaxation rates in TiFe hydrides are complicated by the strong paramagnetism of this material and the presence of ferromagnetic impurities (e.g., TiFe$_2$) that broaden the resonance and contribute an impurity rate R_{1p} to the measured R_1 at low temperatures. In addition, the high hydrogen desorption pressures prevent heating samples above 400 K, making estimates of diffusion parameters difficult. The principal NMR results for TiFeH$_x$ and the closely similar TiCoH$_x$ system [4.219–221] show that $C_e(^1H)$ values for TiFeH$_x$ are much greater than for TiCoH$_x$. This is attributed [4.221] to the larger hyperfine field that arises from polarization of both the Fe 3d and Ti 3d states at E_F and to a relatively reduced $N_d(E_F)$ in TiCoH$_x$. Estimates of E_a for hydrogen diffusion, based on relaxation rate measurements, are summarized in [4.219, 220].

(C) Laves Phase Hydrides

The most studied Laves Phase hydrides have been ZrV$_2$H$_x$ and HfV$_2$H$_x$, and their properties, including NMR results, were reviewed by *Shinar* [4.222]. These systems possess an extended solid solution phase (up to $x \approx 5$) for $T \gtrsim 325$ K, and at least 4 hydride phases. Order-disorder transitions are clearly revealed by discontinuities in the temperature dependence of $R_{1d}(^1H)$ and $R_{1Q}(^2D)$ in both systems [4.223]. Hydrogen motion is sufficiently fast even at 100 K that reliable determination of the electronic parameter C_e requires measurements down to 4.2 K and at high resonance frequencies. In the high-temperature phases the $R_{1d}(^1H)$ data are reasonably described by the BPP model with a single activation energy, E_a being nearly constant in HfV$_2$H$_x$ over the entire range $0.5 \leq x \leq 4.0$ [4.223]. For the low-temperature ordered phases a double-peaked distribution of E_a values is required [4.223].

Skripov et al. [4.224–226] have utilized ^{1}H, ^{2}D, and ^{51}V spectra and relaxation rates to study TaV$_2$H$_x$. This system is especially interesting because clear evidence of localized hydrogen motion at low temperatures appears in the R_1 vs T behavior of all three nuclear spins, as shown in Fig. 4.16 for $R_1(^{51}V)$ at several x-values in TaV$_2$D$_x$ [4.225]. The low-temperature peak results from modulation of the ^{51}V quadrupole interaction by the localized motion of deuterium. Although the ^{51}V R_{1Q} maxima due to long-range $H(D)$ diffusion are essentially the same, as expected, the low-temperature maxima in the deuterides are nearly three times stronger than in the hydrides, leading to the surprising conclusion that the amplitude of deuterium motion at low temperatures is greater than that of hydrogen motion [4.225]. The random bcc alloy system TaV$_2$-$H(D)$ has also been studied in the same manner [4.226]. Long-range diffusion in the alloys is much faster than in the corresponding ordered intermetallics. No evidence for localized low-temperature motion is found; however, as in other alloy hydrides, a distribution of E_a values is required to fit the measurements (Sect. 4.3.1B).

In both cubic and hexagonal ZrCr$_2$H$_x$ the hydrogen hopping rate is already fast enough to cause complete ^{1}H line-narrowing at 78 K and R_{1d}

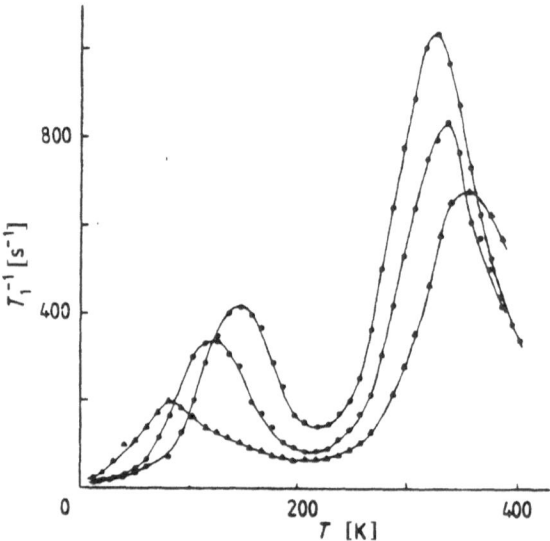

Fig. 4.16. Temperature dependence of the ^{51}V spin-lattice
relaxation rate TaV$_2$D$_{0.50}$(Δ)
TaV$_2$D$_{0.84}$(•) and TaV$_2$D$_{1.08}$(○)
measured at 19.3 MHz. The full
curves are guides to the eye
[4.225]

maxima at $\approx$ 200 K [4.227]. Hydrogen mobility decreases with increasing x, with E_a ranging from 0.29 eV/atom for $x = 4.2$ [4.228] down to 0.06 eV/atom at $x = 0.04$ [4.227] in the hexagonal phases. In ZrTi$_2$H$_x$ the R_{1d}(^{1}H) behavior can be fit with a double-peak distribution of E_a values rather than a continuous E_a distribution of considerable width [4.229]. Measurements of C_e(^{1}H) and κ(1$_H$) in the α and α'-phases of cubic and hexagonal TiCr$_{1.8}$H$_x$(0.63 $\leq x \leq$ 2.85), combined with susceptibility data, were analyzed on the basis of (4.25–27) [4.230]. These results show that $N_d(E_F)$ increases with increasing x and that the electronic structures of the cubic and hexagonal allotropes are quite similar.

Hydrogen diffusivity D has been measured in a number of cubic Laves phase hydrides using the pulsed-field-gradient technique. These include: (a) ZrTi$_2$H$_{3.7}$ (230 $\leq T \leq$ 450 K) [4.231]; (b) ZrV$_2$H$_x$ (0.5 $\leq x \leq$ 5.0) for 245 $\leq T \leq$ 485 K [4.232]; (c) ZrCr$_2$H$_x$ (0.2 $\leq x \leq$ 0.5, and $x = 3.3$) for 130 $\leq T \leq$ 430 K [4.232, 233]; (d) ZrMo$_2$H$_x$ ($x = 0.6, 1.2$) for 264 $\leq T \leq$ 464 K [4.234]; (e) TaV$_2$H$_{1.24}$ for 230 $\leq T \leq$ 445 [4.232]; and (f) HfV$_2$H$_{1.0}$ for 264 $\leq T \leq$ 464 K [4.232]. In addition, hydrogen and deuterium diffusivities have been measured in HfV$_2$H$_x$D$_{4-x}$($x = 1.5, 3.0$) [4.232]. An overview of these results is given in [4.232], including discussion of the dependence of the diffusivity on H concentration x and its relation to the probably site occupancies and jump paths.

In ZrCr$_2$H$_x$ the temperature dependence of D follows an Arrhenius law above $\sim$ 200 K, but shows marked deviations from such behavior below $\sim$ 180 K, suggesting that different mechanisms dominate at high and low temperatures [4.233], as seen in NbH$_x$ and TaH$_x$ [4.27, 157]. The data could be explained consistently within the framework of quantum diffusion, with tunneling transitions between the ground states in neighboring interstices dominating below $\sim$ 200 K and an increasing contribution of tunneling transitions between excited states in neighboring sites above $\sim$ 200 K [4.233].

(D) Other Intermetallic Hydrides

A variety of zirconium intermetallics have been investigated for potential hydrogen-storage applications, and NMR methods have frequently been utilized. A well-documented example of hydrogen motion occuring independently on two sublattices is found [4.235] in the layered structure metallic Zr monohalide hydrides, $ZrClH_{0.5}$ and $ZrBrH_{0.5}$, in which the overall structure consists of infinite four-layer units sequenced x-Zr-Zr-x (x = Cl, Br). Measurements of the temperature dependence of proton lineshapes and widths, and of R_1 and $R_{1\rho}$, showed that hydrogen predominantly occupies alternate chains of T-sites within the metal bilayers, but with a fraction occupying the mostly empty O-site sublattice. Motion occurs first at low temperatures on the O-site sublattice with $E_a = 0.16$ eV/atom, and at higher temperatures on both T and O-sublattices with $E_a \simeq 0.42$ eV/atom. The appearance of a low-temperature secondary maximum in both R_1 and $R_{1\rho}$, together with the observed two stages of motional narrowing, rules out paramagnetic impurities as the origin of this behavior.

Proton relaxation rates have been measured in a number of Zr and Ti-based intermetallics in addition to the Zr-based Laves phases (Sect. 4.5.2C). In most cases, $R_1(^1H)$ was decomposed in the usual manner to obtain $C_e(\frac{1}{H})$, the temperature dependence of τ_d, and hence E_a values, from BPP-type analyses. Examples include Zr_2NiH_x [4.236], and Ti_2CuH_x, Zr_2CuH_x, Ti_2PdH_x and Zr_2PdH_x [4.237, 238 and references therein]. The frequency dependence of $R_1(^1H)$ in Zr_2PdH_x, studied at low temperatures, showed clear evidence of ^1H-metal spin cross-relaxation becoming pronounced at low resonance frequencies [4.239]. These measurements indicate that care must be exercised in determining $C_e(^1H)$ in such systems.

The temperature dependence of $R_1(^1H)$ has been determined over the range $5 < T < 630$ K in the A15 structure intermetallic hydride Ti_3IrH_x, $0.55 \leq x \leq 3.5$ [4.240] and interpreted in terms of R_{1e} and R_{1d} contributions. The behavior of R_{1d} in a similar study of Ti_3SbH_x, $0.6 \leq x \leq 2.4$, to ~ 450 K [4.241] was interpreted in terms of localized hydrogen motion rather than long-range diffusion as in [4.240].

Several ^1H NMR studies have been made of the Fe_2P-type systems $ThNiAlH_x$ and $UNiAlH_x$ [4.242], $YNiAlH_x$ and $ZrNiAlH_x$ [4.243], $CeNiAlH_x$ and $CeCuAlH_x$ [4.244], and $CeNiInH_x$ [4.245]. In most of these superposed broad and narrow resonances appear over some temperature range, suggesting that hydrogen diffusion occurs simultaneously on independent sublattices, as in $ZrClH_{0.5}$. In addition, in $CeNiInH_x$ ($x = 1.0, 1.6$) the broad ^1H resonance displays "Pake doublet" character [see Sect. 4.2.1] in the rigid lattice regime [4.245]. The observed doublet separation, $\Delta B = 55$ kHz, implies a ^1H-^1H spacing of 1.48 Å on the basis of a single close pair of protons, substantially less than the commonly accepted distance for closest approach, ~ 2.1 Å [4.7].

Two stages of motional narrowing were found in Ta_6SH_x [4.246] for temperatures 4.2–280 K, and interpreted in terms of low-temperature localized motion with $E_a \approx 0.01$ eV/atom and long range motion at higher temperatures with $E_a \approx 0.10$ ev/atom.

4.5.3 Random Alloy and Amorphous Systems

(A) Random Alloy Systems

NMR studies of hydrogen in transition metal alloys have been predominantly concerned with alloys of Ti, V and Nb, since these show extended solid solubility ranges. Initially, the work of *Rohy* and *Cotts* [4.1, 247] on R_{1e} in the $V_{1-y}Cr_yH_x$ system (bcc throughout) showed that to a good approximation R_{1e} depended only on the sum $(x + y)$ and not upon the relative amounts of Cr or H. This result supports the rigid-band model in that the H atom gives its electron to the host metal just as an alloying element of the next higher atomic number would do.

Hayashi et al. [4.248–250] utilized ^{1}H and ^{51}V lineshapes and Knight shifts, in conjunction with X-ray diffraction, to study the structure of the Ti-V-H system over a wide range of compositions. These measurements show that hydrogenizing of bcc Ti-V alloys results in several hydride phases with different structures and compositions [4.248]:

$$\beta\text{-Ti-V} \xrightarrow{\;\;H_2\;\;} \alpha\text{-Ti-V-H} + \beta\text{-Ti-V-H}$$
$$\quad \text{bcc} \qquad\quad \text{bcc} \qquad\qquad \text{bcc}$$

$$+ \gamma\text{-Ti- V-H}_2 + \gamma\text{-TiH}_2$$
$$\qquad \text{bcc} \quad \text{fcc} \qquad \text{fcc}$$

The β and γ phases are nominally the monohydrides and dihydrides, respectively, Activation energies for H diffusion in the β-phase were estimated from the temperature dependence of $R_1(^1H)$ [4.248]. The resulting graphs of $\ln \tau_d$ *vs* T^{-1} are strongly curved, and the experimental maximum R_{1d} values are much weaker than those estimated from the calculated second moments. These departures from conventional BPP behavior are expected in random alloy systems (Sect. 4.3.1B).

PFG measurements of $D(^1H)$ have also been made on bcc $\beta\text{-Ti}_{1-y}V_yH_x$, $y = 0.04$ and 0.08, and $0.39 \leq x \leq 0.70$, at temperatures up to ~ 800 K [4.251]. These yielded Arrhenius behavior in the single-phase region, from which values of E_a were derived. Measurements of R_1 on the same sample [4.251] were consistent with purely electronic relaxation. No indication of anomalous relaxation, R_{1x}, appeared even at 800 K.

Baden et al. [4.252] employed the single-foil technique to measure $\kappa(^1H)$ and χ (Sect. 4.4) of bcc alloys $(\alpha, \alpha')\text{-Nb}_{1-y}Ti_yH_x$ $(0 \leq y \leq 0.5; 0 \leq x \leq 1)$ at 180 °C. The molar susceptibility χ_m was found to depend only on the valence electron concentration in agreement with an augmented plane wave band structure calculation and application of the rigid band model. On the other hand, $\kappa(^1H)$ could not be explained within the framework of the rigid band model and is not simply related to the valence electron concentration. A similar study was made of $Ta_{1-y}Nb_yH_x$ alloys [4.253].

Hydrogen diffusion has been studied in bcc Nb-V alloys on the basis of ^{1}H linewidths [4.254] $R_1(^1H)$ [4.47, 255] and PFG $D(^1H)$ [4.68, 256] measurements. This system holds particular interest because V and Nb form an infinite range of solid solutions in the bcc phase and because the terminal solubility of H is greatly enhanced over that in V and Nb [4.254, 257]. In addition, the temperature at which hydride phase precipitation occurs is sharply depressed [4.47] so that in $Nb_{0.5}V_{0.5}H_x$, for example, no indication of the hydride phase proton NMR is found, even at 4.2 K, for $x = 0.21$ [4.47, 255].

The behavior of the $R_1(^1H)$ measurements [4.47] in intermediate composition alloys, $V_xNb_{1-x}H_y(0.25 \leq x \leq 0.75)$, differs significantly from that in either the V–rich or Nb–rich compositions. In this composition range, trapping effects are not evident, and the temperature dependence of $R_1(^1H)$ is well represented by a Gaussian distribution of E_a values. In contrast, the behavior of $R_1(^1H)$ in $V_{0.1}Nb_{0.9}H_{0.19}$ could not be represented by an E_a distribution. Rather, it was concluded that a transition occurs from long-range lattice motion $(T \gtrsim 290K)$ to localized motion (tunneling) between states adjacent to V atoms $(T \lesssim 270K)$ with an activation energy of ≈ 14 meV/atom [4.47] substantially less than that estimated (90–150 meV/atom) on the basis of linewidth measurements [4.254].

The PFG measurements of $D(^1H)$ [4.256] found that V caused a decrease in D in the Nb-V alloys. An H-V trap binding energy of 80 meV/atom was estimated.

Tracer diffusion coefficients, $D_t = D(^1H)$ measured by the PFG method [4.68, 256], were compared with chemical diffusion coefficients, D_c, from the literature [4.258] by taking into account the thermodynamic factor, f_{therm} (Sect. 4.5.1C), which is obtained from pressure-composition isotherms. Satisfactory agreement between D_t and D_c was obtained [4.259].

Knight shifts and relaxation rates have been used to study the effects of alloy additions of V and Nb on the electronic structure of fcc Ti dihydride. $\kappa(^{93}Nb)$, $R_1(^{93}Nb)$ [4.260], and $\kappa(^1H)$ [4.261], and $R_1(^1H)$ [4.262] were measured in $Ti_{1-x}Nb_xH_{1.94}(0.05 \leq x \leq 0.65)$ over the approximate temperature range 100–300 K. Combined with susceptibility data, the ^{93}Nb results were partitioned into s, d, and orbital contributions [4.260]. A similar analysis of $\kappa(^{51}V)$ and $R_1(^{51}V)$ measurements was made for $Ti_{1-x}V_xH_2$ [4.263]; however, in this case it was found that the rigid band model, which was a useful approximation for Ti-Nb-H, could not account for the composition dependence of the NMR and susceptibility data. Similar measurements on this system [4.264] found that the temperature dependencies of κ and C_e confirmed that the reduction of d-electron states through a Jahn-Teller effect in γ-TiH$_{\sim2}$ is eliminated by V substitution for Ti.

(B) Amorphous Systems

The properties of hydrogen in amorphous alloys and intermetallics, including NMR results, have been reviewed by *Bowman* [4.265]. The absence of long-range lattice structure in amorphous metals is expected to result generally in distributions of site and barrier energies for hydrogen motion. Consequently, as discussed Sect. 4.3.1B, such distributions are reflected in the ^1H relaxation rates R_{1d} and $R_{1\rho}$, in that these are calculated, for example, by folding a distribution of activation energies, $G(E_a)$, with $R_{1d}(E_a, \tau_d)$ [(4.15) and similarly for $R_{1\rho}$]. The distribution function used need not be symmetric, indeed in the example of a- $Zr_3RhH_{3.5}$ shown in Fig. 4.3, an asymmetric distribution, Gaussian on the low-temperature side and Lorentzian at high temperatures, gave the best fit [4.46].

Recent studies of hydrogen in amorphous systems include a- Zr_2FeH and a-$(Ni_{0.5}Zr_{0.5})_{1-y}P_yH_x$. In the case of $a - Zr_2FeH_x$ [4.266] $R_1(^1H)$ measurements, $100 \lesssim T \lesssim 270$ K, indicated values of $C_e^{1/2}$ which is nominally proportional to $N(E_F)$ nearly an order of magnitude greater than in other Zr-based hydrides. However, the difference is not believed to reflect an unusually large $N(E_F)$, but rather a very strong hyperfine interaction with Fe d-states at the Fermi level as previously found in crystalline $TiFeH_x$ [4.221].

For a-$(Ni_{0.5}Zr_{0.5})_{1-y}P_yH_x$, $R_1(^1H)$, $R_2(^1H)$, and $R_{1\rho}(^1H)$ measurements have been reported [4.267, 268] over the temperature range 140–400 K. At the high resonance frequency used (90 MHz), the maximum of R_{1d} was not reached; however, that of $R_{1\varrho}$ was reached. For $y = 0.007$, $x = 0.8$ all three rates were described [4.267] by a Gaussian distribution of E_a values having $\bar{E}_a = 0.32$ eV/atom and $\Delta E \simeq 0.10$ eV/atom, plus an R_{1e} contribution at the lower temperatures. $R_{1\rho}$ measurements only, reported for $y = 0.05$ and 0.07, $x \approx 0.55$, yielded similar $\bar{E}_a$ and ΔE values [4.268].

In the case of a-$Ni_{0.33}Zr_{0.67}H_x$ $(0.25 \le x \le 1.45)$ [4.269], $R_1(^1H)$ and $R_{1\rho}(^1H)$ measurements found the temperature of the motionally induced R_{1d} peak to be a non-monotonic function of x. The motion appears fastest at low and high x values and slowest at $x = 0.85$. PFG measurements of hydrogen diffusivity show that the measured relaxation is due to long-range diffusion and not to localized motion.

An example of Pake doublet structure in an amorphous hydride is reported for the ^1H resonance in a-$Zr_{0.5}Cu_yNi_{0.5-y}H_1$ [4.270] indicating a ^1H-^1H separation of 1.67 Å.

4.6 Summary

The primary focus of this review has been those developments in experimental and theoretical aspects of NMR as applied to metal-hydrogen (M-H) systems that have advanced the utility and capabilities of this technique for such investigations. A brief survey of its applications and accomplishments has also been included.

Experimentally, increasing attention has been given to extending the temperature range over which measurements are made. At low temperatures this has lead to the identification of localized hydrogen motion in hcp Sc and Lu [4.63, 104], for example, and in some intermetallic hydrides [4.225]. More examples of such motion will undoubtedly be discovered. Additionally, the localized motion of H trapped at interstitial impurities at moderately low

temperatures has been identified in Nb-O-H [4.27, 165]. At high temperatures, an anomalous spin-lattice relaxation has been found to occur in most
M-H systems studied.

Similarly, the increased accessibility of spectrometer systems based on
superconducting magnets having field strengths on the order of 8–10 T has
resulted in significantly increased sensitivity by raising the proton resonance
frequency to the 300–400 MHz range and the frequencies of most metal
nuclei to the 20–100 MHz range. The former has been valuable, for example,
in establising the true conduction electron contribution to $R_1(^1H)$ in α-LuH$_x$,
where $R_1(^1H)$ is otherwise dominated by cross-relaxation to the ^{175}Lu spins
[4.63], and the latter has enabled the first measurements of the ^{91}Zr resonance
parameters in Zr dihydride to be made [4.154, 155].

Relaxation rate theory has reached an even greater degree of sophis-
tication than was already noted in [4.1]. In particular, the application of
Monte Carlo (MC) calculational techniques [4.31, 32] has advanced the
determination of the dipolar spectral density functions, $J^{(q)}(\omega)$, for R_{1d}
and $R_{1\rho}$ so that these are now known in numerical form for arbitrary
hydrogen concentrations on the three fundamental interstitial sublattices,
sc, bcc, and fcc [4.32]. In the bcc case the calculations have also included
varying degrees of site-blocking [4.36]. Moreover, a new analytic form
[4.40] can now be used to represent the MC results, making it relatively
convenient for the experimenter to extract reliable values of the mean
dwell time τ_d from R_{1d} data. The next stage in these developments may be
that of determining $J^{(q)}(\omega)$ for the basic sublattices that occur in intermetallic hydride systems.

Several spin relaxation mechanisms which can complicate the conventional resolution of spin-lattice relaxation rate measurements into dipolar
and electronic contributions have been recognized as occurring in M-H
systems. These are: (i) relaxation via the dipolar interaction with the fluctuating electronic moments of paramagnetic impurity ions; (ii) cross-relaxation with quadrupolar metal nuclei of the host lattice; and (iii) an
anomalous relaxation rate at high temperatures, as yet not clearly understood. Because of these additional relaxation channels, and to insure that the
true dipolar and electronic rates are derived, it is important that measurements be made at several resonance frequencies, including high frequencies if
possible, and over a wide temperature range.

Both experimental [4.46, 47] and theoretical [4.45] studies of proton relaxation rates in alloys, amorphous hydrides, and even some intermetallics
have shown that distributions of motional parameters are required to account for the observed temperature and frequency dependencies. Again, the
importance of having measurements over wide ranges of temperature and
frequency, whenever possible, must be emphasized. Unfortunately, in the
case of amorphous hydrides, high temperatures may be difficult to achieve
without encountering recrystallization.

Further improvements have been made in the pulsed field gradient (PFG)
technique with which direct measurements of hydrogen diffusion are made.
These include both more elegant pulse sequences that minimize more effec-

tively the "background" gradients [4.81] and gradient coils and pulse circuitry that yield substantially stronger field gradient pulses, thereby extending the range of measurable diffusivities to as low as 10^{-8} cm^2 s^{-1} [4.86]. The latter improvement has enabled measurements of deuterium diffusion in solid solution in Ta to be made [4.157].

Considerably greater utilization has been made of NMR of host lattice metal nuclei despite the fact that these are generally complicated by electric quadrupole interactions. The relaxation rates of the stationary metal nuclei, although primarily quadrupolar in origin, can monitor hydrogen motion without some of the complexities inherent in the rates of the moving spins themselves. This has especially been the case for ^{45}Sc and ^{51}V [4.104, 225], but applications involving ^{91}Zr and ^{93}Nb will surely follow.

Measurements of Knight shifts and of the conduction electron contribution, R_{1e}, to the spin-lattice relaxation rate have been made for both protons and metal nuclei in an increasingly extensive number of systems. Estimates of the electronic density of states, $N(E_F)$, at both hydrogen and metal sites, have been derived from these measurements and compared with band-structure calculations. As an example, such data for all of the binary dihydride phases (except HfH$_2$) have now been obtained and the results summarized with respect to the band structures [4.91].

NMR measurements of some kind have now been made on every type of M-H system, ranging from the solid solution phases through binary hydrides to alloys, intermetallics, and amorphous metals. In many cases, the unique properties of deuteron NMR have also been utilized, and in a few instances triton NMR has been studied as well. The magnetic resonances of nearly all metal nuclei that occur in M-H systems have been exploited to some degree. These accomplishments, together with advances in experimental methods and underlying theory, testify to the present mature status of nuclear magnetic resonance in the study of metal-hydrogen systems.

References

4.1   R.M. Cotts: In *Hydrogen in Metals I, Basic Properties,* ed. by G. Alefeld, J. Volkl (Springer, Berlin Heidelberg 1978)
4.2   A. Abragam: *The Principles of Nuclear Magnetism* (Oxford University Press, London 1961)
4.3   J.G. Powles, J.H. Strange: Proc. Phys. Soc. **82**, 6 (1963)
4.4   W.-K. Rhim, A. Pines, J.S. Waugh: Phys. Rev. Letters **25**, 218 (1970)
4.5   R.C Bowman, Jr., W-K. Rhim: J. Magn. Res. **49**, 93 (1982)
4.6   G.E. Pake: J. Chem. Phys. **16**, 327 (1948)
4.7   A.C. Switendick: Z. Phys. Chem. NF **117**, 89 (1979)
4.8   D.R. Torgeson, R.G. Barnes, and R.B. Creel: J. Chem. Phys. **56**, 4178 (1972)
4.9   R.G. Barnes, J.W. Bloom: J. Chem. Phys. **57**, 3082 (1972)
4.10   D.R. Torgeson, R.J. Schoenberger, R.G. Barnes: J. Magn. Res. **68**, 85 (1986)
4.11   R.M. Cotts: In *Proce. Int. Symp on the Electronic Structure and Properties of Hydrogen in Metals,* ed. by P. Jena, C. Satterthwaite (Plenum, New York 1983) p. 451

4.12   Y. Fukai, H. Sugimoto: Advances in Physics **34**, 263 (1985)
4.13   E.F.W. Seymour: J. Less-common Metals **88**, 323 (1982)
4.14   M. Jerosch-Herold, L.-T. Lu, D.R. Torgeson, D.T. Peterson, R.G. Barnes, P.M. Richards: Z. Naturforsch. **40a**, 222 (1985)
4.15   J.-W. Han, C.-T. Chang, D.R. Torgeson, E.F.W. Seymour, R.G. Barnes: Phys. Rev. B **36**, 615 (1987)
4.16   N. Bloembergen, E.M. Purcell, R.V. Pound: Phys. Rev. **73**, 679 (1948)
4.17   T.-Y. Hwang, R.J. Schoenberger, D.R. Torgeson, R.G. Barnes: Phys. Rev. B **27**, 27 (1983)
4.18   H.C. Torrey: Phys. Rev. **96**, 690 (1954)
4.19   P.A. Fedders, O.F. Sankey: Phys. Rev. B **18**, 5938 (1978)
4.20   C.A. Sholl: J. Phys. C: Solid State Phys. **14**, 1479 (1981)
4.21   C.A. Sholl: J. Phys. C: Solid State Phys. **14**, 447 (1981)
4.22   P.A. Fedders: Phys. Rev. B **25**, 78 (1982)
4.23   I.R. MacGillivray, C.A. Sholl: J. Phys. C **18**, 1691 (1985)
4.24   I.R. MacGillivray, C.A. Sholl: J. Phys.: Condens. Matter **1**, L829 (1989)
4.25   O.F. Sankey, P.A. Fedders: Phys. Rev. B **20**, 39 (1979)
4.26   O.F. Sankey, P.A. Fedders: Phys. Rev. B **22**, 5135 (1980)
4.27   R. Messer, A. Blessing, S. Dais, D. Hopfel, G. Majer, C. Schmidt, A. Seeger, W. Zag: Z. Phys. Chem. NF, Suppl.-H2, 61 (1986)
4.28   N. Salibi, R.M. Cotts: Phys. Rev. B **27**, 2625 (1983)
4.29   A.F. McDowell, C.F. Mendelsohn, M.S. Conradi, R.C. Bowman, Jr., A.J. Maeland: Phys. Rev. B **51**, 6336 (1995)
4.30   W.A. Barton, C.A. Sholl: J. Phys. C **13**, 2579 (1980)
4.31   L.D. Bustard: Phys. Rev. B **22**, 1 (1980)
4.32   D.A. Faux, D.K. Ross, C.A. Sholl: J. Phys. C **19**, 4115 (1986)
4.33   C.A. Sholl: J. Phys. C **7**, 3378 (1974)
4.34   D.A. Faux, C.K. Hall: J. Phys. C **21**, 3967 (1988)
4.35   O.J. Zogal, R.M. Cotts: Phys. Rev. B **11**, 2443 (1975)
4.36   D.A. Faux, C.K. Hall: Z. Phys. Chem. NF **164**, 859 (1989)
4.37   D.A. Faux, C.K. Hall: J. Phys. Condens. Matter **1**, 9919 (1989)
4.38   D.A. Faux: J. Phys. Condens. Matter **3**, 2201 (1991)
4.39   H.C. Torrey: Phys. Rev. **92**, 962 (1953)
4.40   C.A. Sholl: J. Phys. C **21**, 319 (1988)
4.41   P.A. Fedders: Phys. Rev. B **25**, 78 (1982)
4.42   P.M. Richards: Phys. Rev. B **33**, 3064 (1986)
4.43   L. Schimmele, A. Klamt: J. Phys. Condens. Matter **4**, 3405 (1992)
4.44   J. Shinar: J. Less-Common Metals **104**, 87 (1984)
4.45   P.M. Richards, J. Shinar: J. Phys. F **17**, 1659 (1987)
4.46   J.T. Markert, E.J. Cotts, R.M. Cotts: Phys. Rev. B **37**, 6446 (1988)
4.47   L.R. Lichty, J. Shinar, R.G. Barnes, D.R. Torgeson, D.T. Peterson: Phys. Rev. Lett. **55**, 2895 (1985)
4.48   D. Wolf: *Spin Temperature and Nuclear Spin Relaxation in Matter* (Clarendon, Oxford 1979) Chap. 12
4.49   S.W. Kelly, C.A. Sholl, E.F.W. Seymour: Z. Phys. Chem. NF **164**, 883 (1989)
4.50   S.W. Kelly, C.A. Sholl, E.F.W. Seymour: J. Less-Comm. Met. **172–174**, 572 (1991)
4.51   W.A. Barton: J. Phys. C **15**, 5123 (1982)
4.52   R.G. Barnes, F. Borsa, M. Jerosch-Herold, J.-W. Han, M. Belhoul, J. Shinar, D.R. Torgeson, D.T. Peterson, G.A. Styles, E.F.W. Seymour: J. Less-Comm. Met. **129**, 279 (1987)
4.53   H.T. Weaver: J. Magn. Res. **15**, 84 (1974)
4.54   W.A. Barton, E.F.W. Seymour: J. Phys. C **18**, 625 (1985)
4.55   S.W. Kelly, C.A. Sholl: J. Phys. Condens. Matter **4**, 3317 (1992)
4.56   T.-T. Phua, B.J. Beaudry, D.T. Peterson, D.R. Torgeson, R.G. Barnes, M. Belhoul, G.A. Styles, E.F.W. Seymour: Phys. Rev. B **28**, 6227 (1983)
4.57   M. Belhoul, G.A. Styles, E.F.W. Seymour, T-T. Phua, R.G. Barnes, D.R. Torgeson, D.T. Peterson, R.J. Schoenberger: J. Phys. F **15**, 1045 (1985)

4.58 J.-W. Han, D.R. Torgeson, R.G. Barnes: Phys. Rev. B **42**, 7710 (1990)
4.59 T.-T. Phua, D.R. Torgeson, R.G. Barnes, R.J. Schoenberger, D.T. Peterson, M. Belhoul, G.A. Styles, E.F.W. Seymour: J. Less-Common Metals **104**, 105 (1984)
4.60 M. Belhoul, R.J. Schoenberger, D.R. Torgeson, R.G. Barnes: J. Less-Comm Met **172–174**, 411 (1991)
4.61 L.R. Lichty, J.-W. Han, D.R. Torgeson, R.G. Barnes, E.F.W. Seymour: Phys. Rev. B **42**, 7734 (1990)
4.62 Y.-S. Hwang, D.R. Torgeson, R.G. Barnes: Solid State Commun. **24**, 773 (1977)
4.63 D.R. Torgeson, J.-W. Han, C-T. Chang, L.R. Lichty, R.G. Barnes, E.F.W. Seymour, G.W. West: Z. Phys. Chem NF **164**, 853 (1989)
4.64 D.B. Baker, M.S. Conradi, P.A. Fedders, R.E. Norberg, D.R. Torgeson, R.G. Barnes, R.C. Bowman Jr.: Phys. Rev. B **44**, 11759 (1991)
4.65 J.T. Markert, R.M. Cotts: Phys. Rev. B **36**, 6993 (1987)
4.66 R.G. Barnes: Z. Phys. Chem. NF **164**, 841 (1989)
4.67 J.-W. Han, D.R. Torgeson, R.G. Barnes, D.T. Peterson: Phys. Rev. B **44** 12353 (1991)
4.68 J.-W. Han, L.R. Lichty, D.R. Torgeson, E.F.W. Seymour, R.G. Barnes, J.J. Billeter, R.M. Cotts: Phys. Rev. B **40**, 9025 (1989)
4.69 M. Hampele, G. Majer, R. Messer, A. Seeger: J. Less-Comm. Met. **172–174**, 631 (1991)
4.70 K.J. Barnfather, E.F.W. Seymour, G.A. Styles, A.J. Diannoux, R.G. Barnes, D.R. Torgeson: Z. Phys. Chem. NF **164**, 935 (1989)
4.71 U. Stuhr, D. Steinbinder, H. Wipf, B. Frick: Europhys. Lett. **20**, 117 (1992)
4.72 U. Stuhr, H. Wipf, B. Frick: J. Less-Comm. Met. **172–174**, 678 (1991)
4.73 G. Majer, W. Renz, A. Seeger, R.G. Barnes: Z. Phys. Chem. NF **181**, 731 (1993)
4.74 R.G. Barnes, J.-W. Han, D.R. Torgeson, D.B. Baker, M.S. Conradi, and R.E. Norberg: Phys. Rev. B **51**, 3503 (1995)
4.75 R.G. Barnes, M. Jerosch-Herold, J. Shinar, F. Borsa, D.R. Torgeson, A.J. Lucas, G.A. Styles, E.F.W. Seymour: Phys. Rev. B **35**, 890 (1987)
4.76 D.B. Baker, N.L. Adolphi, M.S. Conradi, P.A. Fedders, R.E. Norberg, R.G. Barnes, and D.R. Torgeson: Phys. Rev B **46**, 184 (1992)
4.77 R.M. Cotts: J. Less-Comm. Met. **172–174**, 467 (1991)
4.78 D.B. Baker, M.S. Conradi, R.E. Norberg, R.G. Barnes, D.R. Torgeson: Phys. Rev. B **49**, 11773 (1994)
4.79 M. Bogdan, K.R. Jeffrey, R.L. Armstrong: J. Chem. Phys. **98**, 6154 (1993)
4.80 B. Michel, O. Kanert, B. Gunther: Acta Metall. Mater **42**, 3409 (1994)
4.81 R.M. Cotts, M.J.R. Hoch, T. Sun, J.T. Markert: J. Magn. Res. **83**, 252 (1989)
4.82 R.F. Karlicek, Jr., I.J. Lowe: J. Less-Comm. Met. **73**, 219 (1980)
4.83 R.F. Karlicek, Jr., I.J. Lowe: J. Magn. Res. **37**, 75 (1980)
4.84 J.E. Tanner: J. Chem. Phys. **52**, 2523 (1970)
4.85 A.F. McDowell, P.E. Mauger, R.M. Cotts: J. Less-Comm. Met. **172–174**, 624 (1991)
4.86 G. Majer, W. Renz, A. Seeger, R.G. Barnes: Z. Phys. Chem. **181**, 731 (1993)
4.87 W.D. Knight: Phys. Rev. **76**, 1259 (1949)
4.88 J. Korringa: Physica (Utrecht) **16**, 601 (1950)
4.89 G.C. Carter, L.H. Bennett, D.J. Kahan: *Metallic Shifts in NMR* (Pergamon, Oxford 1977)
4.90 A. Narath: In *Hyperfine Interactions*, ed. by A.J. Freeman, R.B. Frenkel (Academic, New York 1967) p. 287
4.91 O.J. Zogal, B. Nowak: Z. Phys. Chem. NF **181**, 577 (1993)
4.92 M. Gupta, L. Schlapbach: In *Hydrogen in Intermetallic Compounds I*, ed. by L. Schlapbach Topics Appl. Phys., Vol. 63 (Springer, Berlin, Heidelberg 1988) p. 139
4.93 M.E. Stoll, T.J. Majors: Phys. Rev. B **24**, 2859 (1981)
4.94 R.E. Taylor, T. Taki, B.C. Gerstein: Phys. Rev. B **23**, 5729 (1981)
4.95 W. Baden, A. Weiss: Z. Naturforsch. **38a**, 459 (1983)
4.96 H.E. Schone, E.F.W. Seymour, G.A. Styles, C.A. Sholl: J. Less-Comm. Met. **118**, 201 (1986)

4.97 E.F.W. Seymour, C.A. Sholl: J. Phys. Condens. Matter 1, 8529 (1989)
4.98 G.A. Styles, E.F.W. Seymour, K.J. Barnfather, C.A. Sholl, H.E. Schone: J. Less-Comm. Met. 129, 345 (1987)
4.99 D. Khatamian, C. Stassis, B.J. Beaudry: Phys. Rev. B 23, 624 (1981)
4.100 D.L. Anderson, R.G. Barnes, S.O. Nelson, D.R. Torgeson: Phys. Lett. A 74, 427 (1979)
4.101 H.T. Weaver: Phys. Rev. B 6, 2544 (1972)
4.102 R.G. Barnes, D.R. Torgeson, T.J.M. Bastow, G.W. West, E.F.W. Seymour, M.E. Smith: Z. Phys. Chem. NF 164, 867 (1989)
4.103 L.R. Lichty, R.J. Schoenberger, D.R. Torgeson, R.G. Barnes: J. Less-Comm. Met. 129, 31 (1987)
4.104 L.R. Lichty, J.-W. Han, R. Ibanez-Meier, D.R. Torgeson, R.G. Barnes, E.F.W. Seymour, C.A. Sholl: Phys. Rev. B 39, 2012 (1989)
4.105 R.G. Barnes: J. Less-Comm. Met. 172–174, 509 (1991)
4.106 G.W. West, E.F.W. Seymour, C.-T. Chang, D.R. Torgeson, R.G. Barnes: Phys. Rev. B 44, 9692 (1991)
4.107 O. Blaschko, J. Pleschiutschnig, L. Pintschovius, J.P. Burger, J.N. Daou, P. Vajda: Phys. Rev. B 40, 1909 (1989)
4.108 I. Svare, D.R. Torgeson, F. Borsa: Phys. Rev. B 43, 7448 (1991)
4.109 D.S. Schreiber, R.M. Cotts: Phys. Rev. 131, 1118 (1963)
4.110 D.L. Anderson, T.Y. Hwang, R.G. Barnes, D.T. Peterson, D.R. Torgeson: J. Less-Comm. Met. 73, 243 (1980)
4.111 E. Boroch, K. Conder, R.X. Cai, E. Kaldis: J. Less-Comm. Met. 156, 259 (1989)
4.112 P. Klavins, R.N. Shelton, R.G. Barnes, B.J. Beaudry: Phys. Rev. 29, 5349 (1984)
4.113 D.G. de Groot, R.G. Barnes, B.J. Beaudry, D.R. Torgeson: J. Less-Comm. Met. 73 233 (1980)
4.114 D.G. de Groot, R.G. Barnes, B.J. Beaudry, D.R. Torgeson: Z. Phys. Chem. NF 114, 83 (1979)
4.115 D.R. Torgeson, L.-T. Lu, T.-T. Phua, R.G. Barnes, D.T. Peterson, E.F.W. Seymour: J. Less-Comm. Met. 104, 79 (1984)
4.116 R.G. Barnes, C.-T. Chang, M. Belhoul, D.R. Torgeson, R.J. Schoenberger, B.J. Beaudry, E.F.W. Seymour: J. Less-Comm. Met. 172–174, 411 (1991)
4.117 J. Shinar, B. Dehner, R.G. Barnes, B.J. Beaudry: Phys. Rev. Lett. 29, 563 (1990)
4.118 O.J. Zogal: Phys. Status. Solidi. (a) 53, K203 (1979)
4.119 R. Goring, B. Schnabel, O.J. Zogal: Phys. Status. Solidi. (a) 59, K147 (1980)
4.120 R.G. Barnes, B.J. Beaudry, R.B. Creel, D.R. Torgeson: Solid State Commun. 36, 105 (1980)
4.121 H.T. Weaver: Phys. Rev. B 5, 1663 (1972)
4.122 O.J. Zogal, Ch. Jager, H. Dohler, B. Schnabel: Phys. Status. Solidi. (a) 82, K153 (1984)
4.123 R.S. Kashaev, A.N. Gil'manov, F.F. Gubaidulin, M.E. Kost: Sov. Phys. Solid State 20, 1 (1978)
4.124 H. Winter, D. Shaltiel, E. Dormann: J. Mag. and Mag. Materials 87, 181 (1990)
4.125 D. Zamir, R.G. Barnes, N. Salibi, R.M. Cotts, T-T. Phua, D.R. Torgeson, D.T. Peterson: Phys. Rev. B 29, 61 (1984)
4.126 A. Raizman, D. Zamir, R.M. Cotts: Phys. Rev. B 31, 3384 (1985)
4.127 O.J. Zogal, Ph. l'Heritier: J. Alloys and Compounds 177, 83 (1991)
4.128 O.J. Zogal: Phys. Status. Solidi. (b) 117, 717 (1983)
4.129 O.J. Zogal, S. Idziak, M. Drulis, K. Niedzwiedz: Phys. Status. Solidi. 167, K55 (1991)
4.130 G. Cinader, M. Peretz, D. Zamir, Z. Hadari: Phys. Rev. B 8, 4063 (1973)
4.131 G.G. Libowitz, J.G. Pack, W.P. Binnie: Phys. Rev. B 6, 4540 (1972)
4.132 O.J. Zogal: J. Less-Comm. Met. 130, 187 (1987)
4.133 O.J. Zogal, K. Hoffmann, W. Petrynski, H. Drulis, B. Stalinski: J. Less-Comm. Met. 101, 259 (1984)
4.134 C. Korn, D. Zamir: J. Phys. Chem. Solids 31, 489 (1970)
4.135 A. Schmolz, F. Noack: Ber Bunsenges. Phys. Chem. 78, 339 (1974)

4.136 C. Korn: Phys. Rev. B **17**, 1707 (1978)
4.137 R. Goring, R. Lukas, K. Bohmhammel: J. Phys. C. **14**, 5675 (1981)
4.138 C. Korn, S.D. Goren: Phys. Rev. B **33**, 64 (1986)
4.139 C. Korn: Phys Rev. B **28**, 95 (1983)
4.140 R.C. Bowman, Jr., E.L. Venturini, B.D. Craft, A. Attalla, D.B. Sullenger: Phys. Rev. B **27**, 1474 (1983)
4.141 C. Korn, S. Goren: J. Less-Comm. Met. **104**, 113 (1984)
4.142 R.C. Bowman, Jr., B.D. Craft: J. Phys. C. **17**, 1477 (1984)
4.143 R.C. Bowman, Jr., B.D. Craft, J.S. Cantrell, E.L. Venturini: Phys. Rev. B **31**, 5604 (1985)
4.144 C. Korn, S.D. Goren: Phys. Rev. B **33**, 68 (1986)
4.145 J-W. Han, R.G. Barnes: J. Korean Phys. Soc. **28**, 199 (1995)
4.146 L.D. Bustard, R.M. Cotts, E.F.W. Seymour: Phys. Rev. B **22**, 12 (1980)
4.147 L.D. Bustard, R.M. Cotts, E.F.W. Seymour: Z. Phys. Chem. NF **115**, 247 (1979)
4.148 D.R. Torgeson, D.T. Peterson, R.G. Barnes, G. Majer, W. Renz, A. Seeger: American Physical Society March Meeting, San Jose, CA (1995)
4.149 G. Majer, W. Renz, R.G. Barnes: J. Phys. Condens. Matter **6**, 2935 (1994)
4.150 J.G. Couch, O.K. Harling, L.C. Clune: Phys. Rev. B **4**, 2675 (1971)
4.151 Y. Fukai, N. Okuma: Jpn. J. Appl. Phys. **32**, L1256 (1993)
4.152 W.A. Oates, H. Wenzl: Scripta Met. Mater. **30**, 851 (1994)
4.153 K. Niedzwiedz, B. Nowak, O.J. Zogal: J. Alloys and Compounds **194**, 47 (1993)
4.154 O.J. Zogal, B. Nowak, K. Niedzwiedz: Solid State Commun. **80**, 601 (1991)
4.155 O.J. Zogal, B. Nowak, K. Niedzwiedz: Solid State Commun. **82**, 351 (1992)
4.156 B. Nowak, O.J. Zogal, K. Niedzwiedz: J. Alloys and Compounds **4**, 53 (1992)
4.157 M. Hampele, G. Majer, R. Messer, A. Seeger: J. Less-Comm. Met. **172–174**, 631 (1991)
4.158 P.E. Mauger, W.D. Williams, R.M. Cotts: J. Phys. Chem. Solids **42**, 821 (1981)
4.159 T. Eguchi, S. Morozumi: J. Jpn Inst. Met. **41**, 795 (1977)
4.160 J.E. Kleiner, E.H. Sevilla, R.M. Cotts: Phys. Rev. B **33**, 6662 (1986)
4.161 E.H. Sevilla, R.M. Cotts: J. Less-Comm. Met. **129**, 223 (1987)
4.162 H.C. Bauer, J. Volkl, J. Tretkowski, G. Alefeld: Z. Physik. B **29**, 17 (1978)
4.163 U. Potzel, J. Volkl, H. Wipf, A. Magerl: Phys. Status. Solidi. (b) **123**, 85 (1984)
4.164 Yuh Fukai, K. Kubo, S. Kazama: Z. Phys. Chem. NF **115**, 181 (1979)
4.165 R. Messer, D. Hopfel, C. Schmidt, A. Seeger, W. Zag, R. Lässer: Z. Phys. Chem. NF **145**, 179 (1985)
4.166 W. Baden, A. Weiss: Ber. Bunsenges. Phys. Chem. **87**, 479 (1983)
4.167 S. Kazama, Y. Fukai: J. Phys. Soc. Jpn. **42**, 119 (1977)
4.168 S. Kazama, Y. Fukai: J. Less-Comm. Met. **53**, 25 (1977)
4.169 W. Baden, P.C. Schmidt, A. Weiss: J. Less-Comm. Met. **104**, 99 (1984)
4.170 D. Zamir: Phys. Rev. **140**, A271 (1965)
4.171 N. Salibi, B. Ting, D. Cornell, R.E. Norberg: Phys. Rev. B **38**, 4416 (1988)
4.172 Y. Fukai, S. Kazama: Acta Metallurgica **25**, 59 (1977)
4.173 S. Hayashi, H. Hayamizu, O. Yamamoto: Solid State Commun. **41**, 743 (1982)
4.174 R.C. Bowman, Jr., A. Attalla, W.E. Tadlock, D.B. Sullenger, R.L. Yauger: Scripta Met. **16**, 933 (1982)
4.175 R.C. Bowman, Jr., A. Attalla, B.D. Craft: Scripta Met. **17**, 937 (1983)
4.176 K. Nakamura: Bull. Chem. Soc. Jpn. **46**, 2588 (1973)
4.177 B. Pedersen, D. Slotfeldt-Ellingsen: J. Less-Comm. Met. **23**, 223 (1971)
4.178 K. Nakamura: Bull. Chem. Soc. Jpn. **46**, 2028 (1973)
4.179 H. Lutgemeier, G.G. Bohn, R.R. Arons: J. Magn. Res. **8**, 80 (1972)
4.180 R.G. Barnes, K.P. Roenker, H.R. Brooker: Ber. Bunsenges. Phys. Chem. **80**, 876 (1976)
4.181 D. Slotfeldt-Ellingsen, B. Pedersen: Phys. Status. Solidi. (a) **25**, 115 (1974)
4.182 K. Nakamura: Bull. Chem. Soc. Jpn. **45**, 3356 (1972)
4.183 K.P. Roenker, R.G. Barnes, H.R. Brooker: Ber. Bunsenges. Phys. Chem. **80**, 470 (1976)
4.184 B. Nowak, O.J. Zogal, H. Drulis: J. Phys. C. **15**, 5829 (1982)

4.185 B. Nowak, M. Minier: J. Phys. C. **15**, 4385 (1982)
4.186 B. Nowak, M. Minier: J. Phys. F. **14**, 1291 (1984)
4.187 R.R. Arons, H.G. Bohn, H. Lutgemeier: J. Phys. Chem. Solids **35**, 207 (1974)
4.188 Y-S. Hwang, D.R. Torgeson, R.G. Barnes: Scripta Met. **12**, 507 (1978)
4.189 B. Nowak, O.J. Zogal, K. Niedzwiedz, M. Tkacz, Z. Zolnierek: Physica B **193**, 102 (1994)
4.190 G. Albrecht, B. Schnabel: Phys. Stat. Sol. **15**, 141 (1966)
4.191 R.E. Norberg: Phys. Rev. **86**, 745 (1952)
4.192 D.A. Cornell, E.F.W. Seymour: J. Less-Comm. Met. **39**, 43 (1975)
4.193 E.F.W. Seymour, R.M. Cotts, W.D. Williams: Phys. Rev. Lett. **35**, 165 (1975)
4.194 P.P. Davis, E.F.W. Seymour, D. Zamir, W.D. Williams, R.M. Cotts: J. Less Comm. Met. **49**, 159 (1976)
4.195 G.A. Styles, E.F.W. Seymour, K.J. Barnfather, C.A. Sholl, H.E. Schone: J. Less-Comm. Met. **129**, 345 (1987)
4.196 H.E. Avram, R.L. Armstrong: Phys. Rev. B **34**, 6121 (1986)
4.197 T. Ito, T. Kadowaki: Phys. Lett. **54A**, 61 (1975)
4.198 S.R. Kreitzman, R.L. Armstrong: Phys. Rev. B **25**, 2046 (1982)
4.199 H. Avram, R.L. Armstrong: J. Phys. C. **17**, L89 (1984)
4.200 D.W. Nicoll, M.M. Pintar: Phys. Rev. B **23**, 1064 (1981)
4.201 A.J. Holley, W.A. Barton, E.F.W. Seymour: In *Electronic Structure and Properties of Hydrogen in Metals*, ed. by P. Jena, C.B. Satterthwaite (Plenum, New York 1983) p. 595
4.202 A. Boukraa, G.A. Styles, E.F.W. Seymour: J. Phys.: Condens. Matter **3**, 2391 (1991)
4.203 R.C. Bowman, Jr., G. Bambakidies, G.C. Abell, A. Attalla, B.D. Craft: Phys. Rev. B **37**, 9447 (1988)
4.204 P. Brill, J. Voitlander: Ber. Bunsenges. Phys. Chem. **77**, 1097 (1973)
4.205 J.D. Will: Ph.D. Nuclear Spin Relaxation of Hydrogen in Thorium Hydrides, Iowa State University (1971), (unpublished)
4.206 J.D. Will, R.G. Barnes: J. Less-Comm. Met. **13**, 131 (1967)
4.207 K.F. Lau, R.W. Vaughan, C.B. Satterthwaite: Phys. Rev. B **15**, 2449 (1977)
4.208 M. Peretz, D. Zamir, Z. Hadari: Phys. Rev. B **18**, 2059 (1978)
4.209 J. Grunzweig-Genossar, M. Kuznietz, B. Meerovici: Phys. Rev. B **1**, 1958 (1970)
4.210 G. Cinader, M. Peretz, D. Zamir, Z. Hadari: Phys. Rev. B **8**, 4063 (1973)
4.211 M. Peretz, D. Zamir, G. Cinader, Z. Hadari: J. Phys. Chem. Solids **37**, 105 (1976)
4.212 G. Cinader, D. Zamir, Z. Hadari: Phys. Rev. B **14**, 912 (1976)
4.213 F.E. Spada, H. Oesterreicher, R.C. Bowman, Jr., M.P. Guse: Phys. Rev. B **30**, 4909 (1984)
4.214 R.G. Barnes, W.C. Harper, S.O. Nelson, D.K. Thome, D.R. Torgeson: J. Less-Comm. Met. **49**, 483 (1976)
4.215 T.K. Halstead, N.A. Abood, K.H.J. Buschow: Solid State Commun. **19**, 425 (1976)
4.216 H. Chang, I.L. Lowe, R.J. Karlicek, Jr.: In *Nuclear and Electron Resonance Spectroscopies Applied to Materials Science*, ed. by E.N. Kaufmann, G.K. Shenoy (Elsevier, Amsterdam 1981) p. 331
4.217 R.C. Bowman, Jr., B.D. Craft, A. Attalla, M.H. Mendelsohn, D.M. Gruen: J. Less-Comm. Met. **73**, 227 (1980)
4.218 F.E. Spada, R.C. Bowman, Jr., J.S. Cantrell: J. Less-Comm. Metals **129**, 261 (1987)
4.219 R.C. Bowman, Jr., G.C. Carter, Y. Chabre, A. Attalla: In *Hydrides for Energy Storage*, ed. by A.F. Andresen, A.J. Maeland (Pergamon, Oxford 1978) p. 97
4.220 R.C. Bowman, Jr., W.E. Tadlock: Solid State Commun. **32**, 313 (1979)
4.221 J.S. Cantrell, R.C. Bowman, Jr.: J. Less-Comm. Met. **130**, 69 (1987)
4.222 J. Shinar: In *Hydrogen Storage Materials*, ed. by R.G. Barnes (Trans Tech, Switzerland 1988) p. 143
4.223 A.V. Skripov, M.Yu. Belyaev, S.V. Rychkova, A.P. Stepanov: J. Phys.: Condens. Matter **3**, 6277 (1991)
4.224 A.V. Skripov, M.Yu. Belyaev, A.P. Stepanov: Solid State Commun. **71**, 321 (1989)
4.225 A.V. Skripov, M.Yu. Belyaev, S.V. Rychkova, A.P. Stepanov: J. Phys.: Condens Matter **1**, 2121 (1989)

4.226 A.V. Skripov, M.Yu. Belyaev, A.P. Stepanov, L.N. Padurets, E.I. Sokolova: J. Alloys and Compounds **190**, 171 (1993)

4.227 A.V. Skripov, M.Yu. Belyaev, A.P. Stepanov: Solid State Commun. **78**, 909 (1991)

4.228 K. Morimoto, M. Saga, H. Fujii, T. Okamoto, T. Hihara: J. Phys. Soc. Jpn. **57**, 647 (1988)

4.229 A.V. Skripov, S.V. Rychkova, M.Yu. Belyaev, A.P. Stepanov: Solid State Commun. **71**, 1119 (1989)

4.230 R.C. Bowman, Jr., J.F. Lynch, J.R. Johnson: Materials Lett. **1**, 122 (1982)

4.231 W. Renz, G. Majer, A.V. Skripov: J. Alloys and Compounds **224**, 127 (1995)

4.232 G. Majer, W. Renz, A. Seeger, R.G. Barnes, J. Shinar, A.V. Skripov: J. Alloys and Compounds **231**, 220 (1995)

4.233 W. Renz, G. Majer, A.V. Skripov, A. Seeger: J. Phys.: Condens. Matter **6**, 6367 (1994)

4.234 W. Renz, G. Majer, A. Seeger (unpublished results)

4.235 T.-Y. Hwang, R.J. Schoenberger, D.R. Torgeson, R.G. Barnes: Phys. Rev. **27**, 27 (1983)

4.236 F. Aubertin, S.J. Campbell, J.M. Pope, U. Gonser: J. Less-Comm. Met. **129**, 297 (1987)

4.237 R.C. Bowman, Jr., J.S. Cantrell, A.J. Maeland, A. Attala, G.C. Abell: J. Alloys and Compounds **185**, 7 (1992)

4.238 R.C. Bowman, Jr., A. Attalla, G.C. Abell, A.J. Maeland, J.S. Cantrell: J. Less-Comm. Met. **172–174**, 643 (1991)

4.239 D.B. Baker, E.-K. Jeong, M.S. Conradi, R.E. Norberg, R.C. Bowman, Jr.: J. Less-Comm. Met. **172–174**, 373 (1991)

4.240 D. Guthardt, D. Beisenherz, H. Wipf: J. Phys.: Condens. Matter **4**, 6919 (1992)

4.241 A.V. Skripov, M.Yu. Belyaev, S.A. Petrova: J. Phys.: Condens. Matter **4**, L537 (1992)

4.242 O.J. Zogal, D.J. Lam, A. Zygmunt, H. Drulis, W. Petrynski, S. Stalinkski: Phys. Rev. B **29**, 4837 (1984)

4.243 B. Bandyopadhyay, K. Ghoshray, A. Ghoshray, N. Chatterjee: J. Phys.: Condens. Matter **2**, 1253 (1990)

4.244 B. Bandyopadhyay, K. Ghoshray, A. Ghoshray, N. Chatterjee: Phys. Rev. B **46**, 2912 (1992)

4.245 K. Ghoshray, B. Bandyopadhyay, Mita Sen, A. Ghoshray, N. Chatterjee: Phys. Rev. B **47**, 8277 (1993)

4.246 Y.-S. Hwang, D.R. Torgeson, A.S. Khan, R.G. Barnes: Phys. Rev. B **15**, 4564 (1977)

4.247 D.A. Rohy, R.M. Cotts: Phys. Rev. B **1**, 2070 (1970); B **1**, 2484 (1970)

4.248 S. Hayashi, K. Hayamizu, O.Yamamoto: J. Chem. Phys. **78**, 5096 (1983)

4.249 S. Hayashi, K. Hayamizu, O. Yamamoto: J. Solid State Chem. **46** 306 (1983)

4.250 S. Hayashi, K. Hayamizu: J. Less-Comm. **161**, 61 (1990)

4.251 E.H. Sevilla, R.M. Cotts: Phys. Rev. B **37**, 6813 (1988)

4.252 W. Baden, P.C. Schmidt, A. Weiss: J. Less. Comm. Met. **88**, 171 (1982)

4.253 K.-H, Richter, A. Weiss: J. Less-Comm. Met. **142**, 301 (1988)

4.254 T. Matsumoto: J. Phys. Soc. Jpn. **42**, 1583 (1977)

4.255 L.R. Lichty, A Nuclear Magnetic Resonance Study of Proton Spin Relaxation Processes in Niobium-Vanadium Alloys; Ph.D. Dissertation, Iowa State University (1988) (unpublished)

4.256 P.E. Mauger: A Study of Hydrogen Diffusion in Tantalum Hydride and Hydrides of Niobium-Vanadium Alloys Using NMR Techniques; Ph.D. Thesis, Cornell University (1982), University Microfilms Int. No. DA 8210779

4.257 D.G. Westlake, J.F. Miller: J. Less-Comm. Met. **65**, 139 (1979)

4.258 D.T. Peterson, H.M. Herro: Metall. Trans. A. **11A**, 645 (1986)

4.259 A.F. McDowell, P.E. Mauger, R.M. Cotts: J. Less-Comm. Met. **172–174**, 624 (1991)

4.260 B. Nowak, O.J. Zogal, M. Miner; J. Phys. C. **12**, 4591 (1979)

4.261 B. Stalinski, B. Nowak: Bull. Acad. Polon. Sci. Ser. Sci. Chim. **25**, 65 (1977)

4.262 B. Nowak, N. Pislewski, W. Leszczynski: Phys. Status. Solidi (a) **37**, 669 (1976)

4.263 B. Nowak, Y. Chabre, R. Andreani: J. Less-Comm. Met. **130**, 193 (1987)

4.264 R.C. Bowman, Jr., W.-K. Rhim: Phys. Rev. B **24** 2232 (1981)
4.265 R.C. Bowman, Jr.: *"Preparation and properties of amorphous hydrides,"* in *Hydrogen Storage Materials*, ed. by R.G. Barnes (Trans Tech Switzerland 1988) p. 197
4.266 R.C. Bowman, Jr., A.J. Maeland, K.M. Unruh, E.L. Venturini, J.J. Rush, J.S. Cantrell: Z. Phys. Chem. NF **163**, 373 (1989)
4.267 K. Tompa, H.E. Schone, A. Werner, I. Pocsik, P. Banki, I. Bakonyi, G. Konczos, A. Lovas: Z. Phys. Chem. NF **163**, 437 (1989)
4.268 K. Tompa, A. Werner, I. Bakonyi, P. Banki, I. Pocsik: J. Less-comm Met. **159**, 199 (1990)
4.269 A.F. McDowell, R.M. Cotts: Solid State Commun. **94**, 529 (1995)
4.270 G. Lasanda, P. Banki, K. Tompa, Solid State Commun. **87**, 665 (1993)

5. Neutron Scattering Studies of Metal Hydrogen Systems

D.K. Ross

With 16 Figures

The use of thermal neutrons to probe condensed matter dates back to the late 1950s when research reactors giving sufficiently high thermal neutron fluxes (10^{17} neutrons/m^2/s) first became available [5.1]. Neutron diffraction was seen to be a potentially useful technique because the wavelengths of thermal neutrons are comparable with atomic separations in solids. Moreover, neutrons were quickly recognised to have advantages over X-rays where differences between the respective scattering powers meant that different information could be obtained. A particularly important example of this was the fact that the neutron, having a magnetic moment itself, would interact with the magnetic structure of the solid. Thermal neutron energies were also appropriate for measuring inelastic scattering – where the neutron exchanges quanta of vibrational energy with the solid – and here some of the first measurements were of the hydrogen vibration frequencies in ZrH_2 [5.2] because zirconium hydride was being developed as a moderator for nuclear reactors and information on the inelastic neutron scattering was needed to predict its efficiency for this purpose. The particular advantage of neutrons as a probe, however, is that they have, simultaneously, appropriate energies and appropriate wavelengths and this means that they can be used to measure spatial correlations in atomic motions. The prime example here is the measurement of phonon dispersion curves, and the development of the triple-axis spectrometer by *Brockhouse* [5.3] was crucial to this application. This property also makes neutrons unique for the study of diffusive processes, as was pointed out by *van Hove* [5.4] and *Vineyard* [5.5]. Here the distinction between incoherent and coherent scattering must be made. The former arises when the scattering power of an element fluctuates about a mean of zero because of spin-dependent or isotope-dependent effects. *van Hove* showed that in this case we can only investigate the motion of a tracer atom – leading, for instance, to measurements of the tracer diffusion coefficient. The latter case arises when the scattering power of an element has a fixed value and the neutron wave functions scattered from different nuclei interfere with each other. Here we can study the dynamics of nuclear density fluctuations and this leads to, for instance, measurements of the chemical diffusion coefficient. In the majority of cases, however, the fluctuations do not average to zero and both types of scattering are present to some degree. Finally, the neutron, being uncharged, interacts weakly in solids when compared with

Topics in Applied Physics, Vol. 73
Wipf (Ed.)
© Springer-Verlag Berlin Heidelberg 1997

many probes and so neutrons are relatively penetrating and can be used with samples of a convenient size.

5.1 The Neutron Scattering Method

5.1.1 Advantages of the Neutron Scattering Technique for Investigating Metal-Hydrogen Systems

Neutron scattering has been very widely used in the investigation of metal-hydrogen systems. The main reason for this is that the proton has a very large scattering cross-section of 80 barns ($1b = 10^{-28}$ m^2) compared with the cross-sections of all other elements which are typically around 4 barns. Moreover, in inelastic measurements, where the hydrogen is vibrating in anti-phase with its surroundings, the inelastic part of the cross-section is weighted inversely with the mass. The cross-section for hydrogen is largely incoherent ($\sigma^{inc} = 79.7$ b, $\sigma^{coh} = 1.8$ b) and so scattering measurements will yield information about the population-averaged dynamics (vibrations and diffusive motions) of individual hydrogen atoms down to quite small concentrations. However, by changing to deuterium, ($\sigma^{coh} = 5.6$ b; $\sigma^{inc} = 2.0$ b), we can use the coherent scattering to investigate spatial correlation effects, such as phonon dispersion curves and chemical diffusion, albeit with rather lower sensitivity. Also, the ease with which one can switch from coherent to incoherent scattering is unique to the case of hydrogen. Such information is particularly important because hydrogen is difficult to detect using more conventional techniques. Also, electromagnetic radiation is strongly attenuated in conducting solids which means that there is a dearth of information about hydrogen's vibrations in such materials using e.m. techniques.

In the present chapter, we discuss both quasi-elastic and inelastic neutron scattering measurements on hydrogen in metallic elements. We summarise the important theoretical ideas and report on experimental work published since the last major review of the field, which covered work up to 1983 and [5.6]. For older work, we refer the reader to reviews by Sköld [5.7], Springer [5.8], and Springer and Richter [5.9].

For a recent review of the application of the technique to intermetallic hydrides, see the chapter by Richter in "Intermetallic Hydrides and Compounds" [5.10] and for a full description of the basic bulk properties of metal hydrogen systems the reader is referred to the recent book by Fukai [5.11]. A general discussion of quasi-elastic neutron scattering can be found in the monograph by Bee [5.12].

5.1.2 Basic Theory of Neutron Scattering

The neutron is an uncharged nucleon with a mass essentially the same as the proton, a spin of a half and a magnetic moment of -1.913 nuclear Bohr magnetons. If we omit consideration of magnetic scattering, which is largely outside the scope of this chapter, we can confine ourselves to consideration of the scattering of neutrons by nuclei via the nuclear interaction. Full discussions of this theory for nuclei bound in solids and liquids can be found in a number of texts [5,13–16] and so the treatment below is considerably simplified, with emphasis being placed on physical principles rather than mathematical rigour.

The probability of a neutron interacting with a nucleus is defined in terms of the cross-section, σ, the rate of reaction per nucleon in unit incident neutron flux. Of the possible types of reaction, we are only interested in scattering events, the cross-section for which is σ_s. In general, a scattering event can result in a change in both the direction and the energy of a neutron, so we define the double differential scattering cross-section $d^2\sigma/dE'd\Omega$, as the rate of scattering from unit incident flux, energy E_0 into unit solid angle about the scattering angle, θ, and unit energy about the final energy E'. From this definition, it follows that integration of the double differential cross section over final energy and then over scattering angle must yield the scattering cross-section

$$\int_0^\infty \int_0^\infty \int_{4\pi} [d^2\sigma(E_0 \rightarrow E', \theta)/dE'd\Omega]d\Omega dE'$$

$$= \int_{4\pi} [d\sigma(E_0, \theta)/d\Omega]d\Omega = \sigma_s(E_0) \ . \tag{5.1}$$

For a fixed nucleus at the origin, this quantity is easily translated into quantum mechanical terms. If the incident neutron flux is represented by a plane wave, $\exp(i\mathbf{k}_0 \cdot \mathbf{r})$, where $\mathbf{k}_0$ is the incident wave vector, then the wave scattered from a nucleus at the origin is given by $(-a/r)\exp(i\mathbf{k}' \cdot \mathbf{r})$ where $\mathbf{k}'$ is the scattered wave vector. Here, a, known as the neutron scattering length, is a measure of the strength of the scattering at a particular nucleus for a given total spin state J $(= I \pm 1/2$, where I is the nuclear spin). The minus sign is conventionally inserted so as to make the value of a positive in the expression for the scattered wave for the vast majority of nuclei. This situation corresponds to potential or hard sphere scattering. In a small fraction of cases, when the system *neutron + nucleus* (the compound nucleus) lies close to a resonant state, a can have a wide variety of values, either positive or negative, depending on how the incident wave matches to the wave function in the compound nucleus. (For recent tabulations, see [5.16, 17]). It can be shown that the scattering from an isolated fixed nucleus is isotropic and that $d\sigma/d\Omega = a^2 = \sigma/4\pi$.

If we now consider the scattering from a set of N nuclei, fixed at positions $\mathbf{r}_i$, the amplitudes of the scattered waves can be added together in a given scattering direction, and hence the angular cross-section is obtained by multiplying by the complex conjugate, yielding

$$d\sigma/d\Omega = (1/N)\left|\sum_i a_i \exp(i\mathbf{Q} \cdot \mathbf{r}_i)\right|^2$$

$$= (1/N)\sum_{ij} a_i a_j \exp[i\mathbf{Q} \cdot (\mathbf{r}_i - \mathbf{r}_j)] \ , \tag{5.2}$$

where $\mathbf{Q} = \mathbf{k}' - \mathbf{k}_0$ is the wave vector transfer in the scattering (i.e., $\mathbf{Q}$ is the momentum given to the neutron). The matrix on the right-hand side of (5.2) can be simplified by averaging over the diagonal terms, yielding $\langle a^2 \rangle$ and over the nondiagonal terms, which, remembering that the values of a are completely uncorrelated with position, can be written as $\langle a \rangle^2 \exp\left[-i\mathbf{Q} \cdot (\mathbf{r}_i - \mathbf{r}_j)\right]$. Now, by reinserting the diagonal terms as $\langle a \rangle^2$, we have

$$d\sigma/d\Omega = [\langle a^2 \rangle - \langle a \rangle^2] + (1/N)\langle a \rangle^2 \sum_{ij} \exp\left[i\mathbf{Q} \cdot (\mathbf{r}_i - \mathbf{r}_j)\right] . \qquad (5.3)$$

Here, the first term, $[\langle a^2 \rangle - \langle a \rangle^2]$, is the incoherent scattering. This is clearly isotropic in this approximation, containing only information about a particular tagged atom. The remaining term is the coherent scattering, which is determined by the relative positions of the atoms taken in pairs.

The expression (5.3) is written for a set of *fixed* nuclei but the separation into incoherent and coherent scattering is quite general and is fundamental to all neutron scattering. Moreover, it is particularly important for hydrogen. If we take the proton, of spin 1/2, it turns out that, because the neutron-proton anti-parallel state ($J = 0$), which is found with a probability 1/4, has a large negative scattering length, while the parallel state, $J = 1$, which is found with a probability of 3/4, has a smaller positive scattering length, so that $\langle a \rangle \approx 0$, the scattering is predominantly incoherent ($\sigma^{inc} = 79.7$ b, $\sigma^{coh} = 1.8$ b). On the other hand, if we take deuterium, we find the scattering is evenly divided ($\sigma^{coh} = 5.6$ b, $\sigma^{inc} = 2.0$ b). Thus, for hydrogen, by exploiting isotope substitution, we can study both self motions and the relative motions of pairs of atoms. Indeed as $\langle a \rangle$ is negative for hydrogen and positive for deuterium, the isotopes can be mixed so as to make $\langle a \rangle_{mix} = 0$. Scattering from tritium is also of interest in some cases. It has a coherent scattering length of 3.1 b and no incoherent scattering.

To conclude these general remarks, we will formally extend the quantum-mechanical picture which has been introduced above to include energy transfers, i.e., by assuming that the nuclei in our set are not rigidly bound but are held together by their mutual interactions. The initial state of the nuclei must therefore be defined by a particular set of quantum numbers which define the Hamiltonian or total energy of the system. In an inelastic collision, one or more of these quantum numbers will be changed and the equivalent amount of energy will be transferred to or from the neutron. This situation is described by Fermi's golden rule [5.18] by means of which the cross-section can be written

$$d^2\sigma/dE'd\Omega = \sum_i \sum_j \sum_n p_n \sum_{n'} (k'/k_0) \int \psi_n(\mathbf{r}) a_i a_j \, \delta(\mathbf{r} - (\mathbf{r}_i \mathbf{r}_j))$$
$$\psi_{n'}^*(\mathbf{r}) \exp[i(\mathbf{Q} \cdot \mathbf{r})] d\mathbf{r} \times \delta(E_0 - E' - E_{n'} + E_n) . \qquad (5.4)$$

Here, the system is assumed to be in thermal equilibrium so that the cross-section has to be averaged over all possible initial states of the system, n, weighted by their thermal occupation number, p_n and summed over all possible final states n'. The incident energy, E_0 is thus $(E_{n'} - E_n + E')$. k_0 and k' are the wave numbers corresponding to E_0 and E'. Atoms i and j

are at positions $\mathbf{r}_i$ and $\mathbf{r}_j$ and their probability density is described by the wave functions ψ_n and $\psi_{n'}$ where the atoms are respectively in the energy levels n and n'. Further discussion of this equation can be found elsewhere, e.g., in *Squires'* book [5.15].

Equation (5.4) is an appropriate starting point for the analysis of scattering from systems with clearly defined quantum levels. For studies of diffusion, however, a more convenient starting point is the representation of the scattering in terms of the van Hove correlation functions. These will be introduced in Sect. 5.2.1.

5.1.3 Sources of Thermal Neutrons

In view of the successes of a variety of neutron scattering techniques in providing unique information on condensed matter in all its forms, a number of nuclear reactors have been built world-wide specifically for this type of research, in particular, the High Flux Beam Reactors (HFBRs). Here, the peak thermal flux is produced in the reflector region that surrounds an undermoderated reactor core. This reflector consists of a large volume of heavy water into which the neutron beam tubes are inserted. The ultimate flux $(\approx 10^{18} n\ m^{-2}s^{-1})$ available from such a reactor is essentially set by the maximum possible rate of heat transfer between the fuel and the coolant (D_2O). To get round this limitation, therefore, a number of pulsed neutron sources have been developed (IPNS at Argonne and LANSCE at Los Alamos in the USA, ISIS at the Rutherford-Appleton Laboratory in the UK and KENS in Japan). The advantage here is that if one is using time-of-flight to measure the neutron energy, one has to pulse the beam from the source and therefore it is the neutron intensity in the pulse that counts, whereas the heat transfer limit is clearly related to the average power density. It is now clear that most types of neutron scattering measurements can be made equally well using time-of-flight techniques – with perhaps the exception of triple axis spectrometry [5.19]. It therefore seems probable that the next significant increase in neutron flux will involve high current proton accelerators. Indeed a US design study for a reactor called the Advanced Neutron Source (ANS) [5.20] has recently been abandoned in favour of pulsed source developments.

5.2 Theory of Quasi-elastic Neutron Scattering

5.2.1 Van Hove Correlation Functions
and the Intermediate Scattering Functions

As mentioned in Sect. 5.1.2, where diffusion is taking place, the neutron does not cause transitions between quantum states. Here, near-elastic or 'quasi-elastic' scattering is observed from atoms that are diffusing sufficiently quickly for a Doppler effect to be seen with the experimentally available resolution. In the present context, we are mainly interested in the diffusion of atoms on a lattice, i.e., in the model usually described as a lattice gas. Quasi-elastic scattering is best described in terms of the van Hove correlation functions [5.4]. From the incoherent inelastic scattering cross section, one first defines the incoherent scattering function, $S^{inc}(\mathbf{Q}, \omega)$, (where $\hbar\omega$ is the energy transfer) and then the self-correlation function, $G_s(\mathbf{r}, t)$ which is the Fourier transform of the incoherent scattering function.

$$d^2\sigma^{inc}/dE'd\Omega = (\sigma^{inc}/4\pi)(k'/k_0)\, S^{inc}(\mathbf{Q}, \omega)$$

$$= (\sigma^{inc}/4\pi)(k'/k_0)(1/2\pi)\iint G_s(\mathbf{r}, t)\, \exp[i(\mathbf{Q}\cdot\mathbf{r} + \omega t)]d\mathbf{r}dt \ .$$

$$(5.5)$$

In an exactly analogous way, one can define the coherent scattering function, $S^{coh}(\mathbf{Q}, \omega)$, and the corresponding total correlation function, $G(\mathbf{r},t)$. In general, the scattering functions are not symmetric in ω and hence the correlation functions have imaginary components. However, if we confine ourselves to situations where there are only small energy transfers, so that the Boltzmann factor can be approximated to unity, $S^{inc}(\mathbf{Q}, \omega)$ and $S^{coh}(\mathbf{Q}, \omega)$ are symmetric and the correlation functions are therefore real. In this case their meaning is clear. $G_s(\mathbf{r}, t)$ is the probability that a nucleus will be at the point, $\mathbf{r}$, at time t, given that the same atom was at the origin at $t = 0$. Similarly, $G(\mathbf{r}, t)$ is the probability of finding any atom at $\mathbf{r}$ at t, given that an atom was at the origin at time $t = 0$. The above restriction to the case of small ω means that this formulation is most useful as a way of describing the diffusional motions of atoms in liquids and solids. Qualitatively, a neutron scattered from a diffusing atom suffers a Doppler broadening and we therefore observe a quasi-elastic broadening of the elastic peak.

It should be mentioned here that, for some purposes, it is convenient to work with the "intermediate scattering functions", $I_s(\mathbf{Q}, t)$ and $I(\mathbf{Q}, t)$. These functions are obtained by Fourier transforming the correlation functions in space only.

5.2.2 Long-Range Tracer and Chemical Diffusion

The simplest model for $G_s(\mathbf{r}, t)$ is obtained from the solution of the diffusion equation for a delta function source at $t = 0$ for times that are long compared with the individual phases of motion of the atom. This solution is a Gaussian in $\mathbf{r}$ with a mean square deviation $\langle r^2 \rangle = 6D_t t$ where D_t is the tracer diffusion coefficient. It is valid for distances that are long compared with the individual steps in the diffusion process so that, after Fourier transforming, we expect it to be valid at small Q. *Vineyard* [5.5] showed that the corresponding form of the scattering function is

$$S^{inc}(\mathbf{Q}, \omega) = (1/\pi)(D_t Q^2)/[(D_t Q^2)^2 + \omega^2] \ .$$

$$(5.6)$$

This solution is a Lorentzian with a half-width at half maximum of $D_t Q^2$. Thus, by measuring the width of the quasi-elastic peak at low Q, we can extract a value of D_t.

The case of coherent scattering is less obvious. There will only be coherent intensity at low Q if the scattering atoms are in a compressible fluid because this allows density fluctuations. This intensity arises because, at $t = 0$, $G(\mathbf{r}, t)$ has excess concentration at the origin, as, by definition, there is an atom there. This excess concentration diffuses away in the same way as for $G_s(\mathbf{r}, t)$ except that the diffusion coefficient here is D_{chem}, the chemical or Fick's law diffusion coefficient because, here, we are dealing with an actual concentration distribution whereas in the incoherent case, one is dealing with the probability distribution of the tagged atom. Thus, as the tagged atom moves away from the origin, it tends to push other atoms ahead of it so that D_{chem} tends to be larger than D_t.

The ability of neutrons to distinguish between these two types of diffusion is an important advantage of the present technique. It is therefore worth emphasising the distinction between the two diffusion constants. This is best done by equating the net atomic flow, as defined by Fick's law, to the flow produced by a mobility M and a chemical potential gradient, $d\mu/dx$. This yields

$$J = -D_{\text{chem}} dc/dx = -cM d\mu/dx \ ,$$

where c is the fraction of lattice states that are occupied. It is easily shown [5.21] that $M = D_t/kT$ so that – in the mean field limit where correlation effects are ignored – we have

$$D_{\text{chem}} = cD_t (d\mu/dc)/(kT) \ . \tag{5.7}$$

In a noninteracting lattice gas, it is easily shown that $d\mu/dc = kT/c(1-c)$, so that $D_{\text{chem}} = D_t/(1 - c)$ and the two diffusion coefficients are identical for $c = 0$.

When correlation effects are introduced, this expression is modified by the ratio of the tracer correlation factor to the mobility correlation factor, f_m/f_t, a quantity that is often referred to as H, the Haven's ratio. The correlation factors can easily be evaluated using Monte-Carlo techniques [5.21].

5.2.3 Incoherent Quasi-elastic Neutron Scattering from a Lattice Gas: The Chudley-Elliott Model

The interpretation of the quasi-elastic scattering from hydrogen diffusing through a metal lattice involves being able to build up a model for $G_s(\mathbf{r}, t)$ and hence calculating $S^{\text{inc}}(\mathbf{Q}, \omega)$. The basic model assumes that in the low concentration limit the hydrogen atoms perform a series of random jumps between interstitial sites in the metal lattice. In the simplest case, these sites themselves form a Bravais lattice – the most familiar example is the lattice of octahedral sites in an fcc lattice which is itself fcc (e.g., Pd/H). The jumps are random in the sense that successive jumps are assumed to be entirely uncorrelated with each other. They are chosen from a set of fixed probabilities for jumps to near neighbour sites with a given mean residence time, τ, on each site. The original analysis was given by *Chudley* and *Elliott* [5.22] using differential equations. Although this is undoubtedly the most general method of solving the problem, an alternative derivation due to *Gissler* and *Rother* [5.23] and closely following the earlier NMR derivation due to *Torrey* [5.24], which is based on a Markov-chain approach, undoubtedly yields greater physical insight into the problem. In this approach, we first note that $T_n(t)$, the probability of n jumps at time t, is just the same as the Poisson-statistics expression for n events at time t where $\tau = 1/\lambda$, i.e.,

$$T_n(t) = [(t/\tau)^n/n!]\exp(-|t|/\tau) \tag{5.8}$$

except that we have introduced the modulus signs, $|\ldots|$ to indicate that the functions $T_n(t)$ are symmetric in time – as they must be because we also take n to be the number of jumps between $-t$ and $t = 0$. Now, if the spatial distribution after one jump is $f(\mathbf{r})$, the distribution after two jumps is the convolution of $f(\mathbf{r})$ with itself, i.e., $f(\mathbf{r})*f(\mathbf{r})$, after three jumps, it is $f(\mathbf{r})*f(\mathbf{r})*f(\mathbf{r})$, etc. These convolutions would be complicated to evaluate in space but we are only interested in their Fourier transform, which is just the Fourier transform of $f(\mathbf{r})$ [i.e., $F(\mathbf{Q})$] raised to the nth power. We can thus write the intermediate scattering function as the sum over nth of the Fourier transform for n jumps times the probability that it has taken n jumps at time t.

$$\begin{aligned} I(\mathbf{Q},t) &= \sum_n [F(\mathbf{Q})]^n [(t/\tau)^n/n!]\exp(-|t|/\tau) \\ &= \exp[-[1 - F(\mathbf{Q})]\,|t|/\tau] \ . \end{aligned} \tag{5.9}$$

It is now easy to obtain $S(\mathbf{Q}, \omega)$ by Fourier transformation in time, i.e.,

$$S(\mathbf{Q},\omega) = (1/\pi\tau)(1 - F(\mathbf{Q}))/\left\{\omega^2 + (1/\tau)^2[1 - F(\mathbf{Q})]^2\right\} \ , \tag{5.10}$$

where, if the jumps are to the m nearest-neighbour sites at distance $\mathbf{l}$ from the origin site, i.e., $f(\mathbf{r}) = (1/m)\sum_l \delta(\mathbf{r} - \mathbf{l})$, then $F(\mathbf{Q}) = (1/m)\sum_l \exp(i\mathbf{Q}\cdot\mathbf{l})$. This expression (5.10) is known as the Chudley-Elliott model [5.22]. It will be seen that the scattering function is again Lorentzian in ω and it is easy to show that, by expanding the expression for the width at low $\mathbf{Q}$, the broadening has the expected form of $D_t Q^2$ where $D_t = l^2/6\tau$. It will also be observed that the quasi-elastic width varies sinusoidal in a way that depends on the lattice geometry but has the interesting property of decreasing to zero at the reciprocal lattice points of the lattice of sites occupied by the diffusing atoms. This property arises because, for these values of Q, contributions from defect lattice sites are in phase and the total occupation of such sites is unity, i.e., invariant in time, therefore giving elastic scattering.

5.2.4 Extension of the Chudley-Elliott Model to Non-Bravais Lattices

It will be clear that the above derivation requires that the geometric environment of each site is the same, i.e., that the sites constitute a Bravais lattice. In extending the model to a non-Bravais lattice, it is best to return to the Chudley-Elliott method and to write down a set of linked differential equations, as was first done by *Blaesser* and *Peretti* [5.25]. The method was applied to the lattice of tetrahedral sites in the bcc. lattice by *Rowe* et al. [5.26] and, subsequently, to the case of octahedral and tetrahedral sites in hcp lattices by *Anderson* et al. [5.27]. The solutions obtained are analogous to the phonon dispersion curves of a crystal in that they involve the solution of a matrix, here defined by the jump probabilities. The number of solutions at a given value of $\mathbf{Q}$ is equal to the number of defect lattice sites/unit cell. Unique solutions correspond to $\mathbf{Q}$ values within the Brillouin Zone of the lattice. The eigenvalues correspond to the widths of the different Lorentzian components of the quasi-elastic peak and the eigenvectors correspond to the weights of these Lorentzians. The sum of the weights is automatically normalised to unity. It is a characteristic of the solution that for $Q \rightarrow 0$, one of the roots has unit weight and width $D_t Q^2$ while the other roots have zero weight, as expected from the known low Q limit.

5.2.5 Incoherent Quasi-elastic Scattering from a Lattice Gas at Finite Concentrations

We turn now to the case of finite concentrations, c, of diffusing atoms on the defect lattice where the diffusing atoms do not interact with each other except through the prohibition of double site occupancy. This approximation is usually referred to as the noninteracting case. Here, the analysis in the previous section breaks down because we now have correlations between successive jumps. These correlations arise because a jump towards an occupied site is prevented. Hence, if an atom jumps to a neighbouring empty site, the site it has left is initially unoccupied while the other sites available for its next jump are occupied with a probability, c. Thus, the atom has an increased chance of jumping back to the original site so that one of the basic assumption of the Chudley-Elliott model [5.22] becomes invalid.

The simplest way of dealing with this complication is to ignore this enhanced probability and simply adjust the mean residence time to allow for site-blocking, i.e., $\tau(c) = \tau(0)/(1 - c)$ where $\tau(c)$ is the mean residence time at concentration c. This amounts to a "mean field" approximation and this correction to the mean residence time is referred to as the "site blocking factor".

The first calculations to include the correlation effects properly were carried out by *Ross* and *Wilson* [5.28] using Monte-Carlo simulation methods. In their approach, $G_s(\mathbf{r}, t)$ was expanded in terms of the number of attempted jumps so that the time dependence was still given by a Poisson expansion. A Monte-Carlo simulation was then used to calculate $f_n(\mathbf{r})$, the spatial distributions after n' attempted jumps and hence $F_{n'}(\mathbf{Q})$ is easily computed numerically to obtain the corresponding scattering function. The resulting quasi-elastic peak can be compared with the Chudley-Elliott result corrected for site blocking. At large ω, i.e., for the first jump, the curves coincide. At lower ω, the computed curve first drops below and then rises above the mean field version. At low Q, the peak is again Lorentzian, but with a narrower width, corresponding to a diffusion coefficient, $D_t(c)$ $= f_t(c)\langle l^2 \rangle / \tau(c)$, where $f_t(c)$ is the tracer correlation factor for long-range diffusion at concentration 'c'. The effects of correlations on long-range diffusion have been thoroughly investigated [5.29] and a variety of analytical solutions are available particularly for $c = 1$. The Monte-Carlo method has also been extensively used, particularly for obtaining the concentration dependence of the correlation factor in various lattices [5.30].

An alternative approach to the calculation of $S(\mathbf{Q}, \omega)$ that can be compared with the simulation is the "encounter model" which has been taken over from the NMR problem [5.31] and applies to the high concentration limit, where it is vacancies that move through the lattice in a true random walk and the protons only move as a result of occasional encounters with these vacancies. In each encounter, the proton may exchange with the vacancy a number of times before the vacancy moves away. We can therefore establish the average results of an encounter – either by calculation or Monte-Carlo simulation and then, by assuming that the encounter can be thought of as being instantaneous as compared to the time between encounters, and that successive encounters can be thought of as being entirely uncorrelated with each other, we can calculate $G_s(\mathbf{r}, t)$ by expanding it in terms of the number of encounters. We can now define the resulting scattering function in the same way as for the Chudley-Elliott model, but using the encounter distribution $f_e(\mathbf{r})$, instead of the single jump pattern, $f(\mathbf{r})$, and the mean time between encounters instead of the mean residence time [5.32]. In the vacancy diffusion limit, therefore, the QENS is again Lorentzian, as in the Chudley-Elliott model, but with a width that is modified from the Chudley-Elliott expression.

The effects of correlations on the incoherent scattering function at high concentrations has also been examined analytically by *Tahir-Kheli* and *Elliott* [5.33] using Green's function methods. Although this approach is very successful, the numerical evaluation of the results is quite complex so that it is not very useful for studying a wide range of cases. An alternative approach has also been described by *Kehr* et al. [5.34] who have defined a waiting time distribution which has enabled them to evaluate the scattering function for arbitrary concentrations based on a simple Monte-Carlo calculation.

The models described above all assume that the protons do not interact with each other, except for the absence of double site occupancy. In practice, there will, of course, be near-neighbour interactions at finite concentrations. If the interactions can be taken to be pairwise, the lattice gas will behave in an analogous way to the Ising model in magnetism. Thus, at sufficiently low temperatures, repulsive interactions will give rise to the formation of particular hydride superlattice phases or to mixtures of phases in particular concentration and temperature ranges while attractive interactions will give phase separation. In contrast, at higher temperatures, the interactions can cause either short-range ordering (repulsions) or clustering amongst the protons (attractions). There is no known exact solution to the Ising model in

3 dimensions but useful approximate solutions can be obtained using the cluster variation method [5.35] or Monte Carlo [5.36]. In most cases, – , e.g., in fcc lattices – the protons will repel each other which means that the occupation probability of the nearest neighbour to an occupied site will be smaller than c. The site blocking factor will therefore be smaller. However, the site jumped to will have a higher energy than the original sites so that the jump rate will be reduced by the corresponding Boltzmann factor. Thus, the effect of the interactions will be to change the residence time but this will be only one of the concentration-dependent influences on the residence time. Correlation effects will also be changed but Monte-Carlo simulations would suggest that this effect would also be small and difficult to demonstrate experimentally [5.37]. Of more significance is the case of tetrahedral sites in the bcc metals because these sites are so close together that nearest and next-nearest neighbour sites are almost completely blocked. This, combined with the fact that tetrahedral sites only have four nearest neighbours, means that the correlation effects can be increased substantially so that, in this case, the quasi-elastic peak is considerably distorted from a Lorentzian shape [5.38].

5.2.6 Diffusion in a Lattice with Traps

In the previous section, we discussed the influence of interactions between interstitials on their diffusion in an infinite lattice. An equally important case is that of the diffusion of isolated protons in an infinite lattice with traps. In practice, traps are formed in the vicinity of interstitial or substitutional impurities. One extreme case is where the trapped proton can move between a number of trapping sites but cannot move away from the impurity atom. The characteristic of this situation is that $G_s(\mathbf{r}, t)$ tends to a time-independent shape as time increases, i.e.,→ $G_s(\mathbf{r}, \infty)$. Now, when we Fourier transform to get the scattering function, the time-independent part gives an elastic component [a delta function at $(E - E') = 0$], the amplitude of which is just the spatial Fourier transform of $G_s(\mathbf{r}, \infty)$, i.e.,

$$S^{inc}(\mathbf{Q}, \omega) = \int G_s(\mathbf{r}, \infty) \exp(i\mathbf{Q} \cdot \mathbf{r}) d\mathbf{r}\, \delta(\omega) + \text{quasi-elastic terms}$$

$$= S_{el}^{inc}(\mathbf{Q}) + S_{qe}^{inc}(\mathbf{Q}) \ . \tag{5.11}$$

The term $S_{el}^{inc}(\mathbf{Q})$ is generally known as the Elastic Incoherent Structure Factor (EISF). Its measurement is straightforward and provides direct information on the spatial extent of the trapping sites. The forms of the quasi-elastic terms depend on the details of the intertrap motion. The normalisation condition for $S_{qe}^{inc}(\mathbf{Q}, \omega)$ makes its integral over ω equal to $1 - S_{el}^{inc}(\mathbf{Q})$.

A very simple example of localised diffusion is the case of simple thermally activated hopping between two equipotential sites separated by a vector $\mathbf{r}_0$ [5.8]. Let the mean residence time on each site be τ and let us expand the self-correlation function in terms of the number of jumps, n. Thus, after odd numbers of jumps, the proton is displaced by $+/-$ $\mathbf{r_0}$ (after averaging over the initial sites) while for even numbers of jumps, it is back at the origin. Now, by applying the expression (5.9) for Markovian random walk diffusion to the present case, we have

$$I(\mathbf{Q}, t) = \exp\left(-t/\tau\right)\left[1 + F(\mathbf{Q})t/\tau + (t/\tau)^2/2! + F(\mathbf{Q})(t/\tau)^3/3! + \ldots\right] ,$$

where

$$F(\mathbf{Q}) = [\exp(i\mathbf{Q} \cdot \mathbf{r_0}) + \exp(-i\mathbf{Q} \cdot \mathbf{r_0})]/2 = \cos(\mathbf{Q} \cdot \mathbf{r_0}) .$$

On summing this series and Fourier transforming in time, we have

$$S(\mathbf{Q}, \omega) = 0.5[1 + \cos(\mathbf{Q} \cdot \mathbf{r_0})]\delta(\omega) + \frac{[1 - \cos(\mathbf{Q} \cdot \mathbf{r_0})](2/\pi\tau)}{[(2/\tau)^2 + \omega^2]} . \qquad (5.12)$$

The first term on the right-hand side is the expected form of the EISF, and the second is a Lorentzian of width $2/\tau$. It will be observed that, in this case, the Q dependence is in the structure factor and not the width. Also, while, for $Q = 0$, the peak is entirely elastic, at $Q = \pi/r_0$, it is entirely quasi-elastic.

This expression is valid where the jumps between the sites are thermally activated, as is the case at high temperatures. At lower temperatures, however, if the sites are sufficiently close for the wave functions of a proton to coexist in both sites, the proton will actually tunnel between the sites. The best established cases of this is for H trapped near an oxygen interstitial in niobium. but, as this is beyond the scope of the present chapter, the interested reader is referred elsewhere (Chaps. 2, 3, and [5.39]).

A more general case for hydrogen diffusion in a lattice with traps is where the hydrogen spends a significant portion of its time in untrapped sites. The basic theory for this case is described by *Kehr* and *Richter* [5.40]. These authors show that, for sufficiently small Q, the scattering function becomes a single Lorentzian, with a width of $D_t Q^2$ where the macroscopic tracer diffusion coefficient is just the weighted mean of the appropriate values for the trapped and untrapped sites. On the other hand, at larger Q (near the zone boundary), we find, as expected, that the scattering function is a superposition of two peaks, the relative areas of which are proportional to the fraction of trapped and untrapped protons at any instant. The width of one of these peaks is as for the Chudley-Elliott model in the lattice without traps while the second has a width related to the trapping time. The general problem of diffusion with a distribution of trapping sites is quite difficult to deal with – especially if we make realistic assumptions about the variation of the barrier height between two sites with different trapping energies [5.41]. It should be noted that the case of a regular lattice with variable energy sites is a useful simplification of the case of hydrogen diffusion in an amorphous metal – being one where detailed analysis may be more profitable.

5.2.7 Coherent Quasi-elastic Scattering from a Noninteracting Lattice Gas at Arbitrary Concentrations

Turning now to the case of coherent quasi-elastic scattering, we should first note that there has been remarkably little consideration of this problem, probably because the significance of its relationship to the Chemical Diffusion Coefficient, and hence to the chemical potential of the lattice gas, as illustrated by (5.7), has not been very widely realised. We would first note that, at zero time for the noninteracting case, the total correlation function, $G(\mathbf{r},t)$, consists of a delta function of unit area at the origin plus delta functions of area c on all the other lattice sites. At later times, this excess concentration at the origin spreads out through the lattice. If we imagine the atom at the origin to be tagged, it is apparent that its probability distribution will spread out more slowly than the total excess concentration, given that the tagged atom is bound to partially replace the untagged atoms on the lattice sites near the origin, effectively pushing the other atoms ahead of it. This illustrates the fact that the tracer diffusion coefficient will always be smaller than or equal to the Chemical or Fick's law diffusion coefficient, D_{chem}, in the noninteracting case where the chemical potential, μ is determined entirely by the configurational entropy, see (5.7). If the atoms repel each other, μ increases with concentration at a more rapid rate than for the noninteracting case. Thus, $d\mu/dc$ is larger than for the non-interacting case and hence so is D_{chem}. Attraction between atoms reduces μ relative to the noninteracting case and hence also $d\mu/dc$ and D_{chem}. In the limiting case, the $\mu(c)$ becomes flat at a critical point and D_{chem} tends to zero (critical slowing down). Anything that affects the term $d\mu/dc$, such as a proportion of trapping sites, will influence the magnitude of D_{chem} relative to D_t.

Let us first consider the case of the noninteracting lattice gas. The expression for $S^{\text{coh}}(\mathbf{Q}, \omega)$ for this case was first given by *Ross* and *Wilson* [5.28]. We can write the differential equation for the rate of change of the excess concentration of the diffusing atoms, i.e., the deviation from the average concentration, c, as $G'(\mathbf{r},t)$, so that.

$$dG'/dt(\mathbf{r},t) = \left[\frac{1}{m\tau(0)}\right]\left\{\sum_l [G'(\mathbf{r}+\mathbf{l},t)+c][1-c-G'(\mathbf{r},t)]\right.$$

$$\left. -[G'(\mathbf{r},t)+c][1-c-G'(\mathbf{r}+\mathbf{l},t)]\right\} - \frac{[G'(\mathbf{r},t)+c]}{[1-c-G'(\mathbf{r}+\mathbf{l},t)]} ,$$

where m is the number of nearest-neighbor sites at distance, $\mathbf{l}$, from the first site to which the atom is assumed to jump. The first term on the RHS is the rate of jumping from sites $\mathbf{r}+\mathbf{l}$ to site $\mathbf{r}$ given by the jump rate to a particular empty nearest-neighbour site $[1/m\tau(0)]$ times the concentration on the site $\mathbf{r}+\mathbf{l}$ times the chance that the target site is unoccupied, $[1-c-G'(\mathbf{r}, t)]$, while the second term is the jump rate in the opposite direction.

On simplification, this equation yields

$$dG'/dt(\mathbf{r}, t) = \left[\frac{1}{m\tau(0)}\right] \left[\sum_l G'(\mathbf{r} + \mathbf{l}, t) - G'(\mathbf{r}, t)\right] . \tag{5.13}$$

This differential equation is identical with the equation derived by *Chudley* and *Elliott* [5.22] for the self correlation function, except that the boundary condition at $t = 0$ is $(1 - c)\delta(\mathbf{r})$ and, of course, that the solution is valid for all concentrations. If we now solve the equation by the Chudley-Elliott method, and add in the Fourier transform of the constant concentration, c, on each site which just gives Bragg scattering, we find that, for this case,

$$S^{\text{coh}}(\mathbf{Q}, \omega, c) = c(1 - c)S^{\text{C-E}}(\mathbf{Q}, \omega) + c^2 \left[\frac{(2\pi)^3}{B}\right] \sum_\tau (\mathbf{Q} - \tau) , \tag{5.14}$$

where we have now defined $S^{\text{coh}}(\mathbf{Q}, \omega, c)$ per lattice site and where τ is one of the reciprocal lattice vectors of the interstitial site lattice. Here, $S^{\text{C-E}}(\mathbf{Q}, \omega)$ is the Chudley-Elliott model, expression (5.10), for the incoherent quasi-elastic scattering of a lattice gas in the low concentration limit.

This result shows that the coherent quasi-elastic scattering from a lattice gas is a quasi-elastically broadened version of the Laue diffuse scattering from a disordered lattice [$c(1 - c)$]. Moreover, it has a quasi-elastic width that is the same as the incoherent case at zero concentration. This shows that, as expected, D_{chem} and D_t are identical at low concentrations but, more surprisingly, that D_{chem} for a noninteracting lattice gas is independent of concentration. Moreover, it shows that D_{chem} is given in terms of a correlation factor of unity in the noninteracting case [5.27]. This result does not seem to have been previously appreciated but, as was pointed out by *Kutner* [5.42], it is a direct result of the configurational entropy contribution which completely determines the chemical potential in the noninteracting case and shows that Fick's law is just a special case of the Nernst-Einstein equation. Thus, because the internal energy is concentration-independent in this case, we can write, as above (5.7)

$$D_{\text{chem}} = cD_t(d\mu/dc)/kT \quad \text{where} \quad \mu = dS/dc \quad \text{and} \quad S = k \ln W .$$

Here, $W = N!/(cN)!\{(1 - c)N\}!$ because all possible configurations have the same energy so that, using Stirling's approximation, $d\mu/dc = kT/\{c(1 - c)\}$. Now, using the mean field expression for D_t, i.e., $D_t = (1 - c)\langle l^2 \rangle/[6\tau(0)]$, on substitution, we have $D_{\text{chem}} = \langle l^2 \rangle/6\tau(0)$ as required to be consistent with (5.14). It will be noted that, in general, for interacting atoms, there will be a correlation factor in the expression for D_{chem} because a jump down a concentration gradient will induce subsequent correlated jumps. We will refer to this effect as the mobility correlation factor, f_m [5.21] although a variety of other notation has been used in the past. From the above result it is clear that $f_m = 1$ in a noninteracting lattice gas.

5.2.8 Coherent Quasi-elastic Neutron Scattering from an Interacting Lattice Gas

It was shown above that the coherent quasi-elastic broadening at low Q (large distances), i.e., $Q^2 D_{chem}$ is proportional to $Q^2 d\mu/dc$. Hence we might expect that the coherent quasi-elastic scattering at finite Q would depend on the concentration dependence of the spatial Fourier transform of the pairwise interaction energies between the diffusing atoms; this is indeed the case. Let us first consider the short-range order amongst the lattice gas atoms. It was shown by *Clapp* and *Moss* [5.43,44] that, in the mean field approximation, the diffuse scattering due to short range order in a lattice gas is given by the expression

$$S^{CM}(\mathbf{Q}) = \frac{c(1-c)}{1 + c(1-c)V(\mathbf{Q})/kT} ,$$ (5.15)

where $V(\mathbf{Q}) = \sum_l V_l \exp(i\mathbf{Q} \cdot \mathbf{l})$ is the Fourier transform of the pairwise interaction energies, V_l, between atoms on sites separated by the vector $\mathbf{l}$. (A positive V_l increases the total energy of the system and so corresponds to a repulsive interaction). It will be seen that $S^{CM}(\mathbf{Q})$ is periodic in the reciprocal lattice and that it reduces to the Laue expression for sufficiently high temperatures. It will also be noted that for nearest neighbour repulsions, $V(\mathbf{Q})$ will be positive at $Q = 0$ and other reciprocal lattice points but will be negative at the Brillouin Zone (BZ) boundary and hence that $S(Q)$ lies below $c(1-c)$ at $Q = 0$ and peaks where $V(Q)$ has a minimum. *Clapp* and *Moss* showed that a minimum in $V(Q)$ can only occur at a small number of special points in the BZ, i.e., at (100), (1/2 1/2 1/2) and (1 1/2 0) for the fcc case. We now note that if we reduce the temperature, $S^{CM}(Q)$ will diverge when $T_c = -c(1-c)V(\mathbf{Q})/k$. This is a mean-field estimate of the critical temperature, T_c, for the formation of the superlattice, which implies that we will find superlattice peaks at the particular special points where $V(\mathbf{Q})$ has its lowest value. This type of phase transition is second order and an excellent example is provided by the "50 K anomaly" in Pd D_x [5.45, 46].

This mean-field approach can be extended to analysing coherent quasi-elastic scattering from an interacting lattice by using linear response theory. This may be done by analogy with the treatment of similar magnetic systems [5.47], again neglecting any correlation effects in the results. For a Bravais lattice, this analysis shows that, if we neglect the renormalisation of the host lattice phonons due to the presence of the interstitials, the coherent quasi-elastic scattering is given by a Lorentzian with a width of

$$\Gamma'(\mathbf{Q}) = \Gamma(\mathbf{Q})c(1-c)/S^{qe}(\mathbf{Q}) ,$$ (5.16)

where $S^{qe}(\mathbf{Q})$ is the integrated quasi-elastic intensity which is essentially as in (5.15) for $S^{CM}(\mathbf{Q})$ in the mean-field approximation (where we have ignored lattice distortions) and where $\Gamma(\mathbf{Q}) = 1 - F(\mathbf{Q})$ is the quasi-elastic width in the Chudley Elliott model.

This expression shows that the coherent quasi-elastic scattering narrows as compared with the noninteracting case in proportion to the value of $S(\mathbf{Q})$

as predicted in the *Clapp* and *Moss* expression given above (5.15). The result is comparable to "de Gennes narrowing" [5.48] but avoids uncertainties as to whether or not to include inelastic scattering in the calculation of the moments. It also provides a powerful formalism for extending the treatment to multiple sites/unit cell and to interactions between the diffusive and vibrational modes.

It is particularly useful to evaluate D_{chem} from the broadening at low Q. This yields

$$D_{chem} = \langle l^2 \rangle / [6\tau(0)][1 + c(1 - c) \sum_l V_l / kT] \;, \qquad (5.17)$$

which is easily seen to be equivalent to our previous result that $D_{chem}/D_t = c(d\mu/dc)/kT$ if the additional internal energy introduced per extra interstitial (dU/dc) is written in the mean field approximation $c \sum_l V_l$ so that $d\mu/dc$ becomes $[kT/c(1 - c) + \sum_l V_l]$. The ratio, D_{chem}/D_t can in turn, be written $c/S(Q)_{Q \sim 0}$ from (5.15). We thus see that the coherent and incoherent broadening can be quite different at low Q at finite concentration. If the diffusing atoms attract each other, $S(Q)$ peaks at the origin and the coherent quasi-elastic broadening is smaller than the incoherent. In the limit, its width will go to zero at a critical point where $d\mu/dc$ is also zero. On the other hand, if the atoms repel each other, the coherent quasi-elastic broadening is broader than the incoherent, but, because $S(0)$ becomes progressively smaller, the effect becomes increasingly difficult to observe. This is why one cannot measure any coherent quasi-elastic scattering from a normal incompressible liquid at low Q.

It is obviously useful to be able to compare the coherent and incoherent scattering in a single experiment. This requires that $S(Q = 0)$ must not be too large or too small, i.e., $(d\mu/dc)/kT$ must be close to unity so that the isotherms for hydrogen absorption must be neither flat nor vertical! Also, to ensure that $S(Q)$ is reasonably large, c must lie near the middle of the range 0 to 1. Interesting results are likely to be found near a spinodal decomposition at $Q \approx 0$ or for superlattice formation in a second-order phase transition, for Q near the reciprocal lattice values.

5.2.9 The Effect of Lattice Stress on Coherent Quasi-elastic Neutron Scattering

The above analysis has not considered the effects of lattice stress. Because hydrogen expands the lattice and this requires energy, less energy will be required to insert a hydrogen atom if the lattice is already expanded due to a local stress, σ. Here σ is the average of the trace components of the stress tensor. It is thus only strictly a scalar if the site has cubic symmetry. In this approximation, there is an additional contribution to the chemical potential, μ_s, where $\mu_s = -\sigma \Delta V$ and ΔV is the volume expansion caused by one hydrogen atom. We can thus generalise the expression $J = -cM d\mu/dx$ at the mean-field level if we break up the chemical potential into the three components, the configurational part, μ_c, the part due to interstitial-interstitial interactions, μ_i, and the part due to stress, μ_s, so that

$$J = cM\left[\frac{d(\mu_c + \mu_i)}{dc}\right](dc/dx) + \left(\frac{d\mu_s}{d\sigma}\right)\left(\frac{d\sigma}{dx}\right) , \qquad (5.18)$$

which allows us to define a new stress-dependent diffusion constant,

$$\begin{aligned}
D_s &= \left(\frac{cD_t}{kT}\right)\left[\frac{d\mu_c}{dc} + \frac{d\mu_i}{dc} + \left(\frac{d\mu_s}{d\sigma}\right)\left(\frac{d\sigma}{dc}\right)\right] \\
&= \left(\frac{cD_t}{kT}\right)\left(\frac{kT}{c(1-c)} + \sum_i V_i + \Delta V\frac{d\sigma}{dc}\right) , \qquad (5.19)
\end{aligned}$$

where $d\sigma/dc$ represents the variation in the stress with concentration in any particular circumstances.

Thus, from above, we can write

$$D_s = D_{\text{chem}}\left(1 - \Delta V\left(\frac{d\sigma}{dc}\right)\right)\Big/\left(\frac{kT}{c(1-c)} + \sum_l V)_l\right) . \qquad (5.20)$$

Now, the question is, how do we define $d\sigma/dc$ in a meaningful way for neutron measurements. It is clear that we have to consider a concentration wave with a wavelength much greater than the lattice spacing but much less than the macroscopic dimensions of the sample. In this wave, the stress will be determined by the lattice expansion (contraction) caused by the deviation from the average concentration. As the stress will lie parallel to the direction of the concentration variation, it will be determined by $c_{\parallel}$, one of the elastic moduli in Voigt's notation. The full analysis of diffusion in the presence of stresses has been given by a number of authors [5.49–53]. The present case of a local concentration fluctuation is referred to as a bulk mode and it has been shown that the corresponding bulk diffusion coefficient, D_b, is given by [5.54]

$$D_b = D_{\text{chem}}\left[1 + \left(\frac{\Delta V}{V}\right)\left(\frac{P}{kTf_{\text{th}}}\right)\left(\frac{1-B}{c_{11}}\right)\right] , \qquad (5.21)$$

where P is one of the trace components of the dipole moment tensor of the hydrogen atom which is assumed to be isotropic, B is the bulk modulus and f_{th} is the "thermodynamic factor". This shows that coherent Quasi-Elastic Neutron Scattering (QENS) will actually measure D_b if there is a finite volume expansion when we add hydrogen to the lattice.

The point about this measure of D_b is that it differs from D_{chem} determined from Gorsky Effect measurements on macroscopic bodies with *free* boundaries subject to external stress. Here diffusion takes place on a macroscopic scale where the distribution of stress contributing to the flow of interstitial atoms is determined by the existence of stress-free boundaries and where "macroscopic modes" are observed. One might argue that D_b is the more fundamental quantity as the ambiguities in the interpretation only arise when concentration variation over distances comparable with the macroscopic dimensions of the object are involved. However, the important point to be clear about is which quantity is being measured in a particular experiment.

The final point to be stressed in connection with coherent quasi-elastic scattering is that the models discussed above are correct at the mean field level – that is to say that correlation effects are not included although in general they do exist. We can understand how they arise if we imagine atoms diffusing down a chemical potential gradient, remembering that here we are interested in the total movement down the gradient resulting from one random hop. When one atom hops down the gradient, atoms in front have an increased probability of hopping forward and so do atoms behind. These effects are in the opposite sense to the excess probability of the atom itself hopping back. For the case of noninteracting atoms, it happens that these effects completely cancel each other out and the corresponding correlation factor, which, for clarity, we have called the mobility correlation factor f_m, is unity. However, if the atoms do interact, f_m is no longer unity. As in the case of tracer correlation effects, correlations here have the effect of causing the QENS peak to deviate from a Lorentzian shape. We should also note that f_m/f_t is just the quantity usually called the Haven ratio and that, in general, it can only be calculated by Monte-Carlo simulations [5.21].

Some experiments involving these ideas are described in Sect. 5.4 below.

5.3 Experimental Methods of Measuring Quasi-elastic Neutron Scattering

5.3.1 General Features of Quasi-elastic Neutron Scattering Spectrometry

Quasi-elastic neutron scattering involves the simultaneous measurement of neutron energy change and of the angle of scatter in a given collision. There are essentially four possibilities, depending on whether we are using neutrons from a continuous reactor source or an accelerator pulsed source. In most cases, we either first produce a monochromatic pulse of neutrons incident on the sample and then determine the final energy for each neutron by the time-of-flight of the neutron to a set of detectors at accurately known distances from the sample (direct geometry: Sect. 5.3.2) or we determine the incident energy by time-of-flight measurements for a fixed final energy (indirect geometry) where the final energy is generally defined in back-reflection in a crystal monochromator (Sect. 5.3.3). It is also possible to monochromate the incoming neutrons by Bragg reflection and to scan through the final energies with an analyser crystal as in a triple-axis spectrometer but for various reasons, this has not been much used. However, a crystal monochromator and a crystal analyser can be used where both crystals are in back-reflection geometry to achieve high resolution (Sect. 5.3.4) but here the incident energy is scanned either with a doppler drive or by mounting the crystal in a furnace with a linear temperature drive.

The only alternative to these approaches is the neutron spin-echo method (Sect. 5.3.6). Here, the energy transferred to/from a particular neutron – and hence ω – is measured very accurately using the spin-echo technique [5.55] but the actual neutron energy distribution in the incident beam is only determined with the accuracy necessary to give the required resolution in $\mathbf{Q}$.

A second important classification of QENS spectrometers refers to the energy range in which they operate. Most instruments operate in the incident energy range below around 5 meV (cold neutrons) where perfectly black choppers can be constructed using cadmium alloys. Here higher energy neutrons can be eliminated using a beryllium filter ($\lambda > 2d_{max} = 4$ Å, where $2d_{max}$ is twice the maximum Bragg d-spacing in beryllium) or by passing the neutrons down a curved guide tube. Higher energy choppers can be designed but it is difficult to achieve the necessary absolute energy resolution. The only widely used spectrometer designed for quasi-elastic work in this energy range (IN13 at the ILL) uses back reflection in both the monochromator and analyser and employs a temperature drive to scan the monochromator energy. Cold neutron energies can only measure broadenings out to ~ 2 Å^{-1} which for simple lattices is around the edge of the first Brillouin zone. One often wants to go further out in Q than this – particularly for single crystal samples – but IN13 is one of the few instruments that can achieve this.

Three important points about the experimental technique should be emphasised at the outset. Firstly, for single crystal samples, one would like to be able to measure the quasi-elastic peak shape in ω at points along symmetry directions in $\mathbf{Q}$ space. However, it is easily shown that, for a single sample orientation in either direct or indirect geometry, the $\mathbf{Q}$ vector for different detector angles must lie on a semicircle in reciprocal space that passes through the origin so that only one or two detectors will lie on symmetry directions. Such experiments therefore can be prohibitively slow.

The second point is that any neutron scattering experiment offers a compromise between a low signal strength and contamination of the data with multiply scattered neutrons. A common rule of thumb is that a 10% scatterer will have about 1% of double scattering, i.e., 10% of the measured neutrons will have been scattered twice. However, every case has to be considered on its merits and particular care must be taken where the probability of getting once-scattered neutrons is low. For instance, QENS measurements at low $\mathbf{Q}$ will often be seriously contaminated in the wings of the peaks because the single scattered neutrons will be much less broadened than the multiply scattered ones. However, powerful computer codes such as DISCUS [5.56] and MSCAT [5.57] are available to make corrections for multiple scattering by a process of successive approximation. The corrections are particularly critical and complex in polarisation experiments [5.58].

The third point is that it is clearly best to measure quasi-elastic peaks having a width say two to five times the width of the resolution function of the instrument. If the broadening is too small, it is difficult to extract an accurate value of the broadening. If it is too large, the quasi-elastic scattering becomes indistinguishable from the background. For this reason, it is often useful to start an experiment by doing a temperature scan to see at what temperature the broadening is optimum. The question of how to extract the broadening from the data in the optimum fashion is a large subject in itself. Traditionally this was done by constructing a model of the peak shape by convoluting the resolution function with a Lorentzian and minimising χ^2 [5.59] More recently, the advantages of the "maximum entropy" method have been realised as this method provides a rational way of deciding whether the fit is an optimal one or just one that works [5.60].

Finally, we should note the importance of neutron polarisation techniques (Sect. 5.3.5). Here, the incident neutron beam is passed through a polariser where the beam is split into two components – spin up and spin down – and only one of these components is transmitted. After the collision, again only one of the two spin orientations is selected so that we can measure the cross-section for spin-flip or nonspin-flip. If the cross-section is spin-incoherent (Sect. 5.1.2), 2/3 is spin-flip and 1/3 is nonspin-flip whereas the coherent cross-section is entirely nonspin-flip. Thus, it is possible to separate the coherent scattering from the incoherent. This is particularly important for deuterium because the two parts of the cross-section are of comparable importance ($\sigma^{coh} = 5.6$ b, $\sigma^{inc} = 2.0$ b). Hence, we can simultaneously measure the coherent and the incoherent quasi-elastic scattering, which, as we will see, provides important insight into the mechanisms of diffusion (Sect. 5.4.3).

5.3.2 Direct Geometry Time-of-Flight Spectrometers

One of the most widely used direct geometry time-of-flight spectrometers is the IN5 instrument at the ILL and this will be taken as an example of this type. The layout of the instrument is shown in Fig. 5.1.

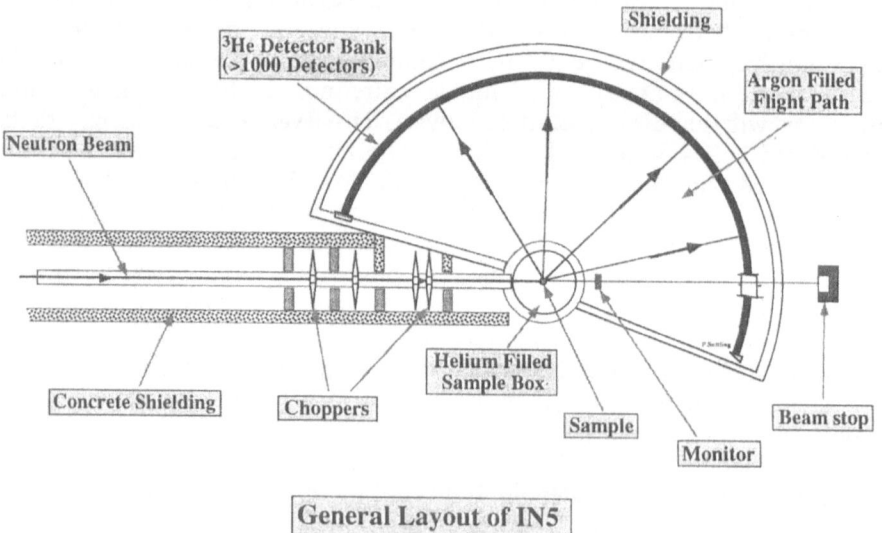

Fig. 5.1. The layout of the IN5 spectrometer at the ILL, Grenoble

Fuller details of this and other ILL spectrometers may be found in *Neutron Scattering Research Facilities at the ILL High Flux Reactor* [5.61]. In this instrument, the monochromatic pulse is produced by a series of disk choppers, [Ref. 5.61, Fig. 49]. The first and the last of these are 6 m apart and define the wavelength that can be selected – in the range from 2–15Å. The second disk eliminates neutrons that would be transmitted by the last rotor in the wrong revolution (higher order contamination) while the third disk operates at a lower repetition rate to control frame overlap. Typical resolution is in the range 10–100 μeV. The scattered neutrons are detected in a very large bank of He^3 proportional counters at 4 m from the sample. The major advantage of this multidisk design is that the incident energy can be varied over a wide range and also a wide range of resolutions can be selected. A second advantage, for quasi-elastic experiments, is that the elastic resolution function is very precisely triangular and this makes the extraction of quasi-elastic broadening functions much more accurate than for resolution functions that spread out in the wings.

5.3.3 Indirect Geometry Time-of-Flight Spectrometers

In principle, indirect geometry could be used for a QENS spectrometer on a reactor but in practice, the double back-scattering design has been preferred. However, this arrangement is widely used on pulsed neutron sources. As an example of the genre, we will describe the IRIS spectrometer on ISIS at the Rutherford Appleton Laboratory, UK. Full details of this and other ISIS instruments can be found in the ISIS Users' Guide [5.62]. The design is illustrated in Fig. 5.2.

IRIS is situated on a guide tube viewing the liquid H_2 moderator (25 K). The range of neutron velocities from this moderator is determined by a disc chopper at 6.4 m from the moderator surface and fast neutron contamination is removed by a slightly curved guide tube. The energy of the incoming neutron is determined by its time of arrival at the sample which is 36.5 m from the moderator. Neutrons scattered from the sample are then reflected in a large array of either graphite or mica crystals in near-back-scattering geometry to a set of scintillation detectors, one viewing each crystal. The disc chopper can select different energy ranges corresponding to different harmonics of the analyser crystals. The graphite (002) reflection gives a final energy of 1.82 meV with a resolution of 15 μeV, the (004) gives 7.28 meV with a resolution of 50 μeV while the mica (002) yields a resolution of 1 μeV. One of the advantages of this instrument, as compared with double back-scattering spectrometers is that the energy range analysed is about 200 times the resolution width compared with about 30 times on IN10 (Sect. 5.3.4). This spectrometer has been in the forefront of tunnelling level measurements but is also extensively used for QENS measurements on H-metal systems.

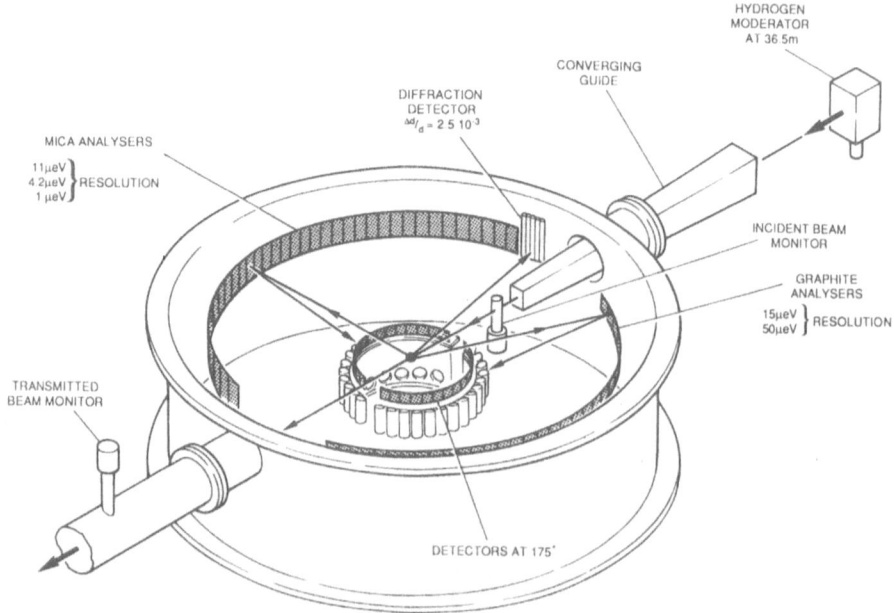

Fig. 5.2. The layout of the IRIS spectrometer on ISIS at the Rutherford-Appleton Laboratory (UK)

5.3.4 Back-Scattering Spectrometers

The principle of the double back-scattering spectrometer was first demonstrated by *Alefeld* et al. [5.63]. In the IN10 design at the ILL (Fig. 5.3, [5.61]), the incident beam in a guide tube is monochromated by back-reflection in one of a series of possible crystals, e.g., Si (111) which produces an incident energy of 2.08 μeV when stationary. The crystal is, in fact, mounted on a Doppler drive which is capable of scanning the back-reflected neutron energy over a range of ± 15 μeV about the stationary value. This back-reflected beam is then diverted out of the guide towards the sample with a graphite crystal. The scattered neutrons are then back-reflected in an array of similar Si (111) crystal planes back onto a set of He^3 detectors. The detected neutrons are then sorted into a histogram according to their incident energy which is determined from the instantaneous velocity of the Doppler drive unit. Recently, the Doppler drive has occasionally been replaced by a crystal in a furnace which is linearly scanned through temperature to enhance the energy range available. IN10 provides excellent resolution, its one disadvantage being that the resolution function has a wide-tailed Lorentzian resolution shape which makes it more difficult to extract a quasi-elastic broadening function.

The other instrument at the ILL based on this principle is IN13. This instrument uses thermal neutrons in order to reach resolutions of 8 μeV or so

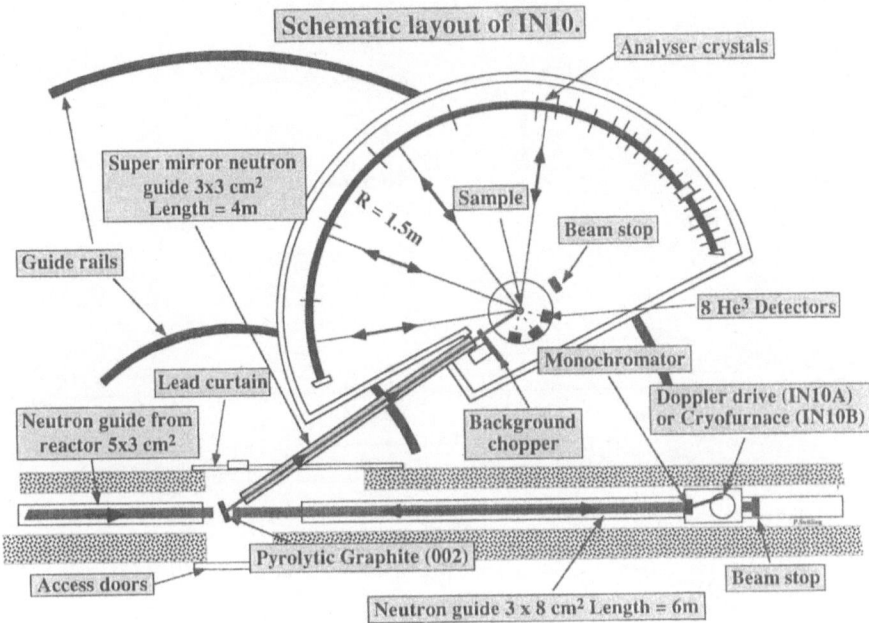

Fig. 5.3. The layout of the IN10 spectrometer at the ILL, Grenoble

out to **Q** values of 5.5 Å^{-1} which is very important if one is interested in short localised jumps. The monochromator crystal can, for instance, be the (422) reflection from CaF_2 crystals (16.45 meV) and, as mentioned above, the scanning is done with a linearly programmed cryofurnace (–196 to 450 °C).

5.3.5 Spin Polarization Analysers

The potential importance of being able to separate spin-flip from nonspin-flip scattering as a way of simultaneously measuring coherent and incoherent QENS has been emphasised above. The only experiment to fully exploit this technique to date [5.58] used the D7 spectrometer at the ILL. This instrument is illustrated in Fig. 5.4. It uses a combination of monochromating graphite crystal and chopper to produce the monochromatic pulse at the sample (energies of 2.5, 3.5 and 8 meV available) and the energy analysis of the scattered neutrons involves the measurement of the time of flight to the detectors (direct geometry), yielding an overall energy resolution of about 100 μeV for the 3.5 meV option. The incident beam is polarised using a "super-mirror" where the magnetisation of the reflecting layer ensures that only neutrons with one polarisation lie within the critical angle for external reflection. There is then a "flipper" coil which, when switched on, flips the

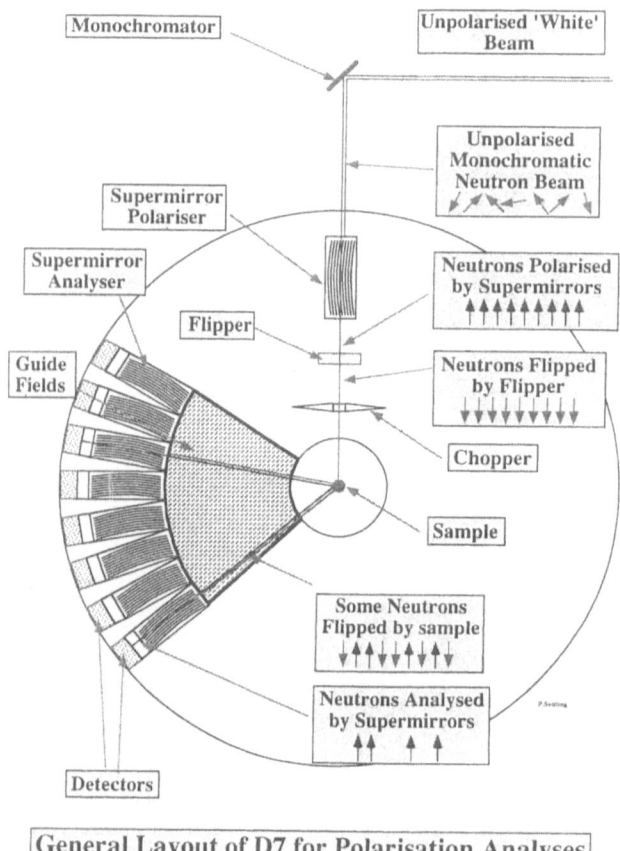

Monochromator

Unpolarised 'White'
Beam

Unpolarised
Monochromatic
Neutron Beam

Supermirror
Polariser

Supermirror
Analyser

Neutrons Polarised
by Supermirrors

Flipper

Neutrons Flipped
by Flipper

Guide
Fields

Chopper

Sample

Some Neutrons
Flipped by sample

Neutrons Analysed
by Supermirrors

Detectors

General Layout of D7 for Polarisation Analyses

Fig. 5.4. The layout of the D7 spectrometer at the ILL, Grenoble

neutron spin. The scattered beam is then analysed for its spin polarisation using a separate super-mirror for each detector angle. The performance of these super-mirrors has been fully described by *Schärpf* [5.64, 65].

5.3.6 The Neutron Spin-Echo Technique

By far the highest energy resolution available from any neutron spectrometer is provided by the spin-echo technique. The principle of this method was proposed by *Mezei* [5.55]. Again it will be described in terms of the IN11 spectrometer at the ILL. The layout of this instrument is shown in Fig 5.5. A beam of neutrons with a narrow band of velocities ($\Delta v/v \simeq 12$–25%) emerge from a velocity selector with a mean wavelength selectable in the range

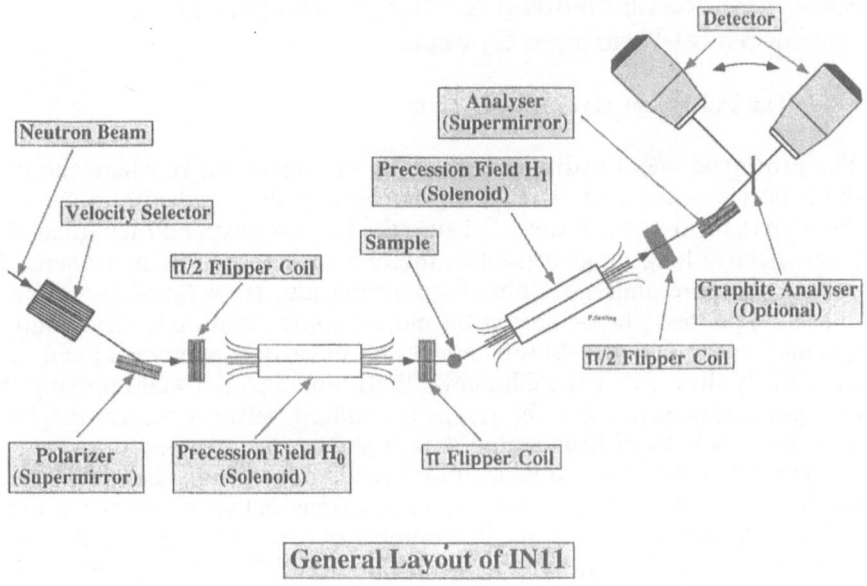

Fig. 5.5. The layout of the Spin-Echo Spectrometer, IN11, at the ILL, Grenoble

4–9 Å, is spin-polarised in the vertical plane by a super-mirror. A flipper coil then rotates its polarisation into the horizontal plane normal to the beam direction. The beam then passes along a tube in which a vertical field causes its spin to precess. It is then scattered in the sample and its spin is subsequently flipped through π radians so that the direction of precession is reversed in the second solenoid. If the scattering were elastic, the beam would emerge from this solenoid with the same polarisation it had on entering the first. This is then followed by a reversing $\pi/2$ coil, a second polarising mirror and a detector. Thus if one of the precession fields is varied slightly, the detected intensity varies sinusoidally – for perfect polarisation it would oscillate from maximum intensity to zero. Now, if the sample scatters quasi-elastically, the time in the second precession field would vary and hence the degree of polarisation would be reduced. If the precession rate is increased in both arms, the resulting polarisation will be further reduced. In fact, the variation of the polarisation with the precession rate yields a Fourier transform of the scattering in time – just what we have defined as $I(Q, t)$ (Sect. 5.2.1). The energy resolution of IN11 is of the order of 10 neV (10^{-9} eV!) – and the instrument is ideal for measuring diffusion coefficients – chemical or tracer – at low Q. To date, it has mainly been used for studies on polymers and magnetic systems but there are considerable opportunities for measurements on hydrogen-metal systems.

5.4 Quasi-elastic Neutron Scattering Measurements on Metal-Hydrogen Systems

5.4.1 The Palladium-Hydrogen System

The prototype metal-hydrogen system is, of course, Pd/H where the palladium lattice is fcc and the hydrogens occupy the octahedral sites which themselves form an fcc lattice. Because the H atoms expand the lattice, there is an effective long-range attractive interaction between the hydrogens. The system therefore undergoes phase separation into the α (gas-like) and β (liquid-like) phases. The α-phase is the closest approximation to the conditions assumed in the Chudley-Elliott model and in early experiments [5.66, 67] it was clearly shown that the Chudley-Elliott model works well implying that, to a good approximation, the jumps are indeed between nearest-neighbour sites. One such set of broadening data is shown in Fig. 5.6.

This work therefore concluded that the hydrogen hops between nearest-neighbour octahedral sites with no correlations between successive jumps. This conclusion has been tested in a number of first principles calculations, in particular, the molecular dynamics calculations of *Gillen* [5.68] and of *Li* and *Wahnstrom* [5.69]. *Gillan* used an empirical pair potential to describe the interatomic interactions and made use of the Born-Oppenheimer separation of the electronic and ionic motions, the so-called adiabatic approximation. *Li* and *Wahnstrom* used the embedded atom model to allow for some essential many-body interactions. This method is now accepted to be very successful in describing a metallic environment. In their simulations using this model, in

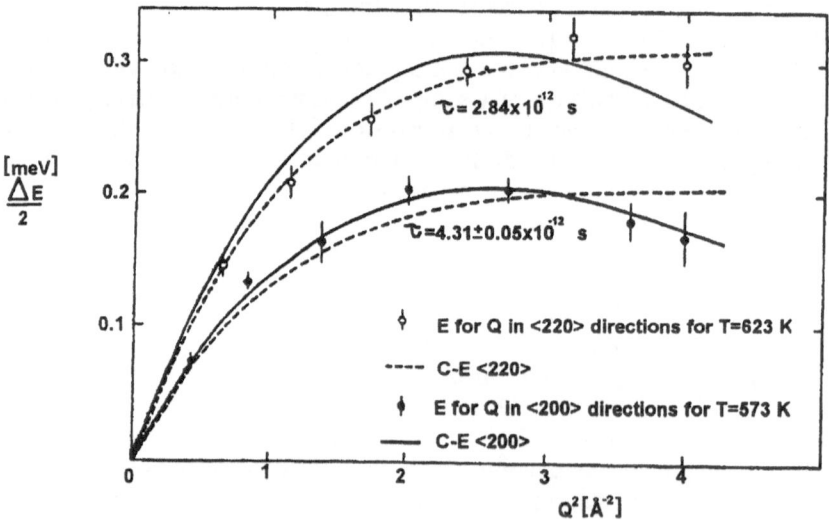

Fig. 5.6. The quasi-elastic broadening of neutrons scattered from hydrogen diffusing in α-Pd/H [5.67]

common with *Gillen*, they found that once a hydrogen had escaped from a site, it tended to move rapidly beyond the nearest-neighbour position, which is clearly in contradiction to experiment, and established that this discrepancy did not seem to be sensitive to the potential model used. *Gillen* had suggested that the problem was due to the inadequacy of the adiabatic approximation in these circumstances arguing that, in a metal, where there is a continuum of electronic energies at the Fermi surface, it is possible for the hydrogen to exchange infinitesimally small amounts of energy in a very large number of interactions. When *Li* and *Wahnstrom* incorporated these effects into their model, the agreement with experiment became excellent.

The situation in the β-phase is more complex, because firstly, the site blocking factor, $(1-c)$ will come into the expression for the tracer diffusion coefficient, causing it to reduce linearly towards zero at the stoichiometric composition. The jump-attempt frequency may also change as the lattice parameter increases with increasing concentration. Secondly, the shape of the quasi-elastic peak will deviate from a Lorentzian shape due to correlation effects, as calculated by *Ross* and *Wilson* [5.28]. Thirdly, we might expect the short-range repulsions between the hydrogens, the existence of which can be inferred from the absorption isotherms [5.11], to influence the mean residence time and the details of the correlation effects. These three effects were found to be rather difficult to separate in practice and the best that could be done [5.36] was to fit the peak with a model that assume an elastic component plus a Lorentzian-broadened component. This model shape was then convoluted with the instrument resolution function before fitting to the data. It was found that this simple shape fits the data with a satisfactory χ^2 after multiple scattering corrections have been applied but that correlation effects were small enough to ignore. The resulting broadening also fitted the Chudley-Elliott model and yielded values of the tracer diffusion coefficient that were in good agreement with NMR measurements at the same temperatures.

5.4.2 Hydrogen in the bcc Metals: The Niobium, Tantalum, and Vanadium Systems

The situation for incoherent QENS from hydrogen in the bcc metals is complicated by the fact that the hydrogen is situated on tetrahedral sites. There are six of these $(0,1/2, 1/4)$ sites/metal atom which are not on a Bravais lattice and so one has to use the generalisation of the Chudley-Elliott model, as described in Sect. 5.2.4 above [5.25, 26]. Now the quasi-elastic peak consists of six Lorentzians. Data are available for low concentrations of hydrogen in vanadium, niobium and tantalum [5.70, 71]. At room temperature, the data were well described by the nearest-neighbour tetrahedral–tetrahedral jumps for all these metals. At higher temperatures, however, it was necessary to assume that there is an increasing probability of jumps to further-neighbour sites although the exact model was not uniquely determined.

There remains one major problem with this analysis, which arises from the fact that the lattice distortion field around a proton in Nb is known to be cubic whereas the tetrahedral site has tetragonal symmetry. In two recent papers, *Dosch* et al. [5.72, 73] have suggested that this is because the hydrogen is in fact undergoing rapid localised jumps between a number of adjacent tetrahedral sites. It can be shown that such jumps would average out the noncubic parts of the distortion field and would explain the shape of the diffuse scattering when deuterium is measured instead of hydrogen. It would also explain the anomalous shape of the Debye-Waller factor. The evidence for delocalisation onto a number of sites is convincing, but it is not quite clear whether a model based on a permanent delocalisation of the wave function as opposed to rapid localised jumps would not explain these results just as well – although it is certainly attractive to link a localised jump time of 10^{-13} to 10^{-14} s to the 5 meV width of the inelastic scattering peak for excitation to the first excited state of the proton wave function.

Further support for this model comes from recent molecular dynamics calculations by *Wahnstrom* and *Li* [5.74, 75, 76]. These authors used established Nb-Nb and Nb-H potentials [5.77, 78] and were able to reproduce the anomalous Debye-Waller factor and the deviation of the quasi-elastic broadening from the simple Chudley-Elliott prediction [70, 71]. Essentially, the MD simulation shows that, on a short time scale, the hydrogen atom moves rapidly between a set of local T sites, and then, over a longer time scale, it moves off through the lattice on normal random walk diffusion. In the MD calculations, the localised T sites are differentiated from the rest of the lattice by a local relaxation. As the localized T sites show cubic symmetry, this explains the well-known cubic symmetry of the lattice distortion [5.79]. Thus, the picture that emerges from the MD calculations essentially supports the Dosch model. The question as to whether the classical MD description is preferable to a quantum model in which the ground-state wave functions on each of the sites overlap so that the proton tunnels rapidly between them is probably not of real physical significance because the effective M-H potential used in the MD calculations is derived from the local mode vibration energies of the proton and this presumably introduces the essential quantum behaviour and hence ensures that the MD calculations reproduce the experimental data.

5.4.3 Quasi-elastic Scattering from the Niobium-Deuterium System

Further insight into the niobium-hydrogen (deuterium) system has come from coherent QENS measurements at higher deuterium concentrations. *Hempelmann* et al. [5.80] and *Cook* et al. [5.58] have separated the scattering from D in Nb into coherent and incoherent parts – in the former case, by fitting a complete model to the peak shape and, in the latter, by using spin analysis. The objective of the experiments was to test the linear response theory for an interacting lattice gas at finite concentrations as described in Sect. 5.2.8 [5.47] i.e., that, where the integrated coherent scattering function, $S(\mathbf{Q})$, peaks, the broadening is reduced relative to the equivalent incoherent case. This effect is best seen for deuterium/niobium ratios around unity where

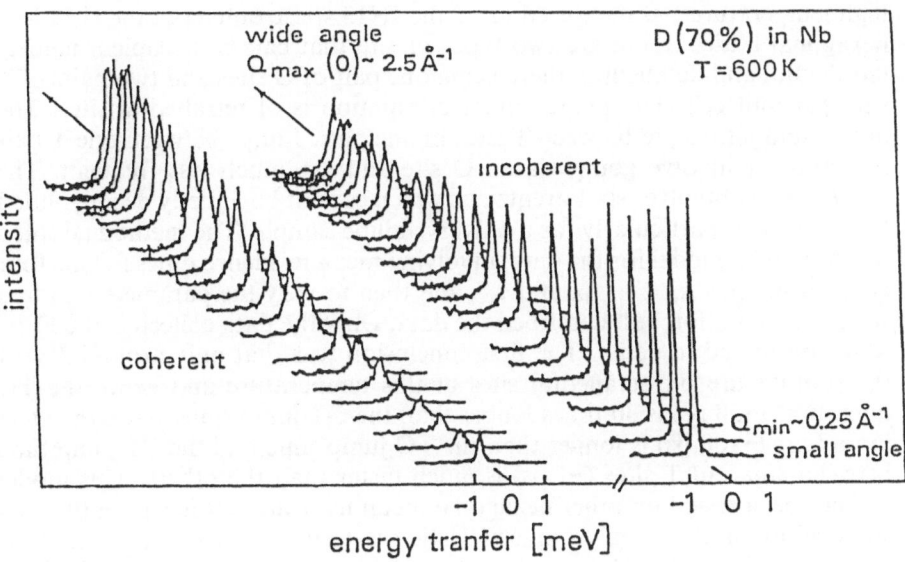

Fig. 5.7. The coherent and incoherent quasi-elastic peaks obtained by spin analysis for a sample of Nb/D using the D7 instrument at the ILL, Grenoble [5.58]

the strong nearest-neighbor repulsions between the deuterium atoms give rise to a superlattice at lower temperatures and a corresponding broader peak in $S(\mathbf{Q})$ above the ordering temperature. The truth of this result can even be seen in the raw data from the spin analysis experiment which is shown in Fig. 5.7. Comparing the coherent with the incoherent, it can be seen that the former increases in intensity and decreases in width for peaks towards the top of the scan where $S(\mathbf{Q})$ peaks. At lower concentrations, around $NbD_{0.5}$, which were also measured in [5.80], the Nb/D system undergoes phase separation below 100 °C due to the long-range attractive interactions between the deuterons caused by lattice distortion. This means that, just above the transition temperature, $\mu(c)$ becomes flatter ($d\mu/dc$ tends to zero) so that D_{chem}, as deduced from the coherent quasi-elastic broadening at low Q, is reduced relative to D_t as predicted by (5.21). *Wipf* et al. [5.54] pointed out that the extent of the reduction was exactly as expected on the basis of Gorsky effect measurements on this system, given that the concentration fluctuations observed by the neutrons give rise to stresses that are local and are not relieved by free boundaries (Sect. 5.2.9).

5.4.4 Quasi-elastic Scattering from Hydrogen Diffusing in a Simple Hexagonal Lattice

As mentioned in Sect. 5.2.4, the theory of incoherent QENS in a simple hexagonal lattice was first given by *Anderson* et al. [5.27] who applied it to the case of hydrogen diffusion in the α-phase of the yttrium-hydrogen system at high temperatures, as measured using the IN13 spectrometer at the ILL. In a hexagonal lattice, there are two types of site that can be occupied, namely tetrahedral and octahedral, there being one pair of O sites and two pairs of T sites per unit cell. The predominant occupation is of tetrahedral sites. The most rapid jumps are between T sites in one pair. Jumps between one T pair and another involve going via an O site and are much less frequent. The model clearly involves six Lorentzians which cannot be directly deconvoluted from the data, particularly for a polycrystalline sample. The method adopted was to build a model for the data, involving mean residence times for the four types of jump, τ_{OO}, τ_{OT}, τ_{TO} and τ_{TT}, and then to vary the parameters to find which combination best described the data. Only the data collected at 600 °C were considered in detail. The first conclusion was that only around 3% of the protons are on octahedral sites at this temperature and hence the TO jump time (~30 ps) is 30 times longer than the OT jump time. Also, the O–O jump time (~100 ps) is longer than the TO jump time and the TT jump time between adjacent T sites (~1 ps) is much faster than the others. This model has not been tested on other hexagonal metal hydrides but has recently been successfully used to analyse quasi-elastic scattering from the hexagonal intermetallic, NiSb [5.81] and more recently from the intermetallic hydride LaNi₅/H [5.82], which is an important system, being the prototype material for metal hydride batteries.

At lower temperatures, it is only the T-T jump that can be observed using QENS. At these temperatures, the situation for the Y-H system is complicated by the fact that most of the protons form pairs either side of a yttrium atom along the c axis. The observed rapid movements are presumably to be connected with the small residue of unpaired protons. The best published data on this behaviour in fact refers to the α phase of hydrogen in scandium [5.83]. The Q-independent broadening was shown to decrease to about 60 μeV at 100 K and then to increase as the temperature was further decreased. This increase in width is to be expected for a two-state tunnelling system. However, it should be noted that there is still some uncertainty about what is happening here because there are NMR data that indicate that the protons are diffusing much more slowly than is indicated by the neutron measurements [5.84]. This, however, introduces the large subject of tunnelling which is dealt with in Chap. 3. A full account of hydrogen in the extended solid solution ranges αY/H, α-Sc/H, etc, have been recently reviewed [5.85].

5.4.5 Diffusion in Hydrides: Titanium Dihydride and Yttrium Dihydride

The remaining example discussed here is the case of hydrides where, as in Sect. 5.4.4 above, there are two hydrogen sub-lattices, where both can, in principle, be occupied. The position now, however, is complicated by the fact that one of the sub-lattices is nearly full so that correlation effects may have to be taken into account. This situation is conveniently illustrated by the work of *Stuhr* et al. [5.86]. These authors describe QENS measurements on

$TiH_x(1.6 \leq x \leq 1.9)$ and $YH_x(1.80 \leq x \leq 2.09)$ at low Q on IN10, thus directly measuring the tracer diffusion coefficient. In the case of TiH_x, the hydrogens are essentially restricted to the tetrahedral sites. The situation is thus the same as for β Pd/H, except that the lattice is simple cubic rather than fcc. The diffusion coefficient is thus given by

$$D_t = \frac{a^2}{4}(1 - c_t)f(c_t)\Gamma_{tt} \ . \tag{5.22}$$

Here, $a/2$ is the distance between the adjacent t sites that make up the simple cubic lattice, $c_t = x/2$ is the concentration on the tetrahedral sites, $f(c_t)$ is the correlation factor (which tends to 0.6532 for $c_t \to 1$) and Γ_{tt} is the jump attempt rate between adjacent tetrahedral sites. The diffusion coefficient thus decreases towards zero as c_t approaches unity. The measured concentration dependence of the diffusion coefficient is shown in Fig. 5.8. The data are well fitted by (5.22).

The case of yttrium hydride is more complex because the activation energy required to move the hydrogen from the tetrahedral to the octahedral sites is rather smaller than for the titanium case and so the octahedral sites play an important role in the diffusive process. The resulting expression for the diffusion can be written [5.87]

$$D_t = \frac{a^2}{4}\left(\frac{4c_0}{x}\Gamma_{oo} + \frac{2c_t(1 - c_t)}{x}\Gamma_{tt} + \frac{2c_t}{x}\Gamma_{to}\right) , \tag{5.23}$$

where Γ_{oo} is the jump attempt rate between a pair of O sites and c_o is the concentration on O sites. Here, the ratio Γ_{to}/Γ_{ot} is given by the equation

$$\frac{\Gamma_{to}}{\Gamma_{ot}} = \frac{c_0(1 - c_t)}{c_t} = \exp\left(\frac{-\Delta E}{k_b T}\right) , \tag{5.24}$$

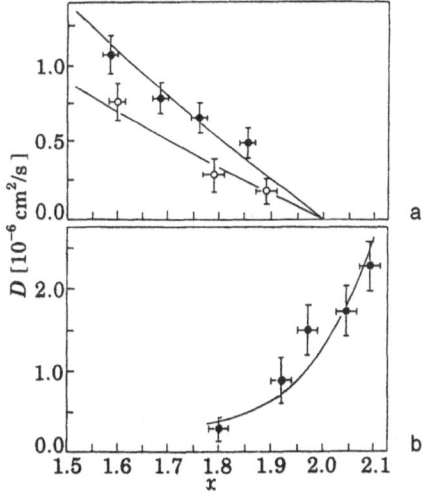

Fig. 5.8a,b. The tracer diffusion coefficient, D_t, plotted against the hydrogen concentration x. (a) Data for TiH_x (•) and TiD_x (○) obtained at 550 °C and 540 °C, respectively. The lines are fits to (5.22). (b) Data for YH_x at 450 °C. The line shows a fit according to (5.23)

where ΔE is the energy difference between the tetrahedral and the octahedral sites. When these two equations are combined, the concentration dependence shown in Fig. 5.8 is obtained. As in the case of the simple hexagonal lattice above, there are not sufficient data to determine all the parameters exactly but *Völkl* et al. [5.87] were able to conclude that the activation energy was in the range 0.35–0.45 eV. It should be noted that (5.23) does not include correlation effects which would contribute to the second and third terms in the bracket. Also, data at higher Q would, in principle, provide a more detailed test of the model.

5.5 Theory of Inelastic Neutron Scattering

When a neutron interacts with a hydrogen atom in a solid, it can transfer the scattering system into a new quantum state with a probability as given by (5.4), the neutron suffering a corresponding energy change. In general, we will be considering cases where the neutron loses energy and the system gains energy. The measurements therefore yield values for the energy differences between quantum levels of the system and also the corresponding transition probabilities. Given that we are interested in hydrogen, the simplest measurements will be incoherent, i.e., the initial and final wave functions in (5.4) will refer only to the same nucleus.

We therefore need to describe the hydrogen atoms by appropriate wave functions. There are two possible starting points. We may either regard the protons as being isolated in the metal matrix – in which case their excited states involve localised defect modes – or we can start with a periodic hydride lattice and then express the energy interchanges in terms of the creation or annihilation of phonons. In the latter case, we would be interested in using coherent scattering from deuterium in single crystal samples to measure phonon dispersion curves and would generally be concerned with the optical phonons where the deuterium lattice is vibrating in antiphase to the metal lattice. However, one of the main interests in recent work in this area has been in the anharmonic shape of the hydrogen-metal potential and this is certainly most easily investigated starting from the localised picture; this is therefore the point from which we will start the present discussion.

5.5.1 Incoherent Inelastic Neutron Scattering from an Atom in a Harmonic Potential: The Einstein Oscillator

The simplest analytic model for an isolated proton in a metal assumes that it is situated in a potential well centred at one or other of a limited set of interstitial sites; either octahedral (O) or tetrahedral (T) sites in fcc systems or tetragonally distorted O or T sites in bcc or hexagonal lattices. If the potential is expanded using Taylor's theorem, the first term is a constant and the second term must be zero at the centre of the site. The third term is the harmonic term and can be separated into independent terms in the x, y and z

directions. If the higher terms in the expansion can be neglected, the wave function of the proton can be represented as the product of three independent 1-dimensional Hermite polynomials where the proton energy levels are equally spaced in energy $[E_n = (n + 1/2)\hbar\omega]$. Fermi's golden rule (5.4) can therefore be used to calculate the cross-sections explicitly. This is straightforward and is given, for instance, in *Turchin* [5.13] or *Lovesay* [5.14]. The resulting expression can be written as

$$S(\mathbf{Q}, \omega) = \exp[-2W(\mathbf{Q})] \exp(\hbar\omega/2kT)$$
$$\times \sum_{n,m,l=-\infty}^{\infty} I_l(s_1)I_m(s_2)I_n(s_3)\delta(\hbar\omega - l\hbar\omega_1 - m\hbar\omega_2 - n\hbar\omega_3) ,$$

$$(5.25)$$

where l, m and n refer to the number of quanta transferred in each of the three Cartesian directions, ω_1, ω_2 and ω_3 are the corresponding harmonic frequencies, and the $I(s)$ are modified Bessel functions where

$$s_i = \frac{\hbar Q^2 i}{2m_H \omega_i} \operatorname{csch}(\hbar\omega_i/2kT) , \tag{5.26}$$

where m_H is the mass of the hydrogen atom. This equation implies that if the wave vector transfer, $\mathbf{Q}$, has a component Q_i in a particular Cartesian direction, i, then there is a finite probability that the proton will be transferred to any energy level of the corresponding wave function that is compatible with energy conservation. Moreover, if the cross-section is measured as a function of Q, it yields the Fourier transform of the product of the initial and final wave functions times Q^2 for one quantum transfer, Q^4 for two quantum transfers, etc. Thus, the technique can yield both the energy levels and the shape of the wave functions. This is also true in more complicated situations and accounts for the importance of being able to measure the Q-dependence of the scattering, particularly if single-crystal samples are available.

5.5.2 Perturbation Theory Analysis of Anharmonic and Anisotropic Effects in Inelastic Neutron Scattering from Hydrogen in Metals

Because the proton wave function is not localised, it samples the potential in a volume close to the centre of the interstitial site. Hydrogen, being the lightest nucleus, samples the potential further from the centre of the site than would any other nucleus (apart from a muon). Its wave functions may therefore be affected by higher terms in the Taylor's expansion of the potential well. The allowed set of such terms depends on the site symmetry. For octahedral sites in a cubic system, the potential can be written – for terms up to the quartic – as

$$(x, y, z) = c_2(x^2 + y^2 + z^2) + c_4(x^4 + y^4 + z^4) + c_{22}(x^2y^2 + y^2z^2 + z^2x^2) .$$

$$(5.27)$$

Now, we can obtain a reasonable estimate of the wave functions associated with this potential using perturbation theory. The solution has been given for a cubic potential by *Eckert* et al. [5.88] who obtained a set of perturbed energy levels and the corresponding wave functions where

$$E_{lmn} = \{(\hbar\omega_0/2) + \beta(j^2 + j + 1/2) + \gamma[(2m + 1)(2n + 1) \\ + (2n + 1)(2l + 1) + (2l + 1)(2m + 1)]\}$$

with $j = l + m + n$. This equation yields a set of energy transfers, ε_{hkl}, from the ground state, E_{000}, given by

$$\varepsilon_{100} = \varepsilon_{010} = \varepsilon_{001} = \hbar\omega_0 + 2\beta + 4\gamma \quad \text{(fundamental terms)} , \qquad (5.28)$$

$$\varepsilon_{200} = \varepsilon_{020} = \varepsilon_{002} = 2\hbar\omega_0 + 6\beta + 8\gamma \quad \text{(first harmonic)} ,$$

$$\varepsilon_{110} = \varepsilon_{011} = \varepsilon_{101} = 2\hbar\omega_0 + 4\beta + 12\gamma, \quad \text{(combination vibration)}$$

where these parameters are related to the parameters in the potential expansion by the expressions

$$\omega_0 = \sqrt{\frac{2\,c_2}{m}}, \quad \beta = \frac{3\hbar^2 c_4}{4\,mc_2}, \quad \gamma = \frac{\hbar^2 c_{22}}{8\,c_2 m} . \qquad (5.29)$$

Here, the notation ε_{110} implies that two of the independent Cartesian oscillators have been raised to the next quantum number. For this to happen, $\mathbf{Q}$ has to have components in both Cartesian directions. We now see that the energy levels are no longer equally spaced. For a U-shaped potential, the spacing increases and for a trumpet-shaped potential, the energy levels get closer together. This model works well for the Pd/H system. Thus, in α-Pd/H, energy levels were observed at 69 meV, 138 meV, and 156 meV, yielding the following values for the parameters,: $\hbar\omega_0 = 50$ meV, $\beta = 9.5$ meV and $\gamma = 0$ [5.89]. It should be noted that these anharmonic parameters are very large and render the use of perturbation analysis rather doubtful.

It is a fortunate coincidence that, just as the development of inelastic neutron scattering techniques is beginning to yield accurate information about the shapes and energies of the proton wave-functions, first principles electron band calculations have advanced to the stage that they can provide accurate predictions of the potential well shape and hence these wave-functions. The calculations can be performed for either isolated protons in a metal lattice or for stoichiometric hydrides. The most common approach used to date, the frozen phonon method, is applicable to the case of a hydride phase. This method was pioneered by *Ho* et al. [5.90, 91] for the case of hydrogen in Nb–H. Their work has recently been extended to the case of Pd–H by *Elsässer* et al. [5.92–94]. The starting point is the Born-Oppenheimer assumption that the electron wave functions will always relax to their correct form at any instantaneous position of the ions. The metal atoms are also assumed to be stationary while the proton is moved across the site. The effective potential energy well is therefore obtained by solving the electron orbital problem and calculating the total energy of the system for every position of the proton along the symmetry axes of the potential well (the "frozen phonon" approach). The calculation is performed assuming a super-cell in a periodic lattice and the electronic orbitals are calculated using the local density-functional approximation. An important aspect of the calculations is that the metal lattice atoms are allowed to relax away from the interstitial atom subject to the periodicity of the super-cell assumed in the calculation. The energy levels in the octahedral site were first obtained using a perturbation calculation as described above but this yielded a vibration frequency that was twice the required value and the discrepancy was attributed to the extent of the anharmonicity of this system which is large enough to make perturbation theory inapplicable. The wavefunctions were then recalculated using a simple Fourier technique which, in contrast to perturbation theory, yielded excellent values for the energy levels. This indicates that the conclusions given above, based on perturbation theory, may not be very reliable. *Elsässer* et al. showed that the sixfold-degenerate ε_{200} level is, in fact, split into three levels, a singly-degenerate level, a doubly-degenerate one, and a triply-degenerate one and they assign the peaks at 137 and 156 meV to the first two of these levels. The third is associated with a weak structure at 115 meV. These calculations were done for various stoichiometries. It is notable that the authors were able to reproduce the lattice expansion caused by the increasing hydrogen concentration and to observe the corresponding decrease in the transition energy. Here, one would expect that, at finite hydrogen concentrations, there would be an isotope effect on the lattice parameter because, even for the ground state, the lighter isotope should expand the lattice more. The authors do not indicate whether they expect different lattice parameters for the different hydrogen isotopes, i.e., a genuine isotope effect in the ground state lattice parameter – not just anharmonicity in the potential well – but such information must now be available from this approach. These calculations undoubtedly open up a new era in the modelling of metal-hydrogen systems.

The above approach, of representing the wave-functions in terms of Cartesian coordinates, is satisfactory where the site symmetry involved is at least orthorhombic. For hexagonal systems, however, a different approach has to be adopted, as described by *Bennington* et al. [5.95]. In order to accommodate 120° angles, the wave function has to be expressed in terms of spherical harmonics which initially have a degeneracy of 1,3,6,10, etc. This degeneracy must clearly be the same as in the Cartesian case for spherical symmetry. Here, the Cartesian wave function (0,0,0) has a degeneracy of 1, the functions (1,0,0) have a degeneracy of 3, (1,1,0) and (2,0,0) have a combined degeneracy of six while (1,1,1), (2,1,0) and (3,0,0) have a combined degeneracy of 10, as expected. The appropriate solution to Schrödinger's equation in this case is:

$$\Psi_{lmn}(r, \theta, \varphi) = N_{lmn} r^l \exp\left(-\tfrac{1}{2}\beta r^2\right) L_{(n-l-D/2)}^{l-\tfrac{1}{2}} Y_{lm}(\theta, \varphi) \ , \tag{5.30}$$

where L is an associated Laguerre polynomial, Y_{lm} is a spherical harmonic and where the normalisation constant, N_{lmn}, is given by

$$N_{lmn} = \frac{2 \times 2^{(n+l+1)/2}}{1 \times 3 \times 5 \times \ldots (n+l)} \left(\beta^{2(1+3)}/\pi \right)^{1/4} .$$

These wave-functions can now be used in perturbation calculations similar to those used with Cartesian coordinates but where, here, the perturbation is written

$$\Delta V = \sum_{lm} a_{lm} \, r^l Y_{lm}(\theta, \phi) .$$

This expression is identical to the perturbation term used to calculate the splitting of electronic energy levels in a crystal field. Indeed, it can be used to represent any perturbing field with a minimum number of parameters. The resulting parameters were used to fit the inelastic spectrum for hydrogen in the yttrium-hydrogen alpha phase. In this case, however, there is a further complication, in that the protons tend to form pairs either side of Y atoms along the c axis [5.96]. This implies that there will be some coupling between the protons and that there will therefore be coupled wave functions, either symmetric or antisymmetric, where we can take the perturbing potential to be:

$$V_{(H-H)} = (1/2)K_Z[r_1 \cos(\theta_1) - r_2 \cos(\theta_2)] , \qquad (5.31)$$

where $r_1 \cos \theta_1$ and $r_2 \cos \theta_2$ are the z coordinates of the paired protons. The measured spectrum is shown in Fig. 5.9a and the fitted energy level diagram is shown in Fig. 5.9b.

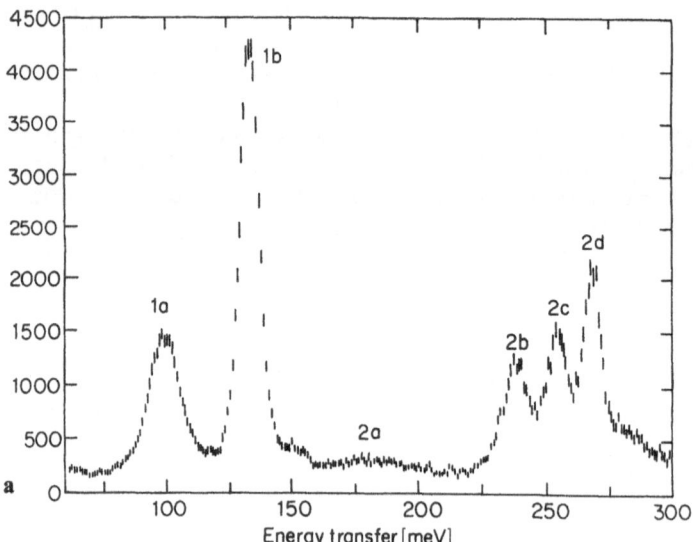

Fig. 5.9. (a) The IINS spectrum measured for α-Y/H at 20 K using the HET spectrometer on ISIS at the Rutherford Appleton Laboratory using an incident neutron energy of 330 meV. **(b)** The energy-level diagram for α-Y/H obtained by fitting the perturbed symmetry-adapted wave functions to the data in **(a)** [5.95]

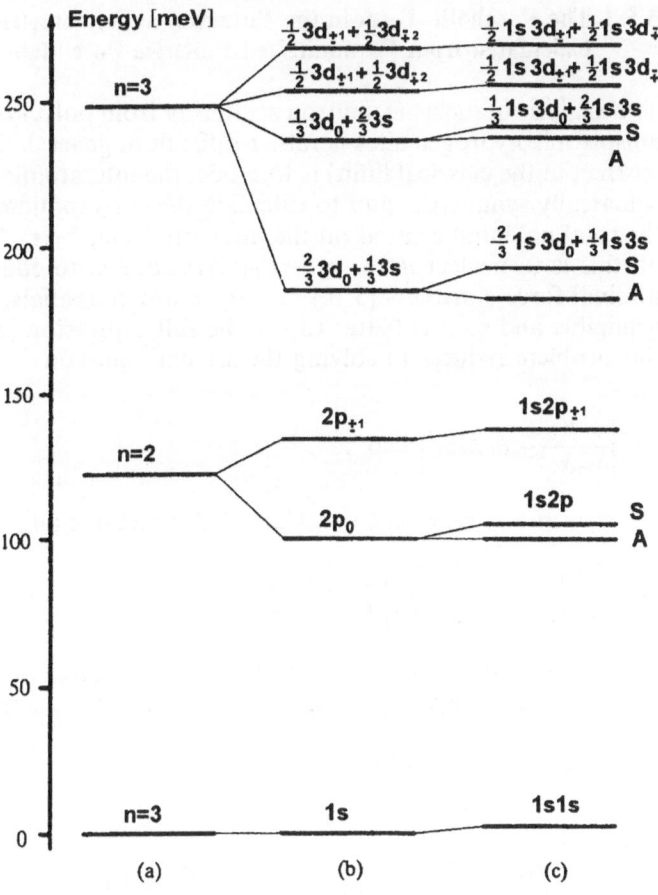

Fig. 5.9b

An important consequence of using perturbed wave-functions is that the Q-dependence of the I-phonon inelastic scattering is no longer of the form $Q^2 \exp(-Q^2 \langle u^2 \rangle)$. Thus, if we can make Q-dependent inelastic cross-section measurements on single crystal samples, it is possible to get very direct confirmation of the shape of the perturbed wave function [5.95]. In fact, a complete measurement of the Q dependence of the intensity of an inelastic level gives a direct measure of the actual shape of the wave functions and hence of the potential energy surface out as far as it is probed by that particular wave function. This approach is much more satisfactory than trying to parameterise the energy levels against the predictions of perturbation theory. The method has, so far, only been applied to hydrogen bonded protons but will undoubtedly add to our knowledge of the hydrogen-metal interaction in future [5.97].

5.5.3 The Parabolic Term in the Potential at Asymmetric Sites Calculated from Paramatrized Pairwise Potentials

The analysis of inelastic neutron scattering from polycrystalline samples with anisotropic hydrogen sites is rather difficult in general. One useful approach (correct in the classical limit) is to model the interatomic potentials, assumed spherically symmetric, and to calculate the corresponding harmonic term in the total potential centred on the interstitial site, $V_T(\mathbf{r})$. The simplest way to do this is to neglect terms in $\partial^2 V_T(\mathbf{r})/\partial r \partial \theta$, i.e., to consider only the long-itudinal force constants [5.10] but, for many potentials, these terms are not negligible and so it is better to use the full expression [5.98, 99]. Essentially this problem reduces to solving the secular equation

$$\left| \frac{\delta^2 V_T}{\delta x_1 \delta x_j} - \delta_{ij} M \omega^2 \right| = 0 \ , \tag{5.32}$$

where x_i and x_j represent two Cartesian directions and where

$$\frac{\partial^2 V_T}{\partial x_i \partial x_j} = \sum_l \left(\frac{\partial V_l(\mathbf{r})}{\partial x_i \partial x_j} \right)_{r=R_l} = \sum_l V_l^{ij} \ , \tag{5.33}$$

where $V_l(\mathbf{r})$ is the potential due to the atom l situated at the distance R_l from the centre of the hydrogen site. Now, in general, V_l^{ij} can be expressed as

$$V_l^{ij} = \Phi_l V_l'' \delta_{ij} + l_l^i l_l^j V_l'' (1 - \Phi_l) \ , \tag{5.34}$$

where

$$V_l'' = \left(\frac{\partial^2 V_l(\mathbf{r})}{\partial r^2} \right)_{R_l} \quad \text{and} \quad \Phi_l = \frac{1}{R_l V_l''} \left\{ \frac{\partial V_l(\mathbf{r})}{\partial r} \right\}$$

and l_l^i are the direction cosines of the lth atom from the mid-point of the potential minimum. This approach has been successfully used to model the hydrogen modes in the Ta_2V_2 site in $TaV_2H_{0.31}$ using Born-Mayer potentials [5.98]. The measured spectrum for this sample fitted with gaussians is shown in Fig. 5.10a and the corresponding contour plot of the potential well calculated from standard B-M parameters is shown in Fig. 5.10b. The predicted energy levels for the potential well in Fig. 5.10b lie close to the measured values, thus giving considerable confidence in the assignment of the levels. Note that the second and fifth gaussians correspond to acoustic phonon side bands.

The use of simple Born-Mayer potentials would suggest that, where the harmonic approximation is good, it should be possible, for a given crystal structure, to relate the hydrogen frequency to the lattice parameter. This certainly works for the tetrahedral sites in the fcc lattice, i.e., for the hydrides with a fluorite structure, where the frequency decreases smoothly with increasing lattice parameter although, as can be seen in Fig 5.11, it is not

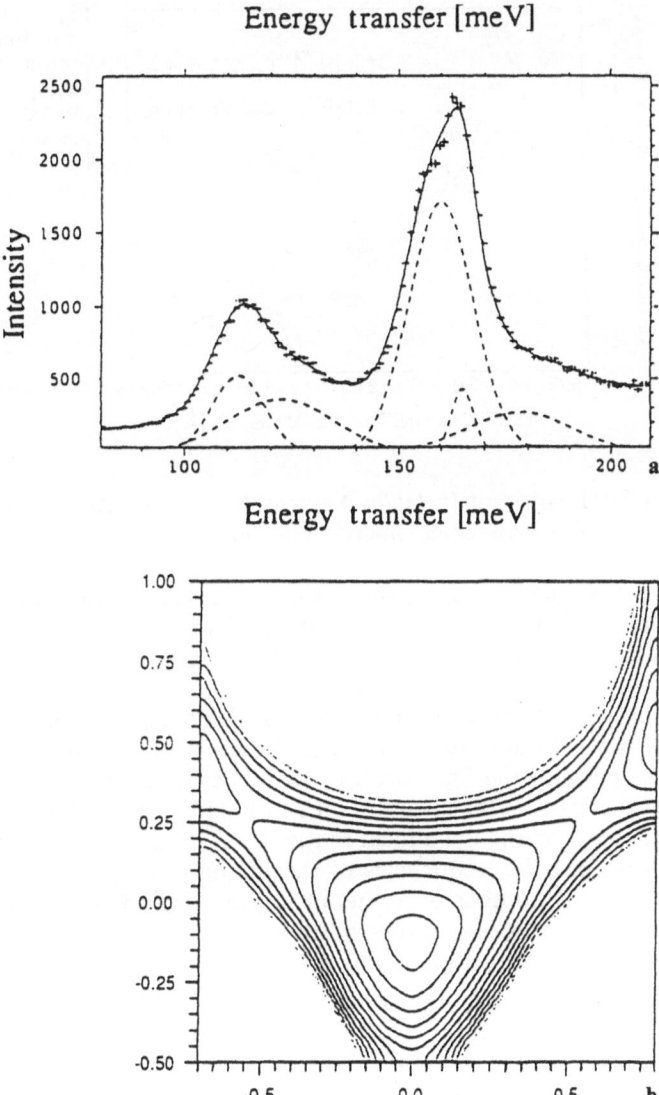

Fig. 5.10a,b. IINS measurements of the local modes of hydrogen in the Ta_2V_2 site in $TaV_2H_{0.31}$. **(a)** The measured spectrum fitted with five gaussians. **(b)** the corresponding contour plot of the potential well obtained on the assumption of Born-Mayer potentials for the hydrogen-metal interactions [5.98]

entirely clear whether an $r^{-3/2}$[5.100] or an r^{-1} [5.101] dependence is most appropriate. For octahedral sites, however, where anharmonicity originating from the scattering of conduction electrons is more important, this kind of correlation does not work. It also fails for pseudo-tetrahedral sites in hexagonal close-packed lattices [5.102].

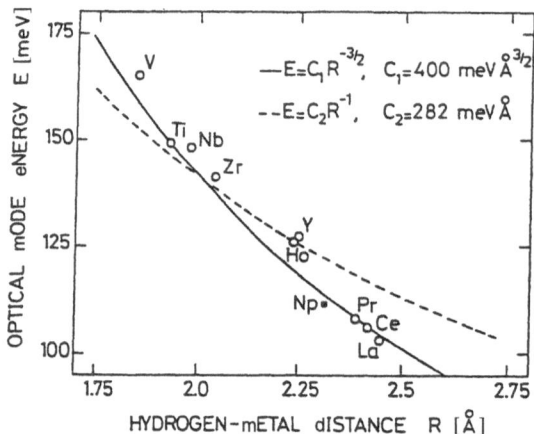

Fig. 5.11. A graph of the measured fundamental frequencies of protons in fcc metal dihydrides having the fluorite structure plotted as a function of lattice parameter. The fitted curves are for $r^{-3/2}$ [5.100] and r^{-1} [5.101]

5.5.4 Incoherent Inelastic Neutron Scattering from Stoichiometric and Non-stoichiometric Hydrides

The above approach, involving the dynamics of protons in individual potential wells, is a valid one at low hydrogen concentrations. For high hydrogen concentrations, however, it is possible that hydrogen–hydrogen interactions are important. In this case, we should view the scattering in terms of phonon dispersion curves where the hydrogen vibration corresponds to one of the optical modes where most of the vibration amplitude is associated with the light proton. In this case, we have to write the atomic displacements in the lattice in terms of plane waves of wave vector $\mathbf{q}$ and we can then derive the dynamical matrix [5.13, 15]. The j eigenvalues of this matrix, $\omega_j(\mathbf{q})$, are the frequencies of the phonon of wave vector, $\mathbf{q}$, where j is three times the number of atoms in the unit cell. The associated eigenvector has components $e_\rho^j(\mathbf{q})$ corresponding to the vector amplitude of vibration of the ρth atom in the unit cell in the jth mode.

From these components, we can evaluate the IINS scattering for neutron energy loss from the hydrogen atoms in a polycrystalline lattice. At small Q, we can ignore the multiphonon terms in the scattering and so the cross-section reduces to the equation

$$\frac{\mathrm{d}^2\sigma}{\mathrm{d}\Omega\mathrm{d}\omega} = \frac{k'}{4\pi k_0}[\langle n(\omega)\rangle + 1]\frac{Q^2}{2\omega}\,\Theta(\omega)\ ,$$

where k' and k_0 are the incident and final wave-vectors of the neutron, respectively and $\Theta(\omega)$ is the amplitude-weighted density of states given by

$$\Theta(\omega) = f(\omega)\sum_\rho \frac{\sigma_\rho^{\mathrm{inc}}}{M_\rho}\exp\left(-Q^2\langle u_\rho^2\rangle\right)\sum_j\left|e_\rho^j(\omega)\right|^2\ . \tag{5.35}$$

Here $\langle n\rangle$ is the phonon occupation number for modes of frequency ω, $f(\omega)$ is the normalised number of phonon modes at frequency ω summed over all phonon modes j and wave vectors $\mathbf{q}$, $\mathbf{e}_\rho^j(\omega)$ is the amplitude of the ρth hydrogen atom in the jth mode averaged over all phonons of frequency, ω, and $\langle u_\rho^2 \rangle$ is the mean square amplitude of displacement of the ρth hydrogen atom averaged over all modes.

Thus, by means of a programme such as PHONON [5.103], the inelastic scattering can be calculated explicitly from a knowledge of the interatomic force constants. It can be shown that if the mass of the metal atom is large enough and all force constants can be neglected except for the one linking the hydrogen with the nearest metal atom, the dispersion curve will be completely flat and the solution will be identical with the simple harmonic oscillator case. In fact, as will be seen in Sect. 5.7 below, there are only a few cases known where the effects of dispersion are clearly present.

5.5.5 Coherent Inelastic Scattering from Metal Deuterides

Turning now to the case of coherent inelastic scattering from deuterium, it should be noted that, in the rare cases where we can obtain single-crystal samples, we can measure the dispersion curves directly. Deuterium is necessary because the coherent cross-section of hydrogen is completely swamped by its incoherent scattering. To get coherent inelastic scattering from a single crystal, we also have to satisfy the momentum conservation condition which introduces the term $\delta(\mathbf{Q}\text{-}q\text{-}\tau)$ into the scattering cross-section, (5.36), where τ is a reciprocal lattice vector of the crystal. Thus we are now directly measuring the frequency of the phonon in the jth mode at phonon wave vector $\mathbf{q}$, i.e., $\omega^j(\mathbf{q})$ (for either acoustic or optical phonons) for a number of fixed values of $\mathbf{q}$ lying along the basic symmetry directions of the crystal. These dispersion curves can be fitted using force constants between neighbouring atoms. It was noted above that for a sufficiently large metal mass, the optical dispersion curve will be completely flat if only the nearest neighbour D-M force constants are involved. In this case, the local mode and phonon pictures become identical as the group velocity of the phonon is zero.

5.6 The Deep Inelastic Neutron Scattering Technique

We close this section with a brief introduction to the "deep inelastic neutron scattering" technique. This technique has been recently developed to a useful level, particularly at the Rutherford Appleton Laboratory, to exploit the high fluxes of epithermal neutrons (energies greater than 1 eV) available from the ISIS pulsed neutron source. The technique involves measuring the recoil energy transferred to a target atom as a function of the angle of scatter of the neutron. Usually the incident energy is determined by time-of-flight and the scattering energy by a resonance absorber using the filter-difference technique, i.e., the scattered neutrons are detected with and without the foil containing the resonant absorber being present in the scattered beam. The usual resonance foils are gold, which has an absorption resonance at 4.9 eV and uranium which has four potential resonances but where so far only the lowest energy one at 6.67 eV has been used. The resulting instrument, the EVS spectrometer, which is still undergoing development, is described in the ISIS Instrument Handbook [5.62].

When inelastic measurements are made in this way, the scattering function shows a peak corresponding to the recoil energy of the target atom, i.e., $h\omega = h^2Q^2/2M$ with a width that is determined by the momentum distribution of the target nucleus before the collision and possibly slightly modified by "final state effects" due to any influence of the lattice on the energy of the recoiling atom. In this high incident energy limit (the impulse approximation), the scattering function can be written in the classical limit as:

$$S_{IA}(Q, \omega) = \int_{-\infty}^{\infty} n(p)\delta(\omega - h(p + Q)^2/2M + hp^2/2M)\mathrm{d}p \ ,$$

where $n(p)$ is the momentum distribution of the target atom before the collision, and M is its mass. This tell us that, if the sample is near absolute zero of temperature and if the atom is not bound in a potential well and therefore has no zero point energy, the scattering function will be a delta function at $\omega = hQ^2/2M$, i.e., the energy loss varies with the angle of scatter and with the target atom mass – the smaller the mass, the larger the energy transfer at the peak. It will be realised that the scattered peaks from H and D are well separated from each other and from the peaks from any other elements present. In these circumstances, the scattered peak width defines the resolution function of the apparatus.

Now, if the target atom has a finite zero point energy, it will also have a finite momentum distribution which will increase in width as the temperature is increased. For a simple harmonic oscillator of frequency ω_0 in its ground state, the delta function becomes a Gaussian. It is convenient to express this Gaussian as a function of the parameter y where

$$y = (M/Q)(\omega - hQ^2/2M)$$

i.e. with the dimensions of momentum. On this scale, the r.m.s. broadening of the peak produced by a zero point energy of $h\omega_0/2$ would be $(Mh\omega_0/2)^{0.5}$. The peak intensity is proportional to the atomic concentration and the free atom cross section.

A recent full exposition of the method of the data reduction techniques has been given in the context of measurements of hydrogen in amorphous carbon [5.104]. The method has also been very fully tested for the case of zirconium hydride [5.105] where the broadening of the peak was found to be well predicted on a harmonic model of the phonon density of states [5.106]. The important question, however, is to determine whether this technique offers any advantage over conventional density of states measurements in the case of hydrogen-metal systems. There seem to be two areas in which the technique is producing interesting results. The first is in the determination of anharmonicity in the isotope dependence of the ground state momentum of H/D without the complications of coherence or indeed of the periodicity of the lattice. Thus, it has been shown that there is a clear anharmonic effect for H/D in palladium [5.107]. The second promising application is in the absolute determination of the amount of hydrogen trapped at lattice defects. At present, using EVS, this can be determined to an accuracy of about 0.05% for hydrogen and 0.2% for deuterium. Analysis to this accuracy would be very difficult by any other means, particularly where high temperatures and good vaccum would be needed to extract the hydrogen. This has, for instance, been recently used to determine the amount of hydrogen trapped in LaNi$_5$ after cycling [5.108].

5.7 Experimental Methods of Measuring Inelastic Neutron Scattering

5.7.1 General Features of Inelastic Neutron Scattering Spectrometry

Neutron inelastic scattering from hydrogen in metals usually involves large energy transfers. Moreover, phonon life-times are largest at low temperature and hence the inelastic scattering is best measured in conditions where $h\omega \gg kT$ and therefore measurements are usually made in neutron energy loss. There are essentially two types of instruments for measuring incoherent neutron scattering: indirect and direct geometry.

Indirect Geometry: Using a fixed low final energy and varying the incident energy (indirect geometry), we can measure along a single locus in Q, ω space where, although Q is small for small energy transfers, it becomes quite large at higher energy transfers. For a rigid lattice and a mass 1 scatterer, this locus, to a good approximation, coincides with the locus of points that, for any vibration mode, give the maximum one-phonon intensity. However, for increasing Q this one-phonon scattering will be increasingly contaminated with multiphonon contributions and these may be difficult to correct for.

Direct Geometry: The alternative approach is to monochromate the incoming beam and to measure the energy spectrum of neutrons observed at a range of scattering angles. For a given value of energy transfer, each angle of scatter will give a different value of Q. For small fractional energy transfers and small angles of scatter, this technique reaches much smaller Q values than does the indirect geometry case. The multiphonon scattering is therefore much less important. Moreover, because the one-phonon term is proportional to Q^2 and the higher phonon terms are proportional to Q^4, Q^6, etc., we can improve the separation of the one phonon term by fitting the Q dependence to the range of Q values measured at a given value of ω. This is essentially the extrapolation technique introduced by *Egelstaff* and *Schofield* [5.109] and, given the Q ranges available in instruments such as MARI described in Sect. 5.6.3 below, it is a very valuable way of improving the determination of $f(\omega)$ from (5.36). This approach is appropriate for harmonic systems but, as mentioned in the previous section, the potential well occupied by the proton will not, in general, be perfectly harmonic, particularly for the excited states of the proton wave function. The actual shape of the Q dependence of the cross-section for fixed ω will thus be of interest because it is determined by the Fourier transform of the wave function in the excited state. Direct geometry instruments provided with a large array of detectors that can measure the Q dependence of the intensity of a particular energy level are therefore of great interest.

Because of their relative novelty and of their major contribution to this kind of work, we will choose two ISIS instruments, TFXA and MARI, as examples of these two techniques [5.62].

5.7.2 TFXA, an Indirect Geometry Time-Focusing Spectrometer on a Pulsed Neutron Source

TFXA, the time-focusing crystal analyser spectrometer, is mounted on the ISIS pulsed neutron source at the Rutherford Appleton Laboratory. Its layout is illustrated in Fig. 5.12. The neutron pulse from the moderator is directly incident on a plane sample mounted normal to the beam. The

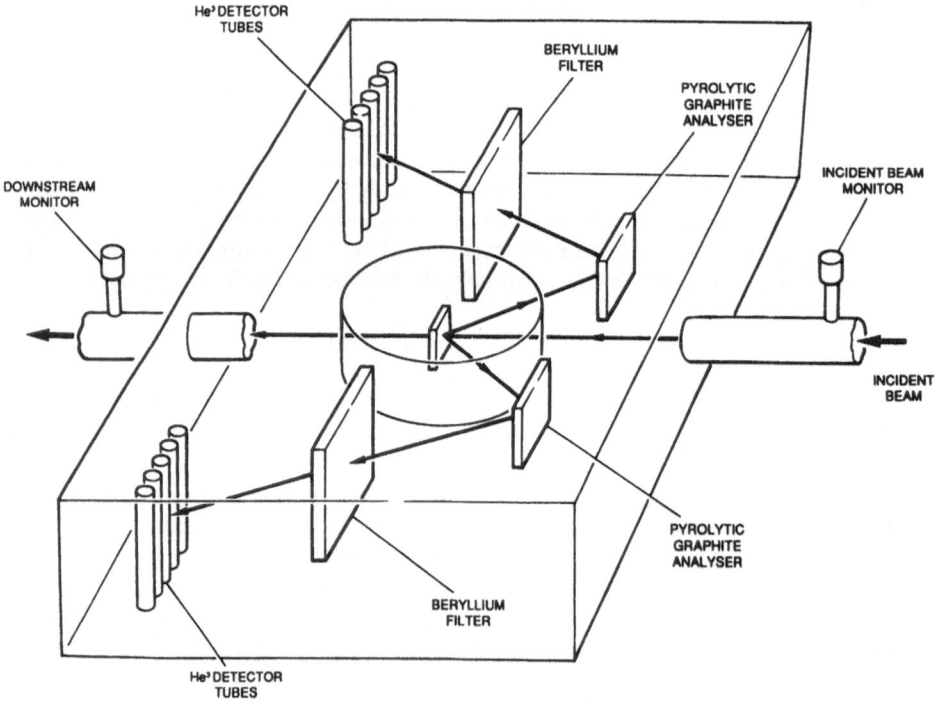

THE TFXA SPECTROMETER

Fig. 5.12. The layout of the TFXA spectrometer at the ISIS pulsed neutron source

scattered neutrons fall on a large flat graphite monochromating crystal, also mounted normal to the beam. Bragg-reflected neutrons are then passed through a beryllium filter to eliminate the higher orders of reflection ($E_f > 5$ meV) and are then detected in He^3 counters, also arranged in a plane normal to the beam. The incident neutron energy is calculated from the measured time-of-flight from moderator to detector less the calculated time-of-flight from sample to detector. It is therefore important to ensure that all neutrons take the same time from sample to detector. It is easily shown that, for this geometry, neutrons with a range of final energies, reaching the detector after Bragg reflection in the analyser crystal, will all have the same flight time – hence "time focusing". It is also possible to provide energy focusing by choosing to put the cylindrical detectors with their axes normal to the scattering plane. Each detector will therefore measure neutrons with a narrow range of Bragg angles and hence final energies. However, by analysing each detector separately, each spectrum can be converted to an energy transfer scale and then summed so that the overall intensity can be maximised without loss of resolution. The two focusing techniques employed improve the resolution at high energy and low energy transfers, respectively, yielding an overall resolution of around 2 % which is remarkably independent of energy transfer. This spectrometer is thus ideal for measuring a full spectral range in one experiment with very high resolution. For metal hydride systems – having relatively rigid lattices – excitations up to 1 eV can be observed.

5.7.3 MARI, a Direct Geometry Chopper Spectrometer on a Pulsed Neutron Source

In a direct geometry time-of-flight spectrometer on a reactor source, the monochromatic incident pulse is selected by means of two choppers a fixed distance, L, apart, the second one opening at a fixed delay D after the first. The system therefore selects a reciprocal velocity of D/L [µs/m]. This is the method used on IN5 at Grenoble, as described in Sect. 5.3.2 above. Alternatively, the monochromation can be performed with a crystal analyser with pulsing by a chopper, as on IN4 at the ILL, Grenoble [5.61]. However, the largest range of incident energies is available from a pulsed source, where, moreover, the neutrons emerge from the moderator in a short time interval, Δt. A monochromatic pulse can therefore be obtained by passing the neutrons through a chopper which is at a fixed distance, L, from the moderator and which is phased so as to open at a fixed time, D, after the neutrons of that energy emerge from the moderator. In the MARI spectrometer (Multi-Angled Rotor Instrument) at ISIS, the slits in the chopper are constructed from a package of aluminium and boron-carbide-fibre-reinforced aluminium sheets which are mounted in an aluminium alloy chopper body. A series of slit packages with different curvatures is available to match different incident neutron energy ranges from 1 eV down to 25 meV. The important features of the spectrometer are that it provides a full range of scattering angles in the

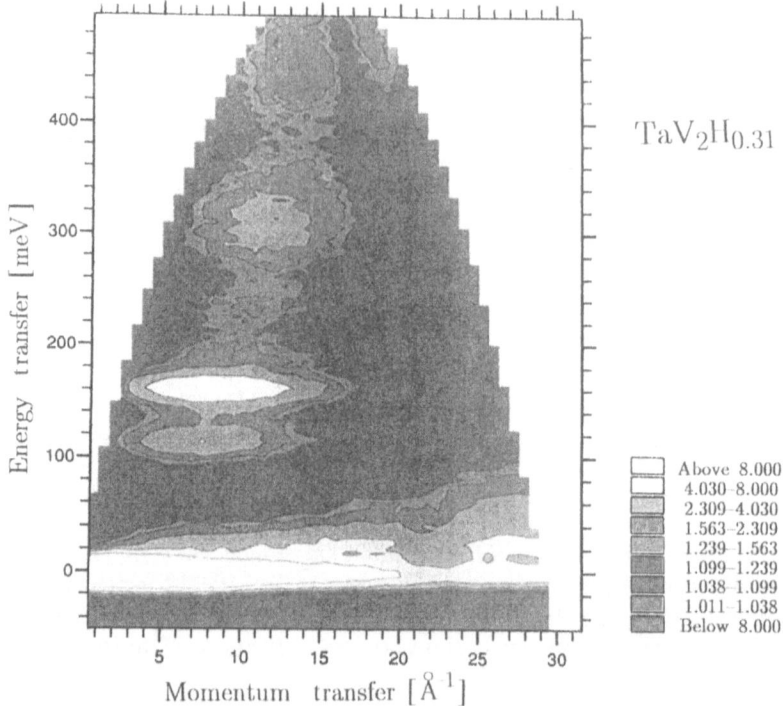

Fig. 5.13. A contour plot of the scattering function $S(Q,\omega)$ obtained from MARI for $TaV_2H_{0.31}$ [5.98]. The Q dependence of each peak is clearly visible and is consistent with the potential well being harmonic as far as it is sampled by the proton in the first excited state

scattering plane and that the corresponding detectors are all at the same distance from the sample. It is thus possible to measure the scattering function in a complete region of Q, ω space with a fixed resolution function. A typical contour plot of the scattering from $TaV_2H_{0.31}$ [5.98] for an incident energy of 525 meV is depicted in Fig 5.13. This shows immediately how the fundamental peaks have maximum intensities at Q values that lie along a locus in (Q, ω) space which represents the values of Q that correspond to the free recoil of an, initially stationary, mass 1 atom that has received energy, $\hbar\omega$.

5.8 Inelastic Neutron Scattering Measurements on Metal-Hydrogen Systems

5.8.1 fcc Hydrides with Hydrogen in the Octahedral Sites

In all the common metal lattices, i.e., the fcc, hcp and bcc metals, a good indication of whether the octahedral or the tetrahedral site is occupied, due to *Hauck* [5.110], is the electropositivity of the metal relative to hydrogen. If the hydrogen is the more electropositive, it will tend to lose its electron to the conduction band of the metal and hence will seek the site with the highest electron density, which, in these lattices, turns out to be the octahedral site. If, however, it is more electronegative, it goes to the tetrahedral site, where the electron density is low. It is noticeable, for instance, that vanadium has about the same electronegativity as hydrogen and in this case, both octahedral and tetrahedral sites are occupied under different conditions. In fact, the electronegativity also correlates with the size of the interstitial site as electropositive metals tend to have large lattice parameters and hence relatively lower zero point energies for the smaller tetrahedral sites. When the sites have about the same energy, as in the case of vanadium, the site chosen in practice will be partly determined by the zero point energy. As this is higher for lighter isotopes and for the smaller (tetrahedral) sites, hydrogen would be expected to prefer the octahedral site more than deuterium. This conclusion is supported by the V/H and V/D phase diagrams as the former shows larger regions in which the octahedral site is occupied [5.111]. Fuller discussions of the isotopic dependence of site occupation in vanadium, taking full account of the anisotropy of the site, can be found elsewhere [5.11]. These qualitative ideas are fully supported by electronic calculations, see [5.92] and references therein.

As mentioned, the characteristic of an octahedral site is that it is generally larger than a tetrahedral site and the harmonic term in a Taylor's expansion of the potential is, in consequence, relatively small, giving a relatively low zero point energy. From observations on the palladium system, the most studied system to date, it is clear that, as expected, the proton vibration frequency decreases with increasing concentration, i.e., increasing lattice constant. In the low concentration limit (lattice parameter 3.96 Å), *Drexel* et al. [5.112] found the peak to be at 66 meV (0.2% H/M) and 63 meV (2.72% H/M). *Rush* et al. [5.89] found 69 ± 0.5 meV also in the α-phase. At the lower end of the β-phase concentration range, values around 60 meV [5.89], 57 meV [5.113], 58 meV [5.114] have been obtained at H/Pd ratios of 0.63, 0.68 and 0.7, respectively. At these nonstoichiometric concentrations, the peak is very broad and shows an intense shoulder that extends up to around 85 meV. Single-crystal, triple axis, measurements on Pd/D indicate that this shoulder is due to longitudinal phonons where strong dispersion is expected due to large H-H interactions between the nearest-neighbour hydrogens [5.115, 116]. The most detailed calculations [5.117] used a supercell of palladium fcc lattice with the octahedral sites randomly-filled with deuterons to the required concentration as a superlattice cell. The nearest-neighbour Pd-Pd and Pd-H force constants were assumed to have specific values throughout the superlattice. However, in practice, it is known that the Ds have strong short-range order and hence the local lattice distortion is bound to change the local values of the force constants depending on the local distribution of empty sites. At higher concentrations, produced by low temperature electrolysis, values of 59, 57.5 and 57.5 meV were obtained at H/Pd ratios of 0.7, 0.85 and 0.93 respectively [5.100]. Recently [5.118], samples of the stoichiometric material with a lattice parameter of 4.095 Å were produced using 40 kbars of hydrogen gas pressures by the High Pressure Group at the Soviet Institute for Solid State Physics at Chernogolovka. The spectra measured on TFXA have a peak at 56 meV, as shown in Fig. 5.14.

Allowing for some slight increase in the peak energy in the highly deformed electrolytic samples, these data show a linear decrease in the local mode frequency corresponding to a linear increase in the lattice parameter with concentration. The other noticeable features of the data for the stoichiometric material are that the peak narrows and the high energy shoulder drops in intensity as the concentration increases, presumably due to the fact that the force constants tend to well-determined values in the stoichiometric case.

The stoichiometric sample also shows a strong first harmonic and indications of the second harmonic. A notable feature of the results was that the shape of the first harmonic peak was strongly dependent on the sample orientation in the neutron beam. This is believed to be due to the fact that the sample had marked preferred orientation such that in one orientation (Q parallel to the (110) direction) it was possible to observe the (1,1,0) peak whereas, in the other (Q parallel to the (100) direction), the scattering was dominated by the (2,0,0) peak. The peak energies could be fitted with anharmonic parameters in a similar manner to the α-phase case, as described in Sect. 5.5.2 above. The anharmonicity constants derived are even greater than for the α phase, as expected because of the larger lattice parameters – so great, in fact, that the use of perturbation theory is very suspect.

It should be noted that the vibration spectrum of the Pd/H system is of particular interest in the high concentration range where the system is superconducting (H/M > 0.75) because of the reverse isotope effect in the

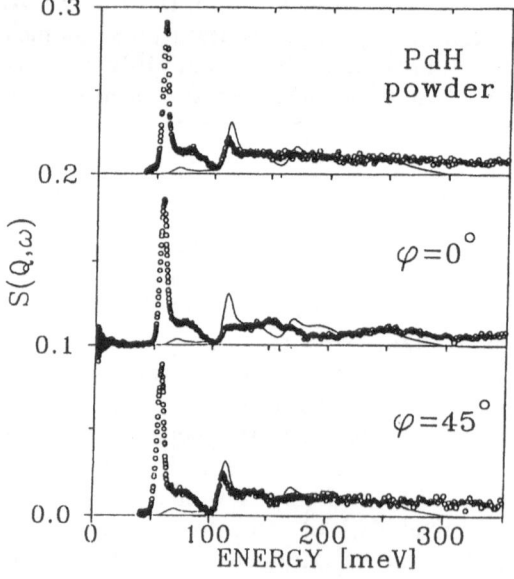

Fig. 5.14. The inelastic spectrum from stoichiometric PdH as measured on TFXA [5.118]. *Top diagram*: PdH$_{1.0}$ powder sample. *Middle diagram*: plate of PdH$_{1.0}$ set normal to the beam. *Bottom diagram*: plate of PdH$_{1.0}$ set at 45° to the beam

measured superconducting transition temperature [5.119]. In most systems, the lighter isotope would have the higher transition temperature but, presumably because of the large amplitude of the hydrogen vibrations and the anharmonicity of the potential well, the reverse is the case here. The transition temperature at a given concentration increases in the order H, D, T. In order to clarify this issue, *Rowe* et al. [5.120] measured the dispersion curves for $PdT_{0.6}$. They used the Born-von Karman model [5.117] to define the frequency at the centre of the Brillouin zone and hence obtained a ratio of force constants given by $(\mu\omega^2)_D(\mu\omega^2)_T$ to be 1.04 ± 0.01. The comparable figure for H:D was 1.12. When all the changes in the standard theory due to the involvement of such light isotopes were taken into account, a plausible expression was obtained for the transition temperature which reproduced the measured transition temperatures but a full explanation of the isotope effect for this case is still awaited.

The only other available metals showing an octahedral site in an fcc lattice are Ni and Rh. Samples of these hydrides have also been produced by the Chernogolovka High Pressure Group [5.121, 122]. They show peaks at 90 meV in $Ni_{1.05}H$ and at 75 meV in RhH. The corresponding lattice parameters are 3.72 Å and 4.01 Å, respectively. It is thus clear that the energy of the fundamental mode in octahedral sites does not correlate with the hydrogen-metal distance in the same way that works for tetrahedral sites [5.100]. This is because, in the tetrahedral case, the short M-H distance ensures that the potential well is determined by the short-range repulsive potential which also determines the metal-metal distance. On the other hand, in the octahedral site, the larger M-H distance means that the detailed nature of the electronic conduction band determines the shape of the potential well.

Some further light is shed on the nature of the hydrogen potential on an octahedral site by recent data on fully ordered polycrystalline $PdAgH_x$ with $x = 0.50$ and 0.86 [5.123]. The lattice structure of this sample consisted of alternate face-centred square planes of Pd and Ag atoms with hydrogen on all the octahedral sites in the Pd layers when $x = 1.0$ and no hydrogen in the Ag layers at any obtainable concentration. For $x = 0.86$, the face-centred square has side 3.997 Å while the equivalent c dimension has increased to 4.27 Å . The inelastic spectrum consists of two peaks centred on 62 and 85 meV which were shown to be associated with modes of vibration polarised in the plane (towards a Pd atom) and normal to the plane (towards an Ag atom), respectively. Again it is seen that an increasing separation gives an increasing frequency, i.e., it does not correlate with the size of the site as in the fluorite structures [5.100]. The authors suggest that this result would imply an H frequency in pure silver of 93 meV (lattice parameter 4.074 Å), reiterating the lack of any correlation with the size of an octahedral site.

Similar results have also been found for Pd_3Mn-H [5.124]. In the ordered alloy, it is found that the hydrogen occupies sites entirely surrounded with palladium atoms and the IINS spectrum from a sample of $Pd_3MnH_{0.4}$ shows a double peak at 76 and 83 meV, attributed to the β- and α-phases, respectively. Again it is found that the vibration energy rises in spite of the increase in the lattice parameter.

5.8.2 bcc Systems: Nb/H, V/H, and Ta/H

In the bcc metals, Nb, V and Ta, which readily dissolve large amounts of hydrogen, the hydrogen normally chooses the tetrahedral site $(1/2, 1/4, 0)$ – except for some phases of V/H at lower temperatures where the proton occupies the octahedral site $(1/2, 0, 0)$, vanadium having a similar electronegativity to hydrogen. Confining ourselves to the tetrahedral sites for the moment, there are four such sites on each face of a cube, giving a composition of MH_6 if all the sites were filled. All the sites do not fill, of course, due to the repulsive interactions between electrons gathered around each proton which effectively block nearest neighbour and next-neighbour sites to protons – the maximum concentration in the bcc lattice thus being about $MH_{1.1}$. Full phase diagrams are given in [5.111]. In the ordered phases, the tetrahedral sites are selectively occupied so as to minimise the number of short-range repulsive interactions. These ordered phases normally show some distortion that mirrors the symmetry of the superlattice but, for many purposes, can be regarded as having the same site geometry as the high temperature disordered (α, α') phase.

The tetrahedron of metal atoms around tetrahedral sites in bcc systems are squashed in the z direction as compared with a regular tetrahedron, although all the M-H distances are the same. This means that on a simple model that assumes longitudinal force constants only, the first excited states should be singly and doubly degenerate, with the singlet level being $1/\sqrt{2}$ the energy of the doublet. This is indeed approximately the case. Thus, for example, from recent room temperature measurements at low concentrations in the α-phase [5.125], it is known that $TaH_{0.037}$ gives peaks at 114 and 154 meV, $NbH_{0.03}$ gives peaks at 107 and 163 meV, while $VH_{0.012}$ gives peaks at 106 and ≈ 170 meV. The main characteristic of these peaks is their considerable width, which increases in the order Ta, Nb, V, with the lower-energy peak always being narrower than the higher. The origin of this broadening is uncertain but is presumably related to the lifetime of the excited state. For the case of Nb, *Dosch* et al. [5.126] have suggested that the self-trapping of the proton due to relaxation of the surrounding metal atoms gives rise to localised quantum states which are separated from each other by a very small threshold energy. The proton can therefore jump between these sites with a mean residence time of 10^{-13} to 10^{-14} s. This is much too fast for the lattice to respond so that adjacent sites can all be relaxed and the corresponding potential wells would be of the same depth. Motion on this time scale might explain the width of the local mode peaks. Lattice relaxation could be imagined to stabilise various groupings of T sites. 3T, 4T and 6T sites are discussed. The 4T site includes the occupation of the O site at $(1/2, 1/2, 0)$ as an extreme case and so this is referred to as the 4T(O) site. This model is used to explain the rapid decay of the noncubic strain components around the interstitial site.

The existence of these self-trapped sites have also recently been discussed in some detail by *Fukai* [5.11], making use of earlier work which approached the problem using model potentials [5.127, 128]. This approach prefers the

4T and 6T sites because these are stabilised using relaxations that preserve the 4-fold and 6-fold symmetry of the respective site cluster. The calculations, using modified Born-Mayer potentials, show that the 6T site is always of higher energy with the 1T and 4T being comparable to each other. These calculations, suggest that the ground-state wave-function extends over the whole cluster of sites, i.e., that the threshold energy between sites lies below the zero point energy. The model due to *Dosch* et al. on the other hand, requires that the ground state is localised on one of the T sites with rapid tunnelling or over-barrier hopping to the other sites. When the temperature is lowered so that the system goes into an ordered superlattice, the reduced symmetry will split the individual site energies and, hence, eliminate the broadening of the transition energy caused by rapid jumping between equivalent sites.

In the ordered phases of Ta/H and Nb/H, the inelastic spectra become much more clearly defined and, using spectrometers on pulsed sources, some eleven excited states have been measured at energies up to 500 meV in ε-Nb/H [5.88, 129]. *Eckert* et al. [5.88] first applied their perturbation-theory approach to explain the observed anharmonicity in the measured spectra and this work was continued up to higher energies by *Ikeda* and *Watanabe* [5.129] who included terms up to z^6 to describe the potential for a 1T site. The latter used six energy levels to define the potential and hence predicted the other five energy levels to within 8%. The potential obtained in this way is, of course, appropriate to the relaxed lattice. An interesting feature of the results is that the harmonic term in the potential is nearly isotropic. The saddle point energies obtained are 350 meV suggesting a wave function localised on one T site. However, the potential is so considerably anharmonic that the use of perturbation theory is very suspect. Indeed, the first principles calculations of *Ho* et al. [5.91], which predict the energy levels by performing a Fourier analysis of the calculated potential surface, appear to make better predictions of the observed energies.

As mentioned above, a more satisfactory way of measuring the shape of the wavefunction, and hence of estimating the potential well shape, is to measure the form-factor of an excited state. Measurements of this kind have been performed on polycrystalline samples of ZrH_2 and of $NbH_{0.3}$ by *Ikeda* et al. using the HET spectrometer at ISIS [5.130]. Because the interstitial sites are randomly orientated, it is really only practical to fit the measured form factor to theoretical expressions. In the ZrH_2 case (fluorite structure), the data fits a harmonic model very precisely. In the $NbH_{0.3}$ case, however, only the 115 meV peak profile coincides with the simple harmonic form. The first excited state in the x, y plane (161 meV) has the predicted shape but only 92% of the intensity while the second excited state in the z direction (220 meV) also has the same shape but only 47% of the intensity of the harmonic prediction. This observation can be explained as follows: The potential well corresponding to the perturbation parameters obtained before [5.129] suggests a threshold at 350 meV between the relaxed site and the nearest-neighbour sites. Thus, the ground state and the 161 meV wave-functions will be confined to the original near-harmonic site while the 220 and 350 meV levels will occupy the nearest-neighbour sites as well. It is easy to show that the effect of this delocalisation will, to a good approximation, involve a reduction in the intensity of the corresponding form-factor. It will now also be clear that another possible cause of the width of the peaks could be the sensitivity of the energy levels to the instantaneous position of the metal atoms undergoing normal thermal motion,. One can assume that further progress in this area will depend on the availability of single crystal samples and of spectrometers designed to measure form factors in 3-D reciprocal space.

As has been mentioned, the V/H system differs from the other two bcc systems in that, in most of the ordered phases, the proton transfers to the octahedral site. The reasons for this difference in behaviour and the mechanisms involved are discussed in some detail by *Fukai* [5.11]. The octahedral site is much more anisotropic than the T site in the bcc system as witnessed by the fact that the fundamental vibration in the x-y plane has an energy of 53.5 meV while the fundamental in the z direction is 220 meV [for the O_z site at $(0,0,1/2)$]. The most detailed observations have been performed by *Hempelmann* et al. [5.131] who managed to produce a single domain crystal of V_2H. Using this, and by careful choice of the Q direction relative to the crystal orientation, they were able to separately excite the vibrations in the x, y and z directions. They noted that the vibrations in the x-y plane appeared in pairs with a separation that was proportional to the energy transfer except for the lowest level. Therefore, instead of using a Taylor's expansion of the potential, they used a separate parabolic potential for each of the x, y and z directions and then added a Gaussian centred on the middle of the site with separate widths in the x and y directions. This model correctly reproduces the energy levels out as far at the fourteenth in the x direction – at 1500 meV! At about the same time, *Ikeda* and *Watanabe* [5.129] reported measurements on polycrystalline $VH_{0.33}$, effectively $VH_{0.5}$, and fitted their data using the perturbation analysis described above. They used six energy levels and were able to predict three more within 10% but, on the face of it, *Hempelmann* et al. using only one more free parameter, were able to predict significantly more energy levels more accurately and to correctly describe the most notable feature of the data, namely the uniformly increasing spacing of the levels. However, it has since been pointed out by *Rush* et al. [5.132] that the *Hempelmann* et al. model does not describe their measurements of the ground state energies for the D and T isotopes. This is not surprising because the added Gaussian terms effectively use two parameters to define the energy of the lowest lying states in the x and y directions, which are the ones actually showing the anharmonicity. Thus the ability to fit the two lowest lying energy levels provides no guide to the shape of the bottom of the potential. Again, it is clear that the most useful information on the shape of this part of the potential would come from measurements of the form factors of the low lying levels in the x-y plane.

5.8.3 Hydrides with the Fluorite Structures

The hydrides showing the fluorite structure have an fcc lattice of metal atoms with hydrogen on the tetrahedral sites at (1/4,1/4,1/4), yielding a dihydride composition when all the sites are filled. In this structure, the tetrahedral sites have full cubic symmetry and the fundamental term in the potential is therefore spherically symmetric. As mentioned above, the characteristic of these systems is that the hydrogen exists essentially as H^- and the potential well is largely due to the filled-shell repulsions between the H^- ions and the metal ions which can be well represented by Born-Mayer potentials. This is the reason that the fundamental modes can be so well correlated with the unit cell parameters [5.100, 101]. The other main feature is that the tetrahedral site is relatively small so that the potential well is accurately parabolic. This means that we expect the energy levels to be uniformly spaced and, for the systems that have been examined in detail, this is indeed the case. Thus, *Ikeda* and *Watanabe* [5.129] have measured the spectrum up to the fifth level in TiH_2 (700 meV) and have found the energies to lie just below values expected in the harmonic model based on the energy of the first excited state. Similar results were obtained by *Kolesnikov* et al. [5.133]. The other interesting feature is the observed structure in the harmonic peaks. This structure was first observed by *Couch* et al. [5.134] in their classic measurements on zirconium hydride. The obvious explanation of this structure, offered by these authors, is that this structure is due to H-H interactions and they calculated a frequency distribution from the predicted dispersion curves The measured structure of the peak, however, is normally seen to be well represented by two superimposed gaussians of nearly equal area whereas calculations based on simple H-H interactions yield a much more complex peak structure. The situation is somewhat complicated by the fact that some of the fluorite dihydrides have a slight tetragonal distortion at low temperatures, probably due to a Jahn-Teller splitting, but it would seem that the splitting of the peak is present whether or not there is a tetragonal distortion. Again, accurate measurements of the inelastic form factors may be useful in clarifying this point – if only to separate off the two-phonon (optic + acoustic) side bands.

This observed fine structure in the measured peaks raises another interesting point – how do we predict the structure of the higher harmonics if there is structure in the fundamental peak due to H–H interactions? If the system were precisely harmonic, structure in the higher levels would be simply obtained by the convolution of the frequency distribution in the fundamental with itself in the usual way. If the potential is anharmonic, however, the picture has to be that two phonons will travel together through the lattice, bound together by the reduction in their combined energy relative to twice the fundamental energy. The theory of these bound biphonons has been developed by *Agranowich* et al. [5.135].

5.8.4 fcc Metals with Both Tetrahedral and Octahedral Sites Occupied

For a number of hydrides with the flouride structure, it is possible to in-
trodouce more than two hydrogen atoms per metal atom. In these cases, the
extra hydrogens are accommodated on octahedral sites, giving a theoretical
maximum stoichiometry of MH_3. The examples of this behaviour are found
amongst the hydrides of the rare earths and yttrium. These systems usually
show a β-phase dihydride and a separate insulating tri-hydride γ-phase with
an hcp unit cell but the β-phase has a wide solubility range. These systems
have increasing interest because they show typical lattice-gas behaviour, i.e.,
a range of superlattice structures, in a similar way to β-Pd/D [5.36] and the
hexagonal solid-solution α-phases of the same elements discussed in the next
section. The current knowledge about these systems has recently been reviewed [5.136]. To date, neutron diffraction and neutron inelastic scattering
have both played an important part in understanding their behaviour.
Neutron diffraction and inelastic scattering from crystal field levels have also
been crucial in the interpretation of the related magnetic structure but these
aspects will not be discussed here. Examples of superlattice formation are
seen in, for example, GdH_{2+x} [5.137], TbD_{2+x} [5.138, 139], DyH_{2+x} [5.140]
and LaH_{2+x}[5.141]. IINS is also very useful in analysing these phases because it can directly measure the ratio of O-to-T site occupation and can also
detect the onset of a superlattice with a reduced site symmetry which causes
the H vibration peaks to split into two components – as first observed for the
tetrahedral site for the case of PrH_{2+x} by *Hunt* and *Ross* [5.114]. Thus, for
instance, in the cases of lanthanum, terbium [5.142] and probably the other
similar systems, the superlattice formed where $x = 0.25$ has I4/mmm symmetry which allows both the tetrahedral sites and the octahedral sites to have
reduced symmetry. Thus the t and o peaks observed at room temperature
split into a more complex concentration-dependent pattern at low temperature. Specifically, at a concentration of $x = 0.25$, both peaks are split into
two components which have intensity ratios of about 2:1 in each case. The
explanation preferred [5.143] differs somewhat. For H on an o site, the next o
site in the c direction is empty whereas the nearest sites in the a and b
directions are occupied. It is argued that the splitting is due to dynamic
$H_0 - H_0$ coupling in the a-b plane. On the other hand, every H_t atom has one
nearest-neighbour o site occupied along a (111) direction, again giving the site
tetragonal symmetry. Here again the vibrations along the tetragonal axis have
a higher frequency than the two-fold degenerate modes normal to this axis.

Again for these systems, first principles electronic structure calculations
are becoming available [5.144, 145].

5.8.5 hcp Metals with Extensive Hydrogen Solubility Ranges

A number of the heavy rare earth metals (Ho, Lu, Tm, Er etc) along with
yttrium and scandium, which all have an hcp structure, show the intriguing
property of dissolving large amounts of hydrogen in tetrahedral sites [5.146]

within a solid solution phase which has a temperature-independent high concentration limit. The vertical phase boundary implies that the solid solution (α-)phase has a very similar entropy to the ordered dihydride phase. Hence this α-phase must have a low entropy, certainly near its high concentration limit. As there are no signs of any superlattice peaks, i.e., there is no long-range order, it is clear that extensive short-range order must be present – particularly at low temperatures. Because considerable quantities of hydrogen are involved and because single crystals are available, this system is ideal for investigation using neutron scattering techniques. Diffuse neutron scattering from deuterated single crystal samples of yttrium confirmed that considerable short-range order is indeed present [5.96]. At high temperatures (above room temperature), the diffuse scattering consists of planes of intensity normal to the c direction with a cosine profile that is entirely consistent with the existence of pairs of hydrogens on tetrahedral sites immediately above and below the yttrium atoms in the c direction. As the temperature is lowered, the diffraction pattern becomes more complex but is still based on modulated planes of intensity normal to c, indicating that these pairs are rearranging themselves into a longer-range structure, predominantly along the c direction, with the pairs repeating themselves approximately every third unit cell. Similar diffuse scattering behaviour has been observed for Lu [5.147] and Sc [5.148]. There have been several attempts to model these complex patterns [5.149, 150] but the direct approach, namely reverse Monte-Carlo simulation, has only recently been performed [5.151].

As was first suggested by *McKergow* et al. [5.96], the origin of these interactions is, almost certainly, electronic. A prominent feature of the Fermi surface of yttrium is that there are planar regions normal to c^* at about 1/3 of the way to the Brillouin zone boundary. *McKergow* et al. suggested that the driving force for the hydrogen ordering would be a splitting of these electron states at the Fermi Surface into those that screen the protons – the energy of which will be lowered – and those that do not – the energy of which will be raised. As these states are at the Fermi surface, in equilibrium, only the former will be filled. These filled states will now form a charge density wave that will determine favourable sites for neighbouring deuteron pairs, particularly in the c direction.

Detailed electron band calculations for Y–H have recently been performed by *Yang* and *Chou* [5.144] using a pseudopotential method, the Local Density Approximation (DLA) and a plane wave basis. They calculated the electronic structure for a series of super-cells with different hydrogen sites. These calculations showed that a structure with pairs either side of a yttrium atom along the c axis where the pairs are repeated every third yttrium atom were the most stable configuration tried. Moreover, an examination of the wave functions that accounted for the lower energy of this configuration did indeed show the splitting of two bands at the Fermi surface, the occupied lower set of which shows electron density above and below the yttrium atom where the pair of hydrogen atoms are located. As the spacing of the paired protons is $(3/4)c$ the wave will also coincide with a pair of hydrogens placed around the third yttrium along the c direction. The other proposed explanation for the ordering of the pairs is that the process is driven by lattice distortion associated with the elastic dipoles which are caused by the hydrogen pairs [5.152] but no detailed model has been given. It is clear that a first principles calculation of this system will show both electronic effects and structural relaxation.

The calculations of *Yang* and *Chou* [5.144] also reveal the potential well in which the hydrogen is located using the frozen phonon method but without relaxing the metal atoms. This is illustrated in Fig. 5.15 where the derived energy levels are also given. The main point to note is the strongly anharmonic potential in the *c* direction, particularly on the side remote from the yttrium atom. Further, the energy to raise the proton to the first excited state in this direction is close to 100 meV as expected. This potential would, of course, be distorted if there were a hydrogen on the other side of either of the yttrium atoms.

A detailed analysis of the observed inelastic scattering spectrum from Y/H was discussed in Sect. 5.5.2 as an example of how to deal with perturbations in a potential well having hexagonal symmetry (Fig. 5.9). It was also pointed out there how the coupling between the protons in a pair gives rise to a splitting of the (001) mode into in-phase and out of phase components. A study of the splitting of this peak as a function of temperature and hydrogen concentration gives a good measure of the fraction of hydrogen atoms that are trapped in pairs at that temperature. This feature of the vibrations has been studied in some detail by *Anderson* et al. [5.153]. Figure 5.16 shows how the inelastic peak at around 100 meV varies with concentration. At 18% H, the peak clearly splits into two roughly equal components, suggesting that virtually all the protons are paired with similar coupling constant. The lower of these two peaks corresponds to in-phase vibration and this is assumed to be the same frequency as for an unpaired proton. The obvious feature of the results is that the high frequency component changes shape drastically as the concentration is reduced, going from a well defined frequency at high concentrations to a broad band at lower

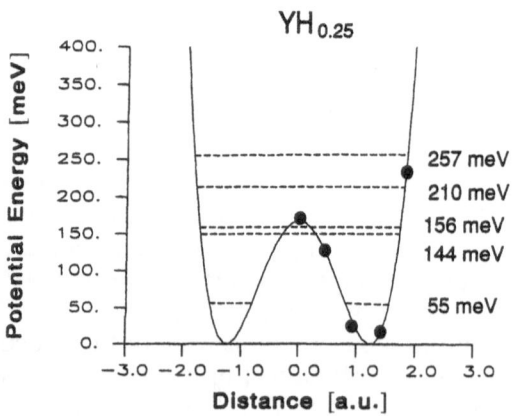

Fig. 5.15. The potential energy well in $YH_{0.25}$ showing the energy levels and the extent of the anharmonicity [5.144]. The black spots represent the positions of the hydrogen atom for which the total energy has been calculated

Fig. 5.16. Vibrational spectra for poly-crystalline YH_x at 80 K as a function of hydrogen concentration [5.153]. The solid lines are only a guide to the eye

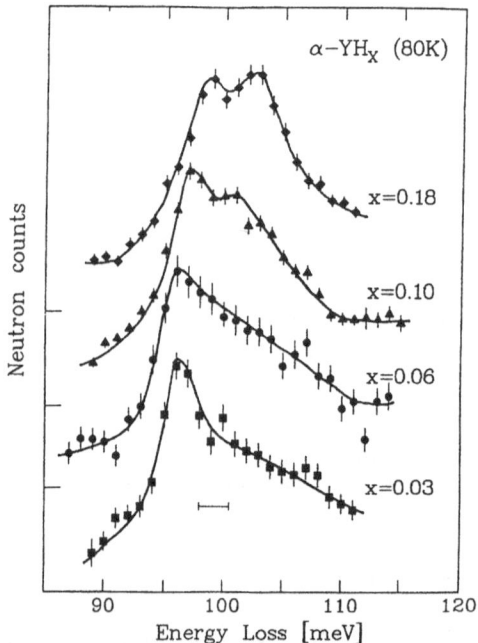

$\alpha-YH_x$ (80K)

x=0.18

x=0.10

x=0.06

x=0.03

Neutron counts

90   100   110   120

Energy Loss [meV]

concentrations. It is suggested that while virtually all the protons are paired at this temperature, it is only at the high concentrations that a regular repeat distance for the pairs is established along the c direction and that this gives rise to a fixed pair-coupling constant. At lower concentrations, there are not enough protons to form the more regular structure and so the electron bands at the Fermi Surface no longer fit the structure so well. Hence, a range of force constants is present which explains the observed peak profiles. For increasing temperatures, however, the basic periodicity remains although it involves increasing disorder and so the components of the peak both broaden with increasing temperature. Thus the detailed structure of the fundamental peak in the c direction is consistent with the general features of the model proposed.

5.9 Conclusions

In the present review, the uses of neutron quasi-elastic and neutron inelastic scattering in the investigation of the properties of hydrogen-metal systems have been described. Because of the large H scattering cross-section, and because of the different significance of coherent and incoherent cross-sections and several other important specific features of the neutron-hydrogen in-teraction, these techniques have been particularly fruitful in the investigation of such systems. The basic understanding of many hydride such comes from neutron methods where other techniques have only been able to provide

subsequent confirmation. A major reason for the power of the technique is that the scattering function measured as a function of Q and ω simultaneously so that models have to be very precise to fit the data. In, contrast, other available techniques often only measure one parameter – e.g., the temperature dependence of T_1 in NMR – and so it is much easier for the experimentalist to be misled.

Future trends are also quite easy to spot. The technique of measuring over wide ranges of Q and ω, as made possible by the development of pulsed sources, and the parallel advance in first-principles calculations have both reached the stage where their results can be compared together in very detailed ways. It thus seems likely that these neutron scattering techniques have a long and rewarding future ahead of them in metal-hydrogen system research.

Acknowledgements. I should like to acknowledge the assistance of all those who let me have reprints before publication. Thanks are also due to Professor W.A. Oates and Dr. S.M. Bennington for their careful reading of the manuscript and for useful suggestions.

References

5.1   G.E. Bacon: *Neutron Diffraction,* 3rd edn. (Clarendon, Oxford 1975)
5.2   W.L. Whittemore: Inelastic Scattering of Neutrons, Proc. Symp., Bombay (1964) IAEA Vienna **2**, 314 (1965)
5.3   B.N. Brockhouse, S. Hauteder, H. Stiller: In Interaction of Radiation with Solids ed. by Strumane et al. (North-Holland 1963)
5.4   L.Van Hove: Phys. Rev. **95**, 249 (1954)
5.5   G.H. Vineyard: Phys. Rev. **110**, 990 (1968)
5.6   T. Springer, D. Richter: In *Methods of Experimental Physics: Neutron Scattering* Vol. 23B, p. 131 (Academic Press, New York 1987)
5.7   K. Sköld: In *Hydrogen in Metals 1* ed. by G. Alefeld, J. Völkl, Topics Appl. Phys., Vol. 28 (Springer, Berlin 1978) Chap. 10
5.8   T. Springer: *Quasi-Elastic Neutron Scattering for the Investigation of Diffusive Motions in Solids and Liquids* Tracts Mod. Phys, Vol. 64 (Springer, Berlin 1972)
5.9   T. Springer, D. Richter: In *Methods of Experimental Physics,* Vol. 23, *(Neutron Scattering)* 131 (Academic Press, New York 1986)
5.10 D. Richter, R. Hempelmann, R.C. Bowman: In *Hydrogen in Intermetallic Compounds II* ed. by L. Schlapbach Topics in Applied Physics, Vol. 67 (Springer, Berlin 1992) Chap. 3
5.11 Y. Fukai: *The Metal Hydrogen System - Basic Bulk Properties* Springer Ser. Mater. Sci., Vol. 21 (Springer, Berlin 1993)
5.12 M. Bee: Quasielastic Neutron Scattering, (Inst. of Phys, Bristol 1988)
5.13 V.F. Turchin: *Slow Neutrons* (IPST, Jerusalem 1965)
5.14 S.W. Lovesey: *The Theory of Thermal Neutron Scattering from Condensed Matter,* (Clarendon, Oxford 1986)
5.15 G.L. Squires: *Introduction to the Theory of Thermal Neutron Scattering,* (Cambridge University Press, Cambridge 1978)
5.16 D. Price, K. Sköld: In *Methods of Experimental Physics: Neutron Scattering,* Vol. 23A, (Academic Press, London 1986)
5.17 V.F. Sears: Neutron News **3**, 26 (1992)
5.18 E. Fermi: Ric. Sci. **1**, 13 (1936)
5.19 C.G. Windsor: *Pulsed Neutron Scattering,* (Taylor and Francis, London 1981)
5.20 R.M. Moon, C.D. West: Physica B **156**, **57**, 522 (1989)

5.21 D.A. Faux, D.K. Ross: J. Phys. C: Solid State Phys. **20**, 1441 (1987)
5.22 C.T. Chudley, R.J. Elliott: Proc. Phys. Soc. **77**, 353 (1960)
5.23 W. Gissler, H. Rother: Physica **50**, 380 (1970)
5.24 H.C. Torrey: Phys. Rev. **92**, 962 (1953)
5.25 G. Blaesser, J. Peretti: In *Proc. Int. Conf. on Vacancies and Interstitials in Metals,* ed. by A. Seeger, D. Schaumander, W. Shilling, J. Drehl (Kernforschungs Anlage, Jülich, Vol. 2, p. 886 (1968)
5.26 J.M. Rowe, K. Sköld, H.E. Flotow, J.J. Rush: J. Phys. Chem. Solids, **32**, 41 (1971)
5.27 I.S. Anderson, J.E. Bonnet, A. Heidemann, D.K. Ross, S.K.P. Wilson: J. Less-Comm. Met. **101**, 405 (1984)
5.28 D.K. Ross, D.L.T. Wilson: In *Neutron Inelastic Scattering*, Vol. 1 IAEA, Vienna, (1977) p. 383
5.29 A.D. Le Claire: Phys. Chem. **10**, 26 (1970)
5.30 G.E. Murch, J. Nucl. Mater. **57**, 239 (1975)
5.31 M. Eisenstadt, A.C. Redford: Phys. Rev. **132**, 635 (1963)
5.32 D. Wolf: Phys. Rev. B. **15**, 37 (1977)
5.33 R.A. Tahir-Kheli, R.J. Elliott: Phys. Rev. B. **27**, 844 (1983)
5.34 W. Kehr, R. Kutner, K. Binder: Phys. Rev. B. **23**, 4931 (1981)
5.35 J.M. Sanchez, D. de Fontaine: Phys. Rev. B. **21**, 216 (1980)
5.36 R.A. Bond, D.K. Ross: J. Phys. F. **12**, 597 (1982)
5.37 I.S. Anderson, D.K. Ross, C.J. Carlile: In *Neutron Inelastic Scattering,* Vol. 2, IAEA, Vienna, (1978) p. 421
5.38 D.A. Faux, D.K. Ross: J. Less-Common. Met. **129**, 229 (1987)
5.39 H. Wipf, D. Steinbinder, K. Neumaier, P. Gutsmiedl, A. Magerl, A.J. Dianoux: Europhys. Lett. **4**, 1379 (1987) and references therein
5.40 K.W. Kehr, D. Richter: Solid State Comm. **20**, 477 (1976)
5.41 W. Kehr, O. Paetzold: Physica A **190**, 1 (1992)
5.42 R. Kutner: Phys. Letts. A **23**, 239 (1981)
5.43 P.C. Clapp, S.C. Moss: Phys. Rev. **142**, 418 (1966)
5.44 P.C. Clapp, S.C. Moss: Phys. Rev. **171**, 754 (1968)
5.45 I.S. Anderson, C.J. Carlile, D.K. Ross: J. Phys. C **11**, L381 (1978)
5.46 R.A. Bond, D.K. Ross: J. Phys. F **12**, 597 (1982)
5.47 S.K. Sinha, D.K. Ross: Physica B **149**, 61 (1988)
5.48 P.G. de Gennes: Physica **25**, 825 (1959)
5.49 J.W. Cahn: Acta Metall. **9**, 795 (1961); **10**, 179 (1962); **10**, 907 (1962)
5.50 M.A. Krivoglaz, Sov. Phys: Solid State **5**, 2526 (1964)
5.51 G. Alefeld: Ber. Bunsenges **76**, 746 (1972)
5.52 H. Wagner, H. Horner: Adv. Phys. **23**, 587 (1974)
5.53 H. Wagner: In *Hydrogen in Metals I: Topics Appl. Phys.* Vol. 28 ed. by G. Alefeld, J. Völkl (Springer, Berlin 1978) p. 5
5.54 H. Wipf, J. Völkl, G. Alefeld: Z. Phys B. Condens. Matter **76**, 353 (1989)
5.55 *Neutron Spin Echo* ed. by F. Mezei, Lecture Notes Phys. Vol. **128**, (Springer, Berlin Heidelberg, 1980)
5.56 M.H. Johnson: AERE Harwell report R-7682 (1974)
5.57 J.R.D. Copley, P. de Kerk, A.A. van Well, H. Friedrikze: Comp. Phys. Commun. **40**, 337 (1986)
5.58 J.C. Cook, D. Richter, O. Schärpf, M.J. Benham, D.K. Ross, R. Hempelmann, I.S. Anderson, S.K. Sinha: J. Phys: Condens. Matter **2**, 79 (1990)
5.59 P.L. Hall, D.K. Ross, I.S. Anderson: Nucl. Inst. Meth. **159**, 347 (1979)
5.60 D.S. Sivia: In *Maximum Entropy and Bayesian* Methods, ed. by P.F. Fougere (Klewer, Dordrecht, Netherlands 1990) p. 195
5.61 *Neutron Research Facilities at the ILL High Flux Reactor,* ed. by B. Maier, available from SCAPRO, Institut Laue Langevin, 156X, F-38042 Grenoble Cedex, France F-38642
5.62 *ISIS User Guide,* available from University Liaison Secretariat, ISIS Facility, RAL, Chilton, Didcot, Oxon, OX11 0QX, U.K
5.63 B. Alefeld, M. Birr, A. Heidemann: Naturwissenschaften **56**, 410 (1973)

5.64 O. Schärpf: Physica B **156/157**, 631 (1989)
5.65 O. Schärpf: Physica B **156/157**, 639 (1989)
5.66 J.M. Rowe, J.J. Rush, L.A. de Graaf, G.A. Ferguson: Phys. Rev. Lett. **29**, 1250 (1972)
5.67 C.J. Carlile, D.K. Ross: Solid State Commun. **15**, 1923 (1974)
5.68 M.J. Gillan: J. Phys. C **19**, 6169 (1986)
5.69 Y. Li, G. Wahnstrom: Phys. Rev. B **46**, 14528 (1992)
5.70 V. Lottner, A. Heim, T. Springer: Z. Physik. B **32**, 157 (1979)
5.71 V. Lottner, J.W. Haus, A. Heim, K.W. Kehr: J. Phys. Chem. Solids **40**, 557 (1979)
5.72 H. Dosch, J. Peisl, B. Dorner: Phys. Rev. B. **35**, 3069 (1987)
5.73 H. Dosch, F. Schmidt, P. Wiethoff, J. Peisl: Phys. Rev. B **46**, 55 (1992)
5.74 G. Wahnstrom, Y. Li: Appl. Phys. Rep. **93–19** (1993); F. Christodoulos, M.J. Gillan: Phil. Mag. B **63**, 641 (1991)
5.75 G. Wahnstrom, Y. Li: Phys. Rev. Lett. **71**, 1031 (1995)
5.76 Y. Li, G. Wahnstrom: Phys. Rev. B **51**, 12233 (1995)
5.77 M.J. Gillen: Phys. Rev. Lett. **58**, 563 (1987)
5.78 F. Christodoulos, M.J. Gillen: J. Phys. Condens. Mat. **3**, 9429 (1991)
5.79 B. von Sydow, G. Wahnstrom: Phys. Rev. B Submitted
5.80 R. Hempelmann, D. Richter, D.A. Faux, D.K. Ross: Z. Phys. Chem **159**, 175 (1988)
5.81 G. Vogl, O.G. Randl, W. Petry, J. Hunecke: J. Phys.: Condens. Matter **5**, 7215 (1993)
5.82 C. Schonfeld, R. Hempelmann, D. Richter, T. Springer, A.J. Dianoux, S.M. Bennington: Phys. Rev. B **50**, 853 (1994)
5.83 I.S. Anderson, N.F. Berk, J.J. Rush, T.J. Udovic, R.G. Barnes, A. Magerl: Phys. Rev. Lett. **65**, 1439 (1990)
5.84 L.R. Lichty, J.-W. Han, R. Ibanes-Meier, D.R. Torgeson, R.G. Barnes, E.F.W. Seymour, C.A. Sholl: Phys. Rev. B **39**, 2012 (1989)
5.85 P. Vajda: In Handbook of Physics and Chemistry of Rare Earths Vol. 20 ed. by K.A. Gschneider Jr., L. Eyring (Elsevier, New York 1995) Chap. 137
5.86 U. Stuhr, D. Steinbinder, H. Wipf, B. Frick: Europhys. Lett. **20**, 117 (1992)
5.87 J. Völkl, H. Wipf, B.J. Beaudry, K.A. Gschneidner Jr.: Phys. Status Solidi B **144**, 315 (1987)
5.88 J. Eckert, J.A. Goldstone, D. Tonks, D. Richter: Phys. Rev. B **27**, 1980 (1983)
5.89 J.J. Rush, J.M. Rowe, D. Richter: Z. Phys. B **55**, 283 (1984)
5.90 K.-M. Ho, H.-J. Tao, X.-Y. Zhu: Phys. Rev. Lett. **53**, 1586 (1984)
5.91 H.-J. Tao, K.-M. Ho, X.-Y. Zhu: Phys. Rev. B **34**, 8394 (1986)
5.92 C. Elsässer, M. Faehnle, K.M. Ho, C.T. Chan: Physica B **172**, 217 (1991)
5.93 C. Elsässer, K.M. Ho, C.T. Chan, M. Faehnle: Phys. Rev. B **44**, 10377 (1991)
5.94 C. Elsässer, K.M. Ho, C.T. Chan, M. Faehnle: J. Phys.: Condens. Matter **4**, 5189 (1992), ibid, **4**, 5207 (1992)
5.95 S.M. Bennington, D.K. Ross, M.J. Benham, A.D. Taylor, Z.A. Bowden, R. Osborn: Phys. Lett. A **151**, 325 (1990)
5.96 M.W. McKergow, D.K. Ross, J.E. Bonnet, I.S. Anderson, O. Schärpf: J. Phys. C **20**, 1909 (1987)
5.97 S.M. Bennington, D.K. Ross: Z. Phys. Chemie **181**, 527 (1993)
5.98 P. Stonadge: Paco; Ph.D thesis, Birmingham University (1993)
5.99 D.K. Ross (unpublished)
5.100 D.K. Ross, P.F. Martin, W.A. Oates, R. Khoda-Bakhsh: Z. Phys. Chemie **114**, 341 (1979)
5.101 Y. Fukai, H. Sugimoto: J. Phys. F **11**, L137 (1981)
5.102 I.S. Anderson, J.J Rush, T. Udovic, M.J. Rowe: Phys. Rev. Letts. **57**, 2822 (1986)
5.103 M. Leslie (private communication)
5.104 J. Mayers, T.M. Burke, R.J. Newport: J. Phys. Cond. Matter **6**, 641 (1994)
5.105 A.C. Evans, D.M. Timms, J. Mayers, S.M. Bennington: Phys. Rev. B **53**, 3023 (1996)
5.106 D.K. Ross, S. Bennington, E.L. Bokhenkov, J. Mayers: ISIS Annual Report (1991) A288

5.107 D.K. Ross, S. Bennington, E.L. Bokhenkov, J. Mayers: ISIS Annual Report (1991) A288
5.108 E. Gray, M. Kemali, J. Mayers, D.K. Ross: J Alloys and Compounds (to be published)
5.109 P.A. Egelstaff, P. Schofield: Nucl. Sci. Eng. 12, 260 (1962)
5.110 J. Hauck, H.J. Schenk: J. Less-Comm. Met. 51, 251 (1977)
5.111 T. Schober, H. Wenzl: In Hydrogen in Metals II, ed. by G. Alefeld, J. Völkl, Topics Appl. Phys., Vol. 29 (Springer, Berlin Heidelberg 1978) p. 241
5.112 W. Drexel, A. Murani, D. Tocchetti, W. Kley, I. Sosnowska, D.K. Ross: J. Phys. Chem. Solids 37, 1135 (1976)
5.113 M.R. Chowdhury, D.K. Ross: Solid State Commun. 13, 229 (1973)
5.114 D.G. Hunt, D.K. Ross: J. Less-Comm. Met. 49, 169 (1976)
5.115 J.M. Rowe, J.J. Rush, H.G. Smith, M. Mosteller, H.E. Flotow: Phys. Rev. Lett. 33, 1297 (1974)
5.116 M.W. McKergow, P.W. Gilberd, D.J. Picton, D.K. Ross, P. Fratzl, O. Blaschko, I.S. Anderson, M. Hagen: Z. Physik. Chem. NF 146, S159 (1985)
5.117 A. Rahman, K. Sköld, C. Pelizzari, S.K. Sinha, H. Flotow: Phys. Rev. B 14, 3630 (1976)
5.118 D.K. Ross, E.L. Bokhenkov, V.E. Antonv, E.G. Pongatovsky, J.C. Li, A.I. Kolesnikov, O. Mose, M.A. Adams, J. Tomkinson: ISIS Annual Report (1991) A288
5.119 R. Stritzker, H. Wuhl: In Hydrogen in Metals II, ed. by (Alefeld, J. Völkl, Topics Appl. Phys. Vol. 29 (Springer, Berlin Heidelberg 1978) p. 241
5.120 J.M. Rowe, J.J. Rush, J.E. Schirbir, J.M. Mintz: Phys. Rev. Lett. 57, 2955 (1986)
5.121 B. Dorner, I.T. Belash, E.L. Bokhenkov, E.G. Ponyatovski, V.E. Antonov, L.N. Pronina: Solid State Commun. 69, 121 (1989)
5.122 D.K. Ross, E.L. Bokhenkov, V.E. Antonov, E.G. Ponyatovskii: ISIS Annual Report (1991) A293
5.123 A.I. Kolesnikov, V.E. Antonov, G. Eckold, M. Prager, J. Tompkinson: J. Phys.: Condens. Matter 5, 7075 (1993)
5.124 J.J. Rush, T.B. Flanagan, A.P. Craft, Y. Sakamoto: J. Phys.: Condens. Matter 1, 5095 (1989)
5.125 A. Magerl, J.J. Rush, J.M. Rowe: Phys. Rev. B 33, 2093 (1986)
5.126 H. Dosch, F. Schmidt, P. Weithoff, J. Peisl: Phys. Rev. B 46, 55 (1992)
5.127 H. Sugimoto, Y. Fukai, Phys. Rev. B 22, 670 (1980)
5.128 A. Klamt, H. Teichler: Phys. Status Solidi (b) 134, 103 (1986)
5.129 S. Ikeda, N. Watanabe: J. Phys. Soc. Jpn. 56, 565 (1987)
5.130 S. Ikeda, M. Furusaka, T. Fukunaga, A.D. Taylor: J. Phys.: Condens. Matter 2, 4675 (1990)
5.131 R. Hempelmann, D. Richter, D.L. Price: Phys. Rev. Lett. 58, 1016 (1987)
5.132 J.J. Rush, N.F. Berk, A. Magerl, J.M. Rowe, J.L. Provo: Phys. Rev. B 37, 7901 (1988)
5.133 A.I. Kolesnikov, V.K. Fedotov, I. Nathanet, S. Khabrylo, I.O. Bashkin, E.G. Ponyatovskii: JETP Lett. 44, 509 (1986)
5.134 J.G. Couch, O.K. Harling, L.C. Clune: Phys. Rev. B 4, 2675 (1971)
5.135 V.M. Agranowich, O.A. Dubrovskii, A.V. Orlov, Solid State Commun. 72, 491 (1989)
5.136 P. Vajda: In Handbook of Physics and Chemistry of Rare Earths Vol. 20 ed. by K.A. Gschneider Jr., L. Eyring (Elsevier, New York 1995) Chap. 137
5.137 P. Vajda, J.N. Daou, J.P. Berger: J. Less-Comm. Met. 172–174, 271 (1991)
5.138 G. André, O. Blaschko, W. Schwarz, J.N. Daou, P. Vajda: Phys. Rev. B 46, 8644 (1992)
5.139 Q. Huang, T.J. Udovic, J.J. Rush, J. Schefer, I.S. Anderson: J Alloys Compounds, in press
5.140 P. Vajda, J.N. Daou: Z. Phys. Chem 179, 403 (1993)
5.141 T.J. Udovic, Q. Huang, J.J. Rush, J. Schefer, I.S. Anderson: Phys. Rev. B 51, 12116 (1995)
5.142 T.J. Udovic, J.J. Rush, I.S. Anderson: Phys. Rev. B 50, 7144 (1994)

5.143 T.J. Udovic, J.J. Rush, I.S. Anderson: J. Phys. Condens. Mat. **7**, 7005 (1995)

5.144 Y. Yang, M.Y. Chou: Phys. Rev. B **49**, 13357 (1994)

5.145 S.N. Sun, Y. Wang, M.Y. Chou: Phys. Rev. B **49** (1994)

5.146 D. Khatamian: J. Less-Comm. Met. **129**, 153 (1987)

5.147 O. Blaschko, G. Krexner, L. Pintschkovius, G. Ernst, P. Vadja, J.N. Daou: Phys. Rev. B **40**, 5605 (1989)

5.148 O. Blaschko, L. Pintschkovius, P. Vadja, J.P. Burger, J.N. Daou: Phys. Rev. B **40**, 5344 (1989)

5.149 O. Blaschko, J. Less-Comm. Met. **172**, 174 (1991)

5.150 J.P.A. Fairclough, D.K. Ross, S.M. Bennington: Z. Phys. Chemiie **179**, S281 (1993)

5.151 J.P.A. Fairclough, PhD thesis, University of Birmingham 1995

5.152 O. Blaschko, G. Krexner, J. Pleschiutschnig, G. Ernst, J.N. Daou, P. Vajda: Phys. Rev. B **39**, 5605 (1989)

5.153 I.S. Anderson, N.F. Berk, J.J. Rush, T.J. Udovic: Phys. Rev. B **37**, 4358 (1988)

6. Hydrogen Related Material Problems

H. Vehoff

With 41 Figures and 4 Tables

Hydrogen degradation of structural materials is a serious problem that has received much attention in the past fifty years. However, most of the research was focused on solving urgent technical problems. The development and selection of appropriate materials with acceptable properties for the chosen application and environment was the prime object of those investigations. Models were developed to rationalize the different effects of hydrogen on the mechanical properties. Still many observed effects seem to be inconsistent. Experiments designed to support theoretical models or to examine these inconsistencies were less frequently conducted.

In this chapter only models of hydrogen degradation of structural materials are summarized that have firm experimental support. Likewise, the status of theories still under discussion is briefly reported. Examples for the behavior of different materials are given in the appropriate sections of the text. They mainly support the discussed models and are neither complete nor meant for the use in design.

In Sect. 6.1 experimental data are summarized which demonstrate effects of hydrogen pressure, temperature and strain rate on material degradation. No efforts are made to give a complete overview of hydrogen effects in different alloy systems. In Sect. 6.2 mechanisms are reviewed which describe direct effects of hydrogen on the mechanical properties, like cavity nucleation and growth, phase transformations, and chemical reactions with second phase particles. Sect. 6.3 addresses questions on the micromechanisms and micromechanics of crack growth. The influence of temperature, hydrogen partial pressure and deformation rate on the nucleation and growth of cracks will be examined in detail. In the last Sect. 6.4 continuum models, dislocation models and atomistic computer simulations of fracture will be discussed. Hydrogen affects bond strength, surface energy, interfacial cohesion, dislocation emission and dislocation velocity, hence the role of hydrogen in fracture must be studied on the micro and meso scale.

6.1 Phenomenology of Hydrogen Damage

Commercial alloys for structural applications are usually designed to combine high strength with a reasonable high-temperature ductility. Hydrogen, however, degrades the ductility of most structural alloys. The influence on

Topics in Applied Physics, Vol. 73
Wipf (Ed.)
© Springer-Verlag Berlin Heidelberg 1997

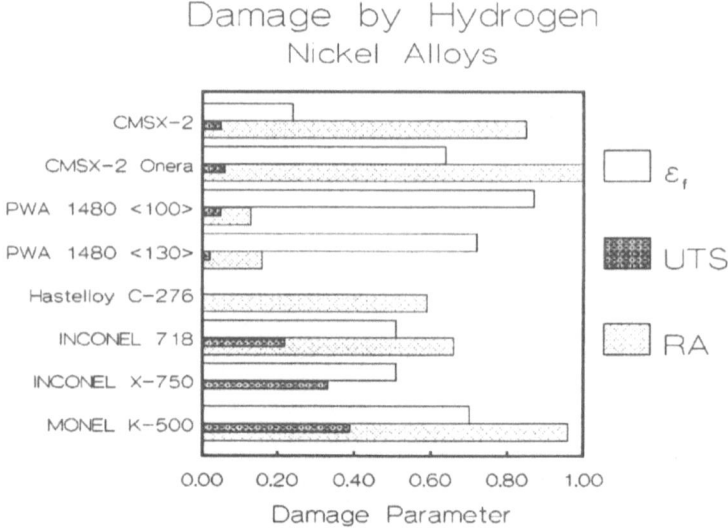

Fig. 6.1. Damage parameter, D, for different single-crystalline and polycrystalline super-alloys

strength is less pronounced and can be nearly avoided by appropriate alloy design. Even modern superalloys, which are most resistant against hydrogen effects, still suffer embrittlement. Figure 6.1 shows a bar diagram which demonstrates the effect of hydrogen on the ultimate tensile strength, UTS, the fracture strain, ε_f, and the reduction of area at fracture after a tensile stress, RA. The damage parameter, for example for UTS, is defined as:

$$D_{UTS} = \frac{UTS^{air} - UTS^{H}}{UTS^{air}}.$$
(6.1)

The maximum damage is defined as one, zero means no damage. Some single crystalline superalloys, like CMSX-2, show only minor hydrogen effects on UTS, but if the ductility at fracture is considered, the relative damage is still nearly one. Similar diagrams can be compiled for other alloy systems [6.1]. Minor changes in the alloy design can drastically alter the degradation of the mechanical properties. An example is given in Fig. 6.2, in which the fracture strain for several Co_3Ti alloys with different minor additives is plotted. The addition of 3 at % Fe clearly improved the ductility at fracture, other additives were less beneficial [6.2]. Hydrogen charging at high temperature and subsequent fast cooling produces hydrogen damage in most materials. In NiAl single crystals, however, the fracture strength is drastically improved by such a heat treatment [6.3]. In this alloy, the solubility of hydrogen is only weakly dependent on temperature [6.3] and the fast cooling rate reduces the strong impurity pinning [6.4]. From the results presented in Fig. 6.3, it follows that the influence of thermal treatment and cooling rate can override any existing hydrogen effects. For the design of alloys, it is therefore of

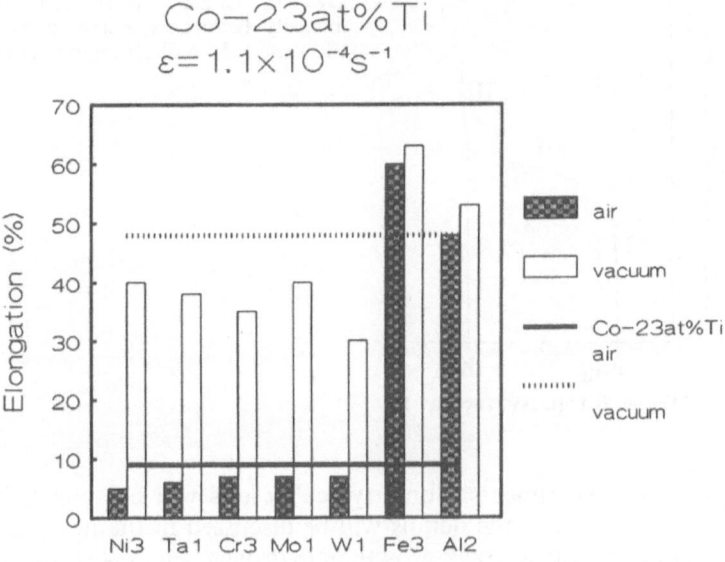

Fig. 6.2. Elongation at fracture in air and vacuum for Co_3Ti alloys with various alloying additions given in at % [6.2]

fundamental interest to understand the interactions between hydrogen and the alloy composition, constitution, impurities, and interface boundaries. These interactions will be discussed in detail in the following sections.

Most structural components contain cracks. The propagation rate of these cracks must be calculated for various loading conditions, to predict the lifetime of structures. Hydrogen can influence the initiation and growth of these cracks. Detailed experiments have shown that the crack growth rate depends in a complicated way on hydrogen charging conditions, temperature and mechanical loading, which cannot be predicted without the help of ap-

Fig. 6.3. Fracture toughness of pre-cracked NiAl single crystals after different heat treatments

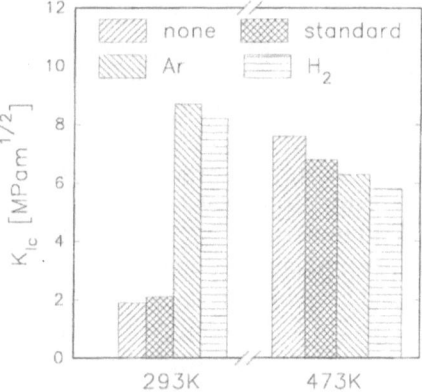

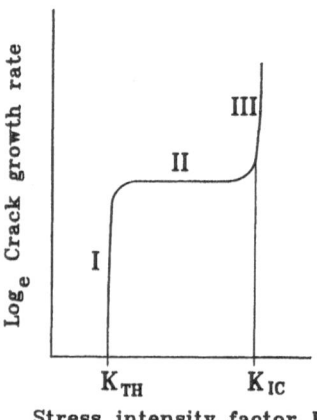

Fig. 6.4. Crack growth rate, da/dt, vs. stress intensity factor, K_I, curve in hydrogen indicating the stage I, II, and III regions of crack growth

propriate experiments. Some typical results will be summarized in the following. However, the details will be discussed in the next sections.

For subcritical crack growth in hydrogen, Fig. 6.4 shows a typical plot of the crack growth rate, da/dt, vs. stress intensity factor, K_I (see Sect. 6.4.1 for the definition of K_I). In air, below the critical stress intensity factor, K_{IC}, no crack growth occurs. In hydrogen, however, the K_{IC} for the propagation of cracks is reduced to K_{TH}. In Fig. 6.4 three regions must be considered: Region I: in this region, the crack starts to propagate and increases in rate very quickly until the rate becomes transport limited by the supply of hydrogen. K_{TH} depends on the equilibrium concentration of hydrogen in the Fracture Process Zone (FPZ) and on the applied stress, as was demonstrated by *Oriani* and *Josephic* [6.5]. Region II: in this region, the crack growth rate is limited by the transport of hydrogen to the FPZ, increasing the crack driving force by increasing K_I has only a minor influence on the crack growth rate since the crack simply blunts. If K_I is further increased, the crack outruns the hydrogen supply and behaves as if it were in air (region III). Hydrogen has a similar influence on the growth rate of fatigue cracks as can be seen in Fig. 6.5 [6.6]. With decreasing loading frequency, the fatigue crack growth rate, da/dN, (more correctly the growth increment per cycle) increases by orders of magnitude compared to the rate observed in vacuum (Fig. 6.5). The various aspects of the interaction between fatigue and hydrogen are discussed in detail in a review article by *Gerberich* [6.7] and will not be further considered here.

The stage II crack growth rate for constant K_I (the plateau in Fig. 6.4) depends on hydrogen pressure and temperature as demonstrated in Fig. 6.6. With increasing temperature the stage II crack growth rate increases, reaches a maximum at around room temperature (dependent on pressure), and drops quickly at higher temperature [6.8]. This behavior will be discussed in detail in Sect. 6.4. There are several other effects of hydrogen (for example hydride formation) which degrade the mechanical properties. Some of them will be discussed in the corresponding sections.

Fig. 6.5. Fatigue crack growth rate, da/dN, as a function of the stress intensity range, ΔK, for various frequencies in vacuum and hydrogen gas (ultra-high-strength steel, [6.6])

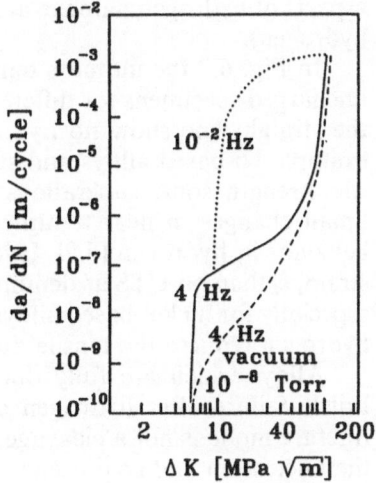

Fig. 6.6. Stage II crack growth rate as a function of temperature for various hydrogen pressures [6.8]

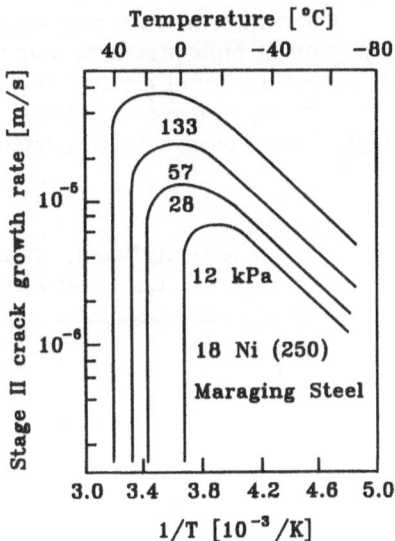

6.2 Degradation of Mechanical Properties Due to High-Pressure Hydrogen

Degradation of the mechanical properties of metals can be observed at high external hydrogen pressures, large internal hydrogen concentrations, supersaturation of hydrogen due to chemical charging or after fast cooling in a hydrogen environment, and at very low pressures and concentrations far below one hydrogen atom/matrix atom. In this section, we will focus on the aspects of hydrogen damage at large internal concentrations (direct effects of hydrogen).

In Fig. 6.7 the ultimate tensile strength, UTS, of hydrogen charged and uncharged specimens for different alloy systems is shown. Alloys which lie on the straight line show no hydrogen effects. For the same alloy system (for example Ni-based alloys and superalloys, open circles), independent of tensile strength some superalloys show pronounced degradation of strength. Small changes in heat treatment and alloying can drastically improve the behavior in hydrogen [6.9]. Hydrogen has a much greater effect on fracture strain, ε_f than on UTS as demonstrated by Fig. 6.8 if compared with Fig. 6.7, especially for nickel-based alloys (open circles). In nearly all structural alloys, hydrogen reduces the tensile ductility.

Alloys which are fully ductile in air can apparently suffer completely brittle failure after hydrogen charging [6.10]. In most cases, the resulting fracture mode is not a cleavage type of fracture, rather the specimens fail by the nucleation and coalescence of micro voids. The growth of these voids is supported by the internal hydrogen pressure which can develop within the voids.

These high internal concentrations are directly achieved by gas-phase charging at high pressures and temperatures with subsequent fast cooling. However, in liquid electrolytes and under electrochemical charging conditions, the hydrogen concentration in subsurface layers can also be extremely high. This is more fully discussed in the following sections.

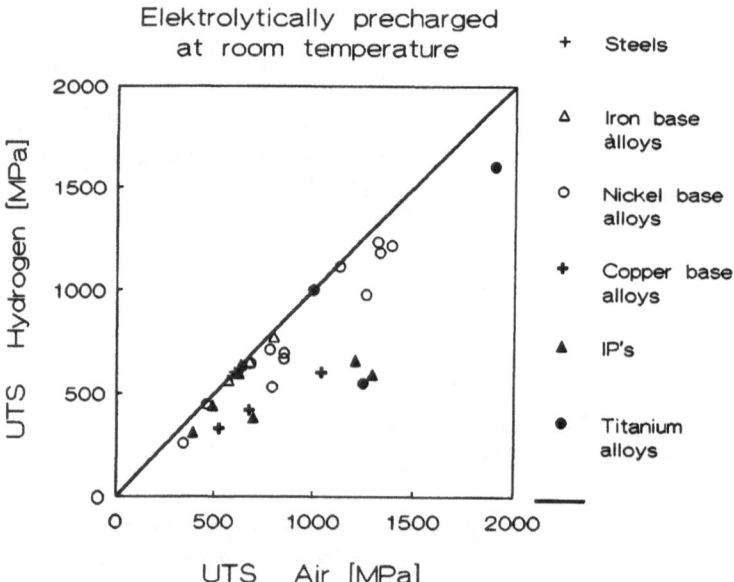

Fig. 6.7. Comparison of the UTS in air and hydrogen for different materials precharged electrolytically at room temperature

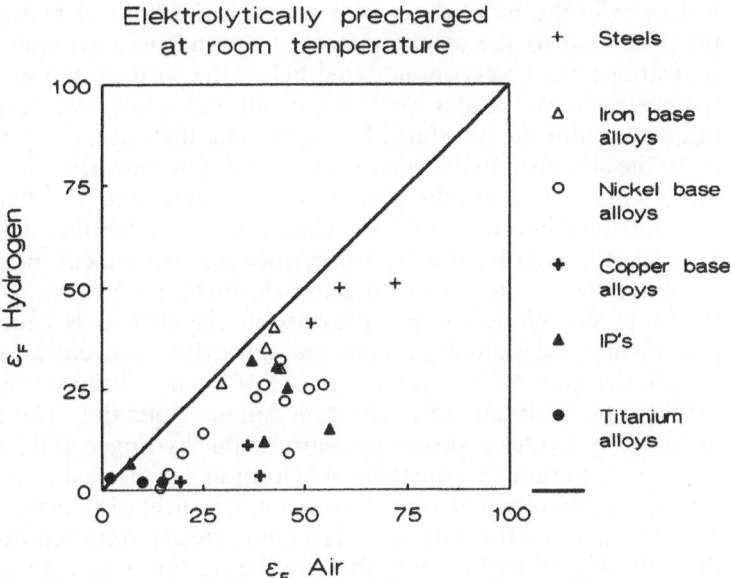

Fig. 6.8. Comparison of the fracture strain, ε_F, in air and hydrogen for different alloy systems for the same charging conditions as in Fig. 6.7

6.2.1 Hydrogen Entry from the Liquid Phase

In the chemical industry as well as in marine structures, engineering alloys are often used in liquid environments. In these environments, besides corrosion, hydrogen embrittlement can occur. At the metal/aqueous interface the following reactions can be found:

(i) Discharge of hydrated protons

$$H_3O^+ + M + e^- \rightarrow MH_{ads} + H_2O . \qquad (6.2)$$

(ii) Recombination of adsorbed hydrogen

$$MH_{ads} + MH_{ads} \qquad \text{a)} \rightarrow H_2 + 2M , \qquad (6.3)$$
$$\text{b)} \leftrightarrow 2MH_{abs} .$$

During these reactions some hydrogen enters the lattice either directly [6.11] or via a precursor state [6.12]. The electrochemical potential of hydrogen at the surface is given by

$$\mu_s = \mu_s^0 + kT \left[\ln\left(\frac{n}{N_s - n}\right) - \ln\left(\frac{n^0}{N_s - n^0}\right) \right] + Ze\Delta\Phi , \qquad (6.4)$$

where n is the number of adsorbed hydrogen ions per unit area, Z the number of elementary charges, $N_s - n$ the number of free sites per unit area, and $\Delta\Phi$ is the applied potential. Equating μ_s with the chemical potential of hydrogen in the bulk, the subsurface concentration of hydrogen can be estimated. Due to the strong potential dependence, extreme hydrogen concentrations can be accommodated below the surface. However, equilibrium is never achieved under charging conditions since hydrogen escapes continuously from the interface. Let us assume that, according to reaction b in (6.3), the adsorbed hydrogen is in local equilibrium with absorbed hydrogen, now, when the charging current is kept constant, the hydrogen surface coverage remains constant, too. The hydrogen subsurface concentration can now be estimated from a typical permeation experiment [6.13, 14].

Consider a thin foil of thickness d, drawn schematically in Fig. 6.9. At the input side either the gas pressure or the current is kept constant by a potentiostat, maintaining a constant subsurface concentration C_{io}. The hydrogen that diffuses out at the exit side of the membrane is electrochemically oxidized by maintaining a constant anodic potential. The anodic current measured provides a direct measure of the hydrogen flux, J, for $i_p = FJ$, where F is Faraday's constant. When input reactions do not determine the rate, the stationary hydrogen flux through a sheet of thickness, d, is given by $J = D(C_{io}-C_d)/d$ for pure materials under steady-state conditions where D is the diffusivity of hydrogen in the ideal lattice (all traps saturated). The input concentration can be calculated when the diffusion coefficient is known. In 0.1 N H_2SO_4 at a charging current between 20 and 170 A/m^2 input fugacities between 10^3 and 10^4 MPa were obtained when As_2O_3 was used as a recombination poison [6.5]. The fugacity, f, describes the deviation of a real gas from ideal gas behavior by maintaining the same functional form of the ideal gas law. These pressures exceed the yield stress of most structural materials.

However, at these high fugacities, f, the pressure which can develop in voids is much lower. This is according to the generalized Sievert's law, $C = (K/\gamma)f^{1/2}$, where γ is the activity coefficient and K is the equilibrium coefficient for the reaction

$$1/2\,H_2(\text{gas}) \leftrightarrow H(\text{solution}) . \tag{6.5}$$

For example, at atmospheric pressure and room temperature the solubility of hydrogen in iron is approximately 0.02 ppm. When a concentration of 3 ppm is charged into iron, according to

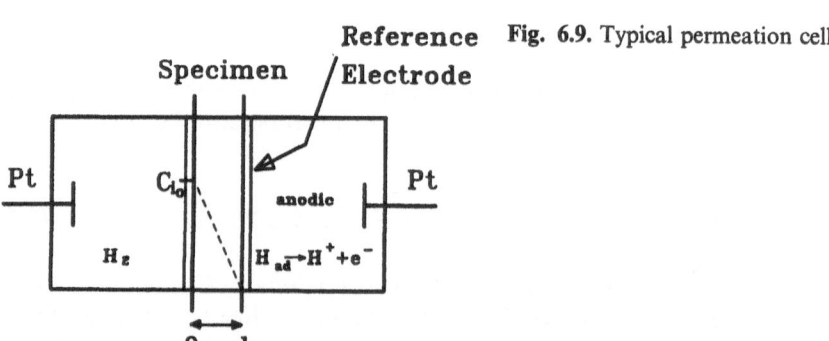

Fig. 6.9. Typical permeation cell

Table 6.1. Fugacity factors for different hydrogen pressures ($T = 300$ K) [6.15]

P/P_N	φ	f/f_N
1 000	2	2 000
2 000	3.9	7 800
5 000	23	115 000
10 000	315	$3.15 \cdot 10^6$
20 000	28 544	$5.7 \cdot 10^8$

$$\frac{C}{C_N} = \sqrt{\frac{f}{f_N}} \tag{6.6}$$

the fugacity of hydrogen inside the void is calculated as 2279 MPa. Since the fugacity coefficient, $\varphi = f/p$, increases drastically with pressure (Table 6.1, [6.15]), inside the cavity this results in a mechanical pressure of 291 MPa only. This is lower than the yield stress of mild steels. Therefore in most cases gas phase charging does not result in critical hydrogen damage.

(A) Hydrogen Entry at Crack Tips

In large offshore structures the propagation of cracks up to critical sizes is the life-controlling failure mechanism. The crack-tip environment in long shallow cracks changes with crack length and must be taken into account for service-life predictions.

The reactions at a crack-tip in an aqueous environment are shown schematically in Fig. 6.10. Chemical reactions can occur at the crack walls as well as at the crack tip and ions must be transported to and from the crack-tip. This alters the local concentration and pH of the crack-tip environment and therefore the potential drop over the crack length. Depending on the crack-tip environment and on the external potential, films can form at the side walls, which again alters the potential drop over the crack length. In the case of hydrogen charging and hydrogen embrittlement the local over-potential at the crack-tip must be known. Some techniques, like the weeping electrode test, allow the measurement of the concentration and pH of the solution in the near-crack-tip region. In most cases, however, the local potential on which the hydrogen activity at the crack-tip depends must be calculated.

Fig. 6.10. Active crack-tip in an aggressive environment indicating the reactions in the crack mouth

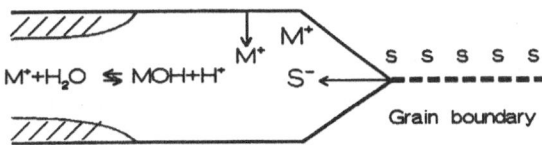

A mathematical model for the crack environment must be solved which describes the flux of dissolved species to the crack tip. In principle, the problem is described by the following set of equations [6.16]:

(i) Mass transport

$$j_i = -M_i \nabla \mu_i \qquad (6.7)$$

with

$$\mu_i = \mu_{i0} + kT \ln a_i + F Z_e \Phi . \qquad (6.8)$$

μ_i is the electrochemical potential for each of the dissolved species, $M_i = D_i c_i / kT$ for a dilute solution, where D_i is the diffusion coefficient.

(ii) Mass conservation of species

$$\frac{\partial C_i}{\partial t} = -\nabla j_i + R_i , \qquad (6.9)$$

where R_i represents the rate of production or depletion of a species. The electrochemical reactions occur at the crack tip and on the crack walls; typical reactions and rate equations are given below:
Anodic dissolution

$$M \Rightarrow M^{n+} + ne^- ; \qquad (6.10)$$

hydrolysis reaction

$$M^{n+} + H_2O \Leftrightarrow M(OH)^{(n-1)+} + H^+ ; \qquad (6.11)$$

cathodic reduction

$$H^+ + e^- \Rightarrow H, \qquad (6.12)$$
$$H_2O + e^- \Rightarrow H + OH^- ; \qquad (6.13)$$

where the fourth reaction represents reduction of water producing hydrogen atoms on the metal surface that can subsequently permeate the steel.
A typical rate equation has the following form:

$$R_{H^+} = k_0 C(H^+) \exp\left(-\frac{\beta F \Phi}{RT}\right) , \qquad (6.14)$$

where k_0 is a rate constant, β the transfer coefficient, F is Faraday's constant, and Φ is the electrode potential.

(iii) Electro-neutrality

$$\sum z_i C_i = 0 . \qquad (6.15)$$

This set of equations was solved by *Turnbull* et al. [6.17] for several aqueous solutions. Their basic findings are summarized in Fig. 6.11 in which the crack-tip potential is drawn as a function of the external potential. With rising or falling external potential the potential at the tip deviates from the

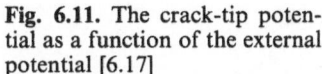

Fig. 6.11. The crack-tip potential as a function of the external potential [6.17]

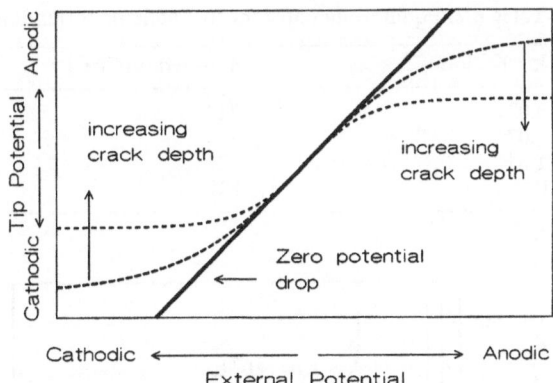

applied potential. When the crack grows the potential at the tip approaches the free corrosion potential and becomes independent of the applied potential. Hence, in long and shallow cracks either hydrogen embrittlement or anodic dissolution can occur depending on the crack-tip environment and the crack growth is independent of the applied potential [6.18, 19].

6.2.2 Hydrogen Attack

In specimens exposed to high-pressure hydrogen at higher temperatures, hydrogen can react with internal particles like oxides and carbides. In steels, methane bubbles can nucleate, grow and merge along grain boundaries forming fissures, which produce a reduction of area in tensile tests. This process is called hydrogen attack. The temperature and pressure range necessary for hydrogen attack depends strongly on the stability of carbides within the steels.

In equilibrium, the internal hydrogen pressure cannot exceed the external pressure, however, the methane pressure which can build up within cavities is given by the equilibrium between hydrogen, carbide and methane according to the reaction

$$Fe_3C + 2H_2 \Rightarrow CH_4 + 3Fe \ . \tag{6.16}$$

Taking unit activity for Fe and Fe_3C, and approximating the fugacity of H_2 by its pressure, for the methane fugacity we obtain

$$f_{CH_4} = K_P(P_{H_2})^2 \ . \tag{6.17}$$

To first order, the pressure of methane increases parabolically with hydrogen pressure. At 327 °C, an equilibrium pressure of 10^3 times the external hydrogen pressure squared builds up within a bubble according to the K_P values given in Table 6.2. The kinetics of this process are still unclear. Tensile tests showed that after an incubation time, t_i, the RA values drop drastically [6.20]. Figure 6.12 shows schematically the temperatures and pressures needed to obtain this drastic RA drop for a given t_i.

Table 6.2. Equilibrium constant for methane formation [6.21]

log K_P [atm]$^{-1}$	Temperature [°C]
5.02	227
3.07	327
1.63	427
0.55	527

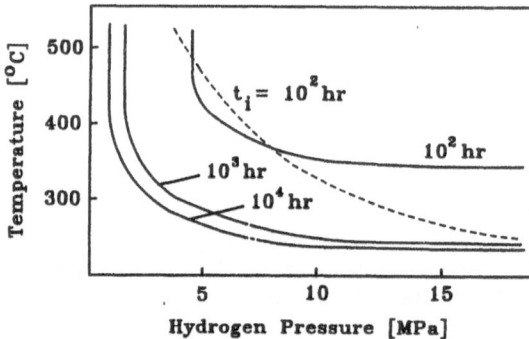

Fig. 6.12. Hydrogen attack incubation times, t_i; the curves limit the temperature and pressure ranges below which no attack occurs during the indicated time

Two regions can be distinguished. In region I, medium temperatures and low pressures, the curves become parallel to the temperature axis. With increasing temperatures lower hydrogen pressures are needed for attack for a given incubation time. Below a critical pressure, attack is virtually independent of temperature. In region II, low temperature and high pressure, the curves become parallel to the pressure axis and attack does not depend on pressure. However, a small increase in temperature results in a drastic decrease in incubation time.

According to *Shewmon* [6.21] this behavior can be understood if the growth kinetics of bubbles are considered more closely. Myriads of small bubbles must grow on the grain boundary to obtain the observed severe reduction in area in tensile tests. TEM observations have shown that these bubbles nucleate and grow on the grain boundary – but rarely on carbides [6.22]. For the growth of methane bubbles, several processes are necessary, any of which may be rate limiting:

(i) $Fe_3C \rightarrow 3Fe + C$ (in solution) ;   (6.18)
(ii) Diffusion of carbon to the bubbles ;
(iii) $C + 2H_2 \rightarrow CH_4$ (*Methane formation*) ;   (6.19)
(iv) Diffusion growth of the cavities driven
     by the internal methane pressure.

Initially, for small cavity sizes, the bubbles grow by diffusion creep which can only occur along the boundary, since the bulk diffusion coefficient of iron is too small in the temperature range in which attack is observed. In region I, it is assumed that reaction three is in equilibrium. According to (6.17), the driving force for bubble growth increases with $(p_{H_2})^2$, but decreases with rising temperature since K_P strongly decreases with temperature. For high temperatures the pressure within the bubbles approaches the equilibrium pressure of hydrogen, and no additional attack occurs.

Attempts were made to calculate the incubation time in region I, assuming diffusion growth driven by the internal methane pressure. The incubation time is assumed to be the time at which the bubbles on the boundary nearly touch which gives the measured reduction in area at fracture. This is considered to be the fracture criterion. Since neither the fugacity nor the initial density of bubbles are known, the model is not presented here (for details see [6.21]). The main results are that the incubation time varies inversely with the methane pressure, hence with the hydrogen pressure squared, and that the temperature dependence is mainly given by the product of the temperature dependence of $1/K_P$ and of $1/D_b$, the inverse temperature dependence of grain-boundary diffusion [6.21]

$$t_i = \frac{\text{const } T}{(p_{H_2})^2 \rho^{3/2}} \frac{1}{K_P D_b \delta} , \qquad (6.20)$$

where ρ = bubble density on grain boundary, δ = grain-boundary thickness and D_b = grain-boundary diffusion coefficient. This solution is given in Fig. 6.12 as a dashed line. In region II (low T, high P) an increase in hydrogen pressure has little effect on the rate of bubble growth since the bubbles grow at a rate limited by the supply of carbon. However, if the temperature is raised at a fixed hydrogen pressure, the bubbles grow much faster since a temperature rise increases the carbon diffusivity, and the mobility of iron atoms moving away from the bubble.

This kind of hydrogen attack is not limited to carbides in iron. Hydrogen can react with oxides or dissolved oxygen forming water bubbles in the material as shown by *Gramberg* [6.23] for copper. This can be detrimental in two aspects: not only bubbles can form but the material loses its strength since the oxide particles which pin dislocations dissolve.

6.2.3 Hydrides

Hydrogen not only reacts with impurities, forming gas bubbles, but it also forms hydrides in many metals. These hydrides, which are often brittle, nucleate preferentially in the stress field of cracks, which reduces the ductility and fracture toughness drastically. In particular, alloys that are candidates for aerospace applications, like the titanium aluminides, suffer under embrittlement due to hydride formation, growth and fracture.

In the typical hydride formers, the elements of the group Vb (Nb, V, Ta), hydrides precipitate readily, for example in Nb at temperatures down to 77 K [6.24]. Hydrides also form in Ni and its alloys. These hydrides are stable under severe charging conditions or high pressure and decompose when the pressure is reduced [6.25]. The main experimental findings of embrittlement due to hydride formation can be summarized as follows [6.26]:

(i) Embrittlement occurs at temperatures that are significantly greater than the solution temperatures of the hydrides.
(ii) The ductility increases as the strain rate increases.
(iii) The fracture surfaces are characteristic for cleavage failure. The cleavage planes correspond to the hydride cleavage plane.

The effects of stress on the nucleation and growth of hydrides were examined systematically by *Matsui* and *Koiwa* [6.27]. They showed that according to the orientation of the hydrides, some hydrides which nucleated without the application of stress disappear by applying stress. Other hydrides which were favorably oriented to the applied stress were observed to grow.

(A) Hydride Formation at Crack Tips

In hydride forming systems embrittlement occurs by hydride formation at stress concentrations followed by cleavage of the hydride. Direct observations of propagating cracks in vanadium by high-voltage electron microscopy showed that the crack propagates by the repetitive formation and cleavage of the hydride at the crack tip [6.28]. Microcracks nucleate within the hydride in front of the main crack, propagate backwards and join the main crack tip, which results in a stepwise growth of the main crack. Hydrides have a positive volume misfit. Therefore the chemical potential of hydrogen is reduced at defects and hydrides can precipitate. The growth of hydrides is autocatalytic, when nucleated, they grow. The resulting volume change is accommodated by the surrounding lattice either elastically or by plastic deformation. This produces a tensile stress at the apex of the growing hydride. Hence new hydrides can form in front of an existing one provided the supply of hydrogen is sufficient and the free energy of the system is decreased by hydride growth [6.29].

In metals with positive energy for hydrogen solution like iron, the regions of large hydrogen concentration are limited to the core of stress singularities and extend at most 1 nm. In front of a crack, however, hydrides precipitate over a length of more than 1 μm [6.30]. Therefore the crack propagates by discontinuous cleavage of the hydride, stops and waits until new hydrides have formed which will be cleaved again.

6.2.4 Formation and Growth of Cavities

In Sect. 6.2.1, it was shown that high hydrogen concentrations can be produced in metals by thermal or electrochemical charging. Hydrogen can accumulate at internal defects like microcracks or cavities which are present in most commercial alloys. Depending on the charging conditions, high pressures can build up within these cavities which then grow and coalesce. The coalescence of these cavities shows up in tensile tests by a reduction of area compared to tests without charging. In low-alloy steels the loss in ductility was found to be hydrogen-enhanced nucleation of microvoids at the carbide-ferrite interface [6.31].

For example in Al-Zn-Mg alloys, the nucleation and growth of hydrogen-filled bubbles was directly observed inside a TEM. The specimens were left in air with 100% relative humidity at 70 °C for up to 96 days. Hydrogen accumulated along the grain boundaries, and during subsequent beam heating inside the TEM, hydrogen-filled bubbles nucleated, grew, and gradually disappeared with time. During growth the molecular hydrogen within the bubbles generated a significant stress which exceeded the yield stress of the surrounding matrix. This stress causes dislocation generation and movement [6.32].

In copper single crystals, which were hydrogen charged above 600 °C at 100 kPa H_2 and then quenched into water, a high density of hydrogen bubbles was found. They were roughly spherical and surrounded by a dense tangle of dislocations and prismatic dislocation loops equally distributed with a density of 10^{10} cm^{-2}. Copper single crystals, prepared to keep the dislocation density at a minimum, showed a bubble density that was four orders of magnitude lower. Hence, hydrogen segregates preferentially into the strain field of dislocations which serve as nucleation sites for hydrogen bubbles [6.33]. In copper polycrystals, however, the bubbles were distributed preferentially along grain boundaries. The growth of the bubbles was diffusion limited by the transport of hydrogen from the surrounding matrix [6.33].

Xie and *Hirth* [6.34] observed softening due to hydrogen in spheroidized steels. With increasing input fugacity the measured softening became irreversible, indicating irreversible damage at fugacities which corresponded to an internal pressure of 270 MPa at internal defects. In notched specimens, hydrogen bubbles and microcracks formed preferentially along shear bands emanating from the notch after straining. This demonstrates again that dislocation tangles or interfaces are necessary for hydrogen precipitation [6.35].

A similar steel with a spheroidized carbide microstructure was strained various amounts after hydrogen charging. After each test longitudinal sections along the center were cut out and examined metallographically, primarily to assess the effect of hydrogen on void initiation and growth. The results showed that hydrogen eased void initiation at carbide/matrix interfaces and accelerated the growth of these voids [6.36].

But the internal pressure mechanism cannot fully explain the observed acceleration of the crack or cavity growth rate in steels. The cracks grow much faster and at lower pressures as predicted by the pressure mechanism. Therefore hydrogen either affects the movement of dislocations as shown by Matsui and co-workers for high purity iron single crystals [6.37], or reduces the cohesion of the interface, or acts by a combination of both. This will be more fully discussed in Sects. 6.3, 4.

6.3 Hydrogen Assisted Crack Growth

6.3.1 Equilibrium Aspects of Hydrogen Damage

The following section describes the adsorption of hydrogen on metal surfaces, and trapping of hydrogen at dislocations and crack tips, since they affect embrittlement and the kinetics of crack growth.

(A) Adsorption of Hydrogen

The binding energy of hydrogen to a free surface relative to a grain or phase boundary is one of the properties that determine the sensitivity of an alloy to hydrogen embrittlement (Sect. 6.4). Therefore measurements of the adsorption equilibria and the kinetics of adsorption are important. In special alloys, the rate of surface reactions controls the crack growth rate. Most structural materials are either iron, nickel or cobalt based. For these materials, a short review is given below of hydrogen adsorption and the role of surface impurities.

A schematic potential diagram of a hydrogen molecule approaching a surface is given in Fig. 6.13. The hydrogen molecule sees a shallow physisorption minimum outside the surface and a deep minimum close to the metal surface corresponding to dissociatively chemisorbed hydrogen. The two states are separated by a potential wall when adsorption is activated. The dissociative chemisorption can be expressed by the following relation:

$$H_2(g) + 2M \rightarrow 2H_{ad} + 2M \rightarrow 2(M - H_{ad}) \ . \tag{6.21}$$

From thermal desorption spectra [6.38] the hydrogen coverage on the surface, adsorption isotherms and the isosteric heat of adsorption can be obtained by solving the Clapeyron equation

$$\left(\frac{\partial \ln p}{\partial (1/T)}\right)_\theta = -\left(\frac{E_{ad}}{R}\right) \ . \tag{6.22}$$

Relevant data for hydrogen on Fe and Ni surfaces are summarized in Table 6.3. These data are used in Sect. 6.3.3 to calculate the hydrogen cov-

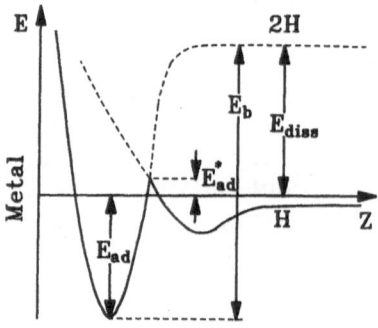

Fig. 6.13. Potential seen by a hydrogen molecule approaching a metal surface

Table 6.3. Data for surface adsorption on metal surfaces

S_0	$\nu_D A_a \left[\frac{m^2}{s \cdot H-ato}\right]$	$E_D(\beta_1)\left[\frac{kJ}{Mol-H}\right]$	$E_D(\beta_2)\left[\frac{kJ}{Mol-H}\right]$	Material	Reference
			41	Fe(Poly)	6.130
		38	50	Fe-{100}	6.39
0.03	1.5×10^{-6}		43	Fe-{100}	6.40
10^{-4}	10^{-9}		30	Fe-{100}	6.40
				$H_2 + O_2$	
			109	Fe (110)	6.39
			88	Fe (111)	6.39
0.06	3×10^{-4}	42	48	Ni-{100}	6.41
0.25	8×10^{-6}		48	Ni-{100}	6.41
	48		48	Ni (100)	6.42
			45	Ni (110)	6.42
			48	Ni (111)	6.42

S_0: Initial sticking probability.
ν_D: Frequency factor of the desorption rate constant.
A_a: Area of (100) surface trapping site. Fe: 8.18×10^{-20} m^2. Ni: 12.4×10^{-20} m^2.
$E_D(\beta_1), E_D(\beta_2)$: Desorption energy of sites β_1 and β_2.

erage, θ, at the crack tip for a stable growing crack. Hydrogen adsorbs on Fe [6.39, 40] and Ni [6.41, 42] surfaces dissociatively. The rate of adsorption is given by:

$$r_a = k_a \cdot (1 - \Theta)^2 \tag{6.23}$$

with $k_a = s_0 A_a Z \exp(-E_a/kT)$ where s_0 is the sticking probability, A_a the area of an adsorption site, Z the collision rate, $Z = p_{H_2}/\sqrt{2\pi mkT}$, and E_a the activation energy of adsorption. The term $(1 - \theta)^2$ describes the number of unoccupied sites for mobile adsorption. For immobile adsorption the number of nearest neighbors plays a role, since a hydrogen molecule needs two adjacent sites for dissociation. In addition, the roughness of the surface and the coverage dependence of the activation energy of adsorption affect the sticking probability. For the description of crack growth, these effects will be neglected to keep the equations simple. However, the surfaces near crack tips are usually rough and activated by the applied stress. Arguments that give diffusion preference over adsorption for the control of crack growth rate might be incorrect, when diffusion is compared with simple adsorption laws only.

The rate of desorption is given by

$$r_d = k_d \cdot \Theta^2 \tag{6.24}$$

with $K_d = \nu_d \exp(-E_d/kT)$, where ν_d is a frequency factor, and E_d is the activation energy of desorption. Subtracting (6.24) from (6.23), the following relationship for the surface coverage is obtained:

$$\frac{d\Theta}{dt} = k_a(1 - \Theta)^2 - k_d\Theta^2 \ . \tag{6.25}$$

This yields for the equilibrium case $r_a = r_d$ the Langmuir isotherm

$$\frac{\Theta}{1 - \Theta} = \sqrt{\frac{k_a}{k_d}} \ . \tag{6.26}$$

Surface additives play an important role in hydrogen adsorption and recombination, and therefore in hydrogen embrittlement. By lowering the recombination rate, a hydrogen recombination poison slows down the rate of H_2 formation and consequently promotes absorption of H-atoms. The effects of different elements that segregate on the surface on hydrogen embrittlement are reviewed by *Marcus* and *Oudar* [6.43]. As an example, the effects of sulfur and phosphorus are considered, since these elements are known to influence embrittlement drastically. On Ni surfaces, total suppression of H_2 adsorption occurs at sulfur saturation. For Ni (100) surfaces, the sticking coefficient follows approximately the relationship $s = s_0(1 - 4\Theta s)$ [6.43] where Θ_s is the sulfur coverage. This indicates that the influence of sulfur spreads over four nearest neighbor nickel atoms. In contrast, the influence of phosphorus is limited to the sites occupied by P atoms [6.44]. The segregation of these elements has a strong influence on the effect of hydrogen on interfacial fracture (Sect. 6.4.1).

(B) Elastic Interactions of Hydrogen with the Metal Lattice

Near the crack tip, the lattice is elastically distorted. Hydrogen atoms, each having a finite partial volume in the lattice, accumulate in the crack-tip region due to the elastic interaction between the stress fields of the hydrogen atom and the crack. Within the elastic model, hydrogen is represented as an elastic distortion occupying the interstices of the lattice. According to *Eshelby* [6.45], the interaction energy between the point defect and the lattice strain can be described by

$$W_H = -V_s \sigma_{ij} \, \varepsilon_{ij} \ , \tag{6.27}$$

where V_s is the volume of the defect site, and σ_{ij} and ε_{ij} are the tensors of the stress and strain field, respectively, produced by internal or external sources in the absence of the defect. This interaction energy must be included into the chemical potential (the lattice part), when the concentration of hydrogen at defect sites such as grain boundaries is calculated within the stress field of the crack. This yields for the chemical potential of the hydrogen atom [6.46]

$$\mu_H = \mu_H^0 + W_{H1} - W_H \ , \tag{6.28}$$

where

μ_H^0: chemical potential of hydrogen at zero stress,

W_{H1}: the change in strain energy of the solid due to the addition of n_H moles of hydrogen,

W_H: work done per mole addition of hydrogen.

For a dilute solution of hydrogen in a sample under uniaxial tension, this yields:

$$C_H^\sigma = C_H^0 \exp\left(\frac{\sigma_{11} V_H}{3RT}\right).$$ (6.29)

(C) Hydrogen-Defect Interaction

The hydrogen concentration in the elastic field of dislocations and other elastic traps can be calculated in similar ways. However, the hydrogen concentration in the dislocation core or at the crack tip core, where elastic calculations predict singular behavior, must be excluded. Comparisons with recent computer simulations (Sect. 6.4.4) show, that the elastic calculations predict correct concentrations within regions of one Burgers vector from the crack tip core. The interaction energy between a spherical particle and the elastic field of an edge dislocation can be approximated by (6.27) [6.47] which yields

$$W_H = \frac{Gb(1+v)V_H \sin\Theta}{3\pi(1-v)r} = \frac{\beta \sin\Theta}{r} .$$ (6.30)

Due to the $1/r$-term in stress, the hydrogen concentration near the core can be so large ($c = 1/4$ at 0.4 nm distance [6.30]) that the Boltzmann description (6.29) must be replaced by a description of the Fermi-Dirac type

$$\frac{c}{1-c} = \frac{c_0}{1-c_0} \exp\frac{-\beta \sin\Theta}{rRT} .$$ (6.31)

Considering a cylinder with a core cut radius of 0.248 nm, the calculations give an excess of hydrogen atoms per unit length of dislocations compared to the lattice without defects of 5×10^{-4}. Thus, the amount of hydrogen in the elastic field, the so-called Cottrell atmosphere, is negligible. Similar calculations can be made for a crack tip, if σ_{11} in (6.29) is replaced by the appropriate stress for a crack [6.30]. For a sharp crack, an elastic excess concentration of 1.2 is obtained at room temperature.

In principle, it can be concluded, that hydrogen concentrations far in excess of the lattice concentration cannot be produced by elastic interaction alone. Rather chemical interaction or trapping at preferred lattice sites (grain boundaries, particle-matrix interfaces, or dislocation cores) give the necessary local concentrations for decohesion phenomena to occur. Trap binding energies can only be measured by permeation techniques (Sect. 6.2.1) or by computer methods, like the embedded atom method (Sect. 6.4.4). In Sects. 6.3.2, 3, (6.31, 25) are used to calculate the effect of trapping on hydrogen embrittlement and on the crack growth rate. However, the defect binding energy in (6.31) is replaced by measured trap binding energies.

6.3.2 Hydrogen Trapping at Crack Tips

In the sections above, it was shown that materials embrittle only at large local hydrogen concentrations. These concentrations are observed at grain and phase boundaries, or at dislocations and crack tips. The transport and the binding of hydrogen to these critical locations must be known in order to describe HE quantitatively. The next sections describe experiments to access these parameters, like the measurement of the size of the Fracture Process Zone (FPZ) and of the mean binding energy of hydrogen to this zone.

Material data like fracture strength, reduction in area, or strain to fracture are needed for comparing the environmental sensitivity of different materials, but they give no information on the local mechanisms that lead to failure. Measurements of the "steady-state" slow crack growth, however, as a function of hydrogen pressure, temperature and loading rate, give more direct information on the local failure mechanism. The most extensive and precise documentation on slow crack growth in hydrogen gas was conducted by *Wei* et al. [6.48]. They correctly pointed out that the different temperature regimes for crack growth (Fig. 6.6) are controlled either by gas phase transport, or adsorption, or diffusion, but no results on the binding energy of hydrogen to local defects in the FPZ are given. Most slow crack growth tests were done on commercial alloys with fracture mechanic testing methods. In these tests, the stress and strain fields near the crack tip are well characterized. However, directly at the crack tip, where the action of hydrogen occurs, neither the stress nor the local microstructure is known.

Therefore, *Vehoff* et al. [6.49, 50] used a different experimental approach. They examined the growth of cracks under fully plastic conditions in FeSi and Ni single and bicrystals with and without hydrogen. In situ crack growth experiments with single crystals under tension-compression loading have shown that the cracks grow by alternating slip [6.51]. Figure 6.14a shows a typical result for a ductile crack tip of a copper single crystal obtained during in situ straining within a SEM. The alternate activation of slip planes directly at the crack is clearly visible in the subpictures 1–6. Figure 6.14b shows a similar V-shaped crack tip in a FeSi single crystal oriented for alternating slip. The active Burgers vectors lie nearly in the side surface, and the active slip planes are perpendicular to the specimen surface. The white lines on the photographs are the corresponding slip bands. The specimen was loaded in constant increments inside the SEM and after each increment a picture was taken. The upper row (subpictures 1–3 in Fig. 6.14b) shows the strengthening of the slip line A, which emanates from the crack-tip apex and runs to the lower right. The corresponding displacement of the right-hand side of the crack towards the lower right is clearly visible, especially if the position of the crack-tip apex is examined in relation to the lowest slip line on the left. In sub-picture 3, work-hardening has stopped the activity on the old slip line, A, and another slip line, B, which emanates from the crack-tip apex and runs to the lower left, is activated and strengthened in subpictures 4–6. Again, the corresponding motion of the left-hand side of the crack towards the lower left is visible. This process goes on with further straining in vacuum.

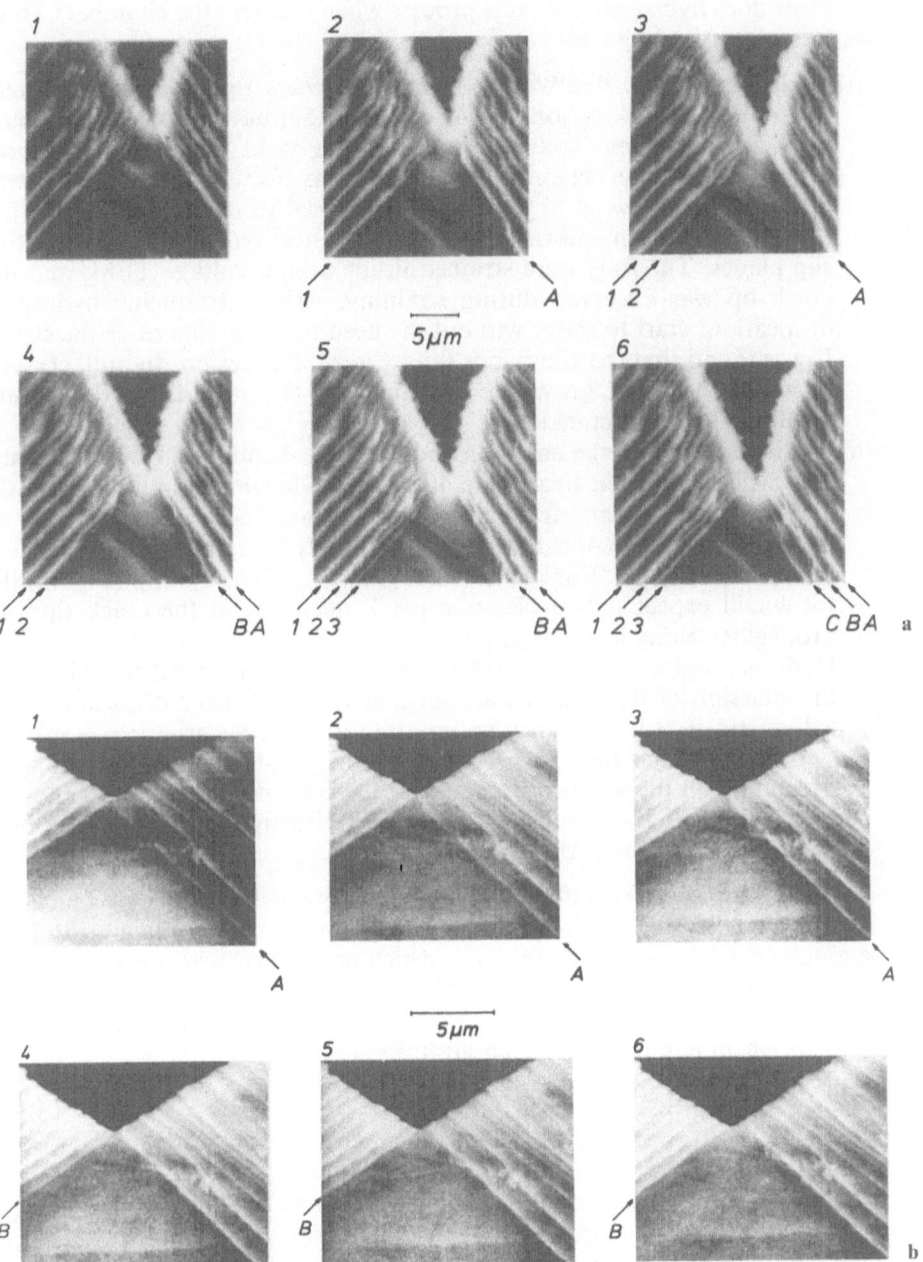

Fig. 6.14. (a) Ductile crack-tip in a Cu single crystal strained in situ in a SEM. **(b)** Ductile crack tip in a FeSi single crystal strained in situ inside a SEM

How does hydrogen alter this process when it enters the chamber? Three processes can be imagined:

(i) Hydrogen is absorbed within the stressed crack-tip zone and facilitates the motion of dislocations. *Matsui* et al. [6.52] have shown that in pure iron single crystals hydrogen reduces the yield stress and enhances dislocation motion when the strain rate was low. Similar evidence was given by *Robertson* et al. [6.53]. They showed in experiments with thin foils that hydrogen embrittles Ni by highly localized rupture along active slip planes. The foils were strained inside a high voltage TEM, and the crack tip was observed during straining. After introducing hydrogen, dislocations start to move without the need to further increase the stress. It was found that the slip bands fail by local decohesion. In bulk crystals in which the crack grows by alternate slip (Figs. 6.14a,b), this would result in slip band cracking.

(ii) Hydrogen impedes the emission and hampers the motion of dislocations. This will, for a given strain rate, increase the local stress at the crack tip and facilitate cleavage fracture. That hydrogen hampers the motion of dislocations was demonstrated experimentally by several groups for less pure iron [6.54, 55]. For a crack that grows by alternate slip (Fig. 6.14b) we would expect that a cleavage crack nucleates at the crack tip and propagates along a cleavage plane.

(iii) Hydrogen reduces the cohesive forces locally, hence bond breaking and the emission of dislocations are facilitated (for an edge dislocation, the ledge term is reduced, since an emitting dislocation produces a surface step at the crack tip). Either cleavage or ductile fracture can happen depending on the detailed processes at the crack-tip. The overall effect is a reduction of the crack-tip angle when cleavage and ductile rupture occur simultaneously along the crack front.

According to *McClintock* [6.56], in fully plastic solids, a crack-tip opening angle, α, smaller than the angle, α_d, between the slip planes, can only be obtained if a second fracture mode, which propagates with the velocity, v_o, and which produces a crack-tip opening angle of zero degrees (cleavage) is superimposed to ductile rupture. For a stable growing crack the resulting crack tip opening angle, α, is then given by:

$$\cot\frac{\alpha}{2} = \cot\frac{\alpha_d}{2} + 2\frac{v_0}{d\delta/dt} \ , \tag{6.32}$$

where $d\delta/dt$ is the crack-tip opening rate of a stable growing crack. Figure 6.15 shows a crack-tip angle, α, which was obtained in FeSi single crystals after straining first in vacuum and then further in hydrogen gas (30 kPa) at a crack-tip opening rate of 100 nm/s. The smaller angle compared to the angle in vacuum is clearly visible. Decreasing the pressure or increasing the crack-tip opening rate increases the crack-tip opening angle continuously (Fig. 6.16). It is important to know, whether the mixture between plastic blunting and cracking happens on an atomistic scale (emission of isolated dislocations mixed with breaking isolated rows of atomic bonds),

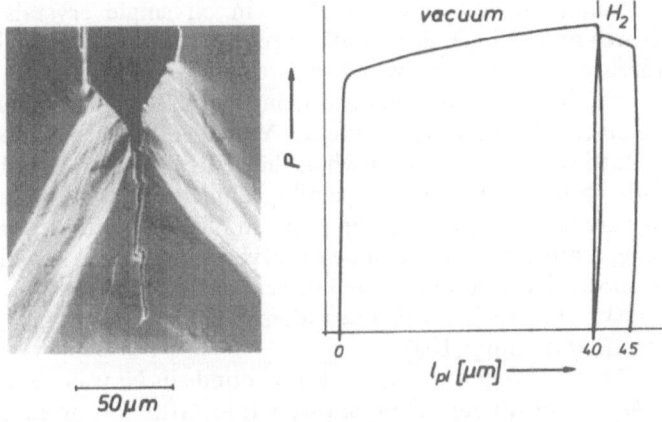

50μm

Fig. 6.15. Crack-tip in FeSi, first strained in vacuum, then further strained in hydrogen gas, and the corresponding load, P, plastic elongation, l_{pl}, curve

on a macroscopic scale (brittle discontinuous crack jumps followed by extensive plastic blunting) or on an intermediate scale. Inspections of secondary carbon replicas of fracture surfaces showed that discontinuous crack jumps must be smaller than 0.1 μm [6.57]. These findings were confirmed by several groups [6.58]. In a recent acoustic emission analysis of fracture in FeSi single crystals it was observed that cracks in hydrogen proceeded along the crack front by the nucleation of submicron microcracks which were embedded in a ductile matrix [6.59].

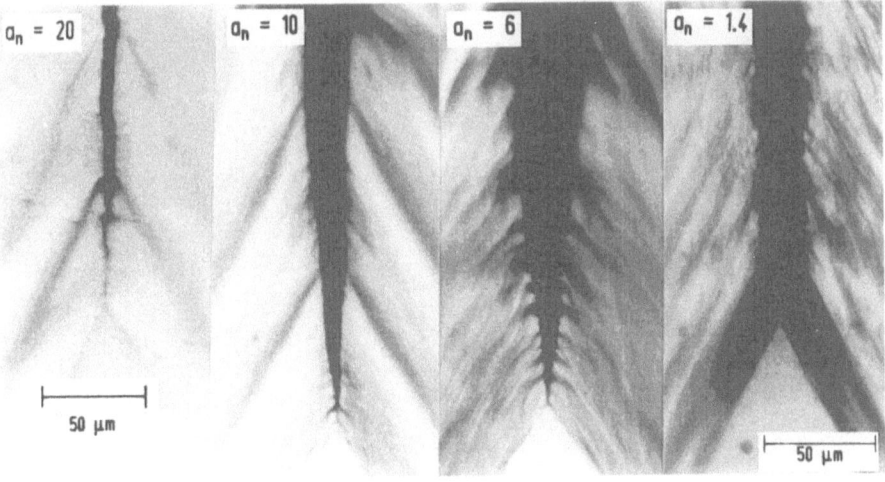

Fig. 6.16. Crack tip opening angles obtained in FeSi single crystals after straining at different hydrogen pressures (100, 10, 0.7 Pa) and in vacuum

Similar results were obtained in Ni single crystals after stable crack growth in hydrogen gas [6.50]. Figure 6.17 shows the typical appearance of a fracture surface in Ni which was obtained after a test at a gas pressure of 60 kPa. The photographs are from matching fracture surfaces of the same specimen. On the surface, ductile, Y-shaped hillocks can be seen which match peak to peak (clearly seen when the central Y-shaped hillock is compared in both photographs). The crystallographic planes forming the surface markings are (111)-planes (the slip plane in Ni). Between these markings, relatively large, featureless areas can be observed which are oriented along (100)-planes as shown by secondary carbon replicas of the fracture surface [6.57]. Obviously, the crack propagates alternately by slip along (111)-planes and by decohesion along (100).

That hydrogen weakens atomic bonds in Ni was clearly demonstrated by *Vehoff* et al. [6.60]. They showed that, after pronounced charging, highly ductile Ni bicrystals fail along the grain boundary by unstable cleavage without any evidence of cavity nucleation and coalescence. This will be further discussed in Sect. 6.4. In this section, we will only keep in mind that the crack proceeds by subsequent bond breaking and dislocation emission and examine the question, how temperature and hydrogen pressure influence this balance under the condition of stable crack growth.

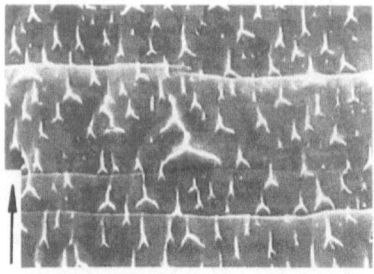
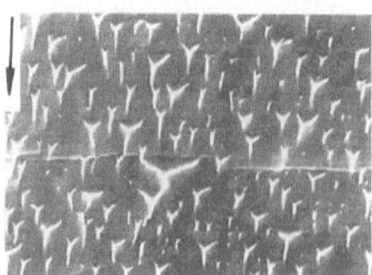

Fig. 6.17. Matching fracture surfaces of a Ni single crystal obtained after fracture in hydrogen gas; the arrows indicate the direction of crack growth

3μm

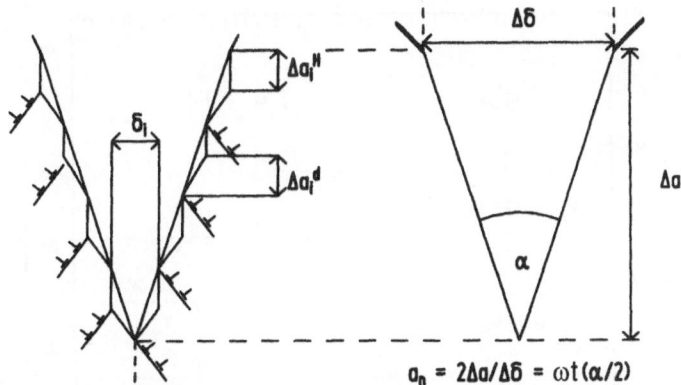

Fig. 6.18. Crack-tip produced by the alternate emission of dislocations and bond breaking

Figure 6.18 shows a schematic drawing of a crack tip obtained by a mixture of alternate slip (Δa_i^d) and local cleavage (Δa_i^H). The local crack increment, Δa_i, is given by:

$$\Delta a_i = \Delta a_i^d + \Delta a_i^H \ . \tag{6.33}$$

Dividing (6.33) by the local crack opening increment, δ_i, yields for the crack tip opening angle, α:

$$\cot\frac{\alpha}{2} = a_n^d + a_n^H = a_n \ , \tag{6.34}$$

where $a_n = 2\Delta a_i/\delta_i$ is the ductile crack growth increment normalized by the crack opening displacement, and a_n^H is given by:

$$a_n^H = \frac{2v_0}{d\delta/dt} \tag{6.35}$$

according to (6.34), ideal ductile crack growth is characterized by $a_n^H = 0$, and $a_n^H > 0$ can be used as a quantitative measure of embrittlement.

(A) Equilibrium Models of Crack Growth

At low crack growth rates, corresponding to region I of the crack growth rate, da/dt, vs. stress intensity, K_I, curve (Fig. 6.4), hydrogen is in local equilibrium at the crack tip. To obtain an estimate of the binding energy of hydrogen to the crack tip or to defects in the near-crack-tip region, we assume that a_n, which is a measure of the ratio of broken bonds, (given by Δa_i), to the number of emitted dislocations (measured by δ_i, the local crack opening increment) is related to the hydrogen coverage at the crack tip in a simple way (Fig. 6.18). By measuring the crack-tip opening angle, α, of a stable growing crack, a_n can be measured directly via the relationship,

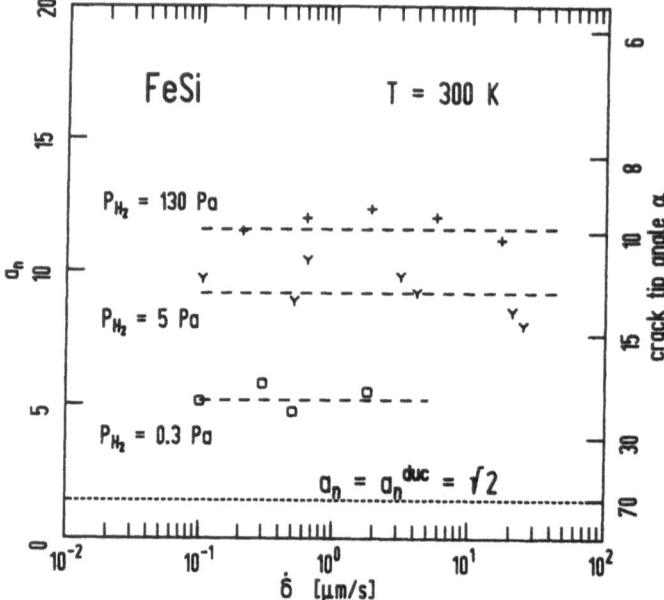

Fig. 6.19. The normalized crack growth rate, a_n, as a function of the crack-tip opening rate for three different pressures

$a_n = \cot(\alpha/2)$. The simplest assumption would be, that a_n is directly proportional to the hydrogen coverage at the crack tip

$$a_n = \beta n , \tag{6.36}$$

where n is the mean coverage of hydrogen at deep binding states near the crack tip. These states can be either surface states, dislocation cores, or even special binding states at the heavily distorted bonds in the near tip region. Computer simulations of *Baskes* and *Daw* [6.61] support the assumption of a special crack-tip state. These calculations predict that hydrogen is more strongly bound to the crack tip than to volume defects, like dislocations, or at free undisturbed surfaces. Further support is given by desorption measurements, which show that hydrogen is more strongly bound to surface steps than to plane surfaces.

Can this assumption be checked experimentally? Combining (6.36) with (6.31), the pressure and temperature dependence of a_n can be described by

$$\frac{a_n^H/\beta}{1 - a_n^H/\beta} = k_0\sqrt{p_{H_2}}\ \exp(U/kT) \tag{6.37}$$

with

$$k_0 = S/N_L$$

for volume traps, where S is the constant of Sievert's law.

In the general case, the binding energy U depends on the stress at the crack tip. To facilitate the comparison with experimental results, (6.37) can be rewritten:

$$\frac{1}{a_n^H} = \frac{1}{\beta}\left(\frac{\exp(-U/kT)}{k_0\sqrt{p_{H_2}}} + 1\right).$$
(6.38)

Plots of $1/a_n^H$ vs $1/\sqrt{p_{H_2}}$ should yield straight lines which intersect the $1/a_n^H$-axis at $1/\beta$, which yields $a_n^H = \beta$ for large pressures.

Experiments on FeSi and Ni single crystals have shown that for low displacement rates a_n is independent of the crack growth rate. Typical results are given in Fig. 6.19 for three different pressures. In this pressure range local equilibrium of the hydrogen at the crack-tip is assumed. Figures 6.20a–c show plots of a_n vs. the hydrogen pressure according to (6.37) and (6.38) for FeSi and Ni single crystals. For both crystal structures the linear relationship between $1/a_n$ and $1/\sqrt{p_{H_2}}$ is fulfilled for a large temperature range (Fig. 6.20b); the straight lines intersect the axis at $1/\beta$ for all temperatures, as proposed by our assumption of proportionality between a_n and β.

According to (6.37), a strong temperature dependence of a_n is expected. Equating pairs of data obtained at different temperatures and pressures, but with the same crack-tip opening angle, hence the same a_n-value, and plotting the data in a semi-logarithmic plot of p_{H_2} vs. $1/T$, a straight line should be

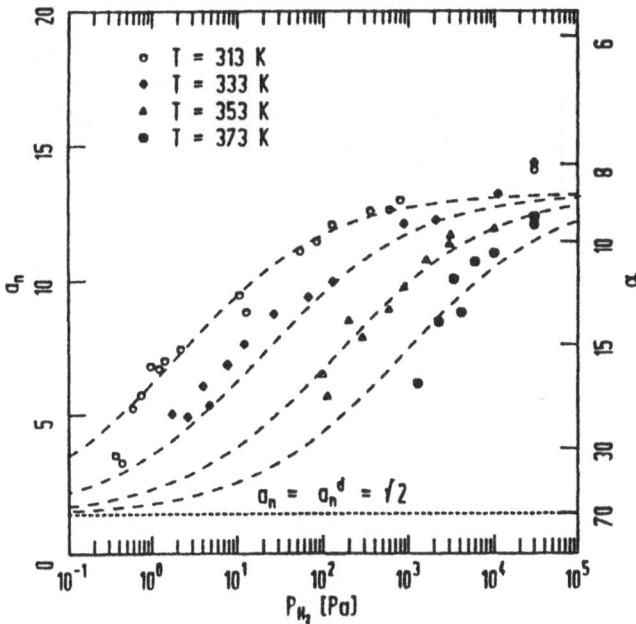

Fig. 6.20a–c. The normalized crack growth rate, a_n, as a function of the hydrogen pressure for different temperatures for FeSi single crystals (**a,b**) and nickel single crystals (**c**)

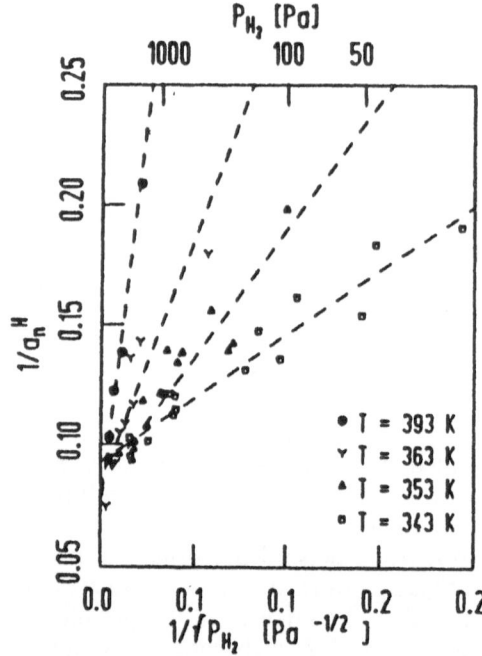

b

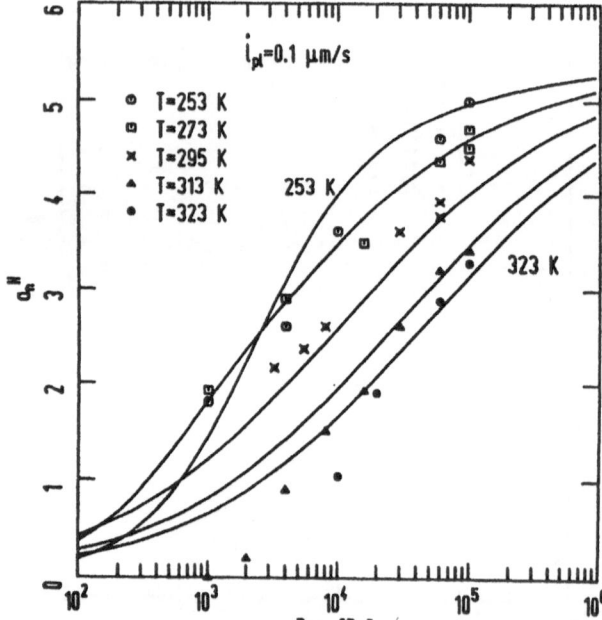

c

Fig. 6.20b,c

obtained. The slope of this line is a measure of the apparent binding energy of hydrogen to the crack tip as can be seen by solving (6.37) for p_{H_2}:

$$p_{H_2} = \left(\frac{a_n^H/\beta}{1 - a_n^H/\beta}\right)^2 \frac{1}{k_0^2} \exp\left(\frac{-2U}{kT}\right) = k_0' \cdot \exp\left(\frac{-2U}{kT}\right). \qquad (6.39)$$

Figure 6.21 demonstrates that (6.39) describes the experimental data satisfactorily. The pressure and temperature dependence of the surface coverage, θ, and the occupation of deep binding states like dislocation cores and interfaces (Sect. 6.3.1) with hydrogen follow similar relationships. In Fig. 6.22, all relevant data for the surface coverage, $x = \theta$, trapping, $x = n$, and fracture, $x = a_n/\beta$ are plotted in a diagram of $\log(\sqrt{p_{H_2}}(1 - x)/x)$ vs. $1/T$. The continuous lines through the data were obtained with the following constants for FeSi: $k_0 = 2.9 \times 10^{-9}$ Pa$^{-1/2}$, $\beta = 12$, $U = 49$ kJ/molH, and for Ni: $k_0 = 1.1 \times 10^{-5}$Pa$^{-1/2}$, $\beta = 5.4$, $U = 22$ kJ/molH.

The different slopes obtained for FeSi and Ni (Fig. 6.22) demonstrate that hydrogen atoms are more strongly bound to a crack-tip in FeSi than to a crack tip in Ni. This is in accordance with the general observation that traps in Ni (f.c.c) are weaker than traps in Fe (b.c.c) [6.62, 63]. When we compare the surface coverage data for Ni with FeSi – lower dashed (Ni) and dotted (FeSi) lines, data from flash desorption measurements [6.39, 42] – it can be seen that for FeSi and Ni similar pressures are needed to fill the surface states. Therefore, if it were the case that embrittlement is controlled by the filling of surface states, we would expect for Ni and FeSi single crystals similar pressure and temperature dependencies. This was not observed.

In Ni, pressures nearly six orders of magnitude higher were needed to fill the crack tip region with hydrogen. In addition, much higher pressures were needed to obtain brittle fracture in Ni than in FeSi, and a definitely lower

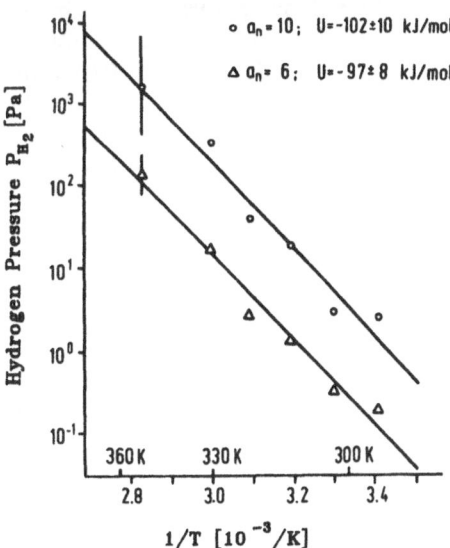

Fig. 6.21. Arrhenius plot of the hydrogen pressure vs. 1/T for two different normalized crack growth rates, a_n

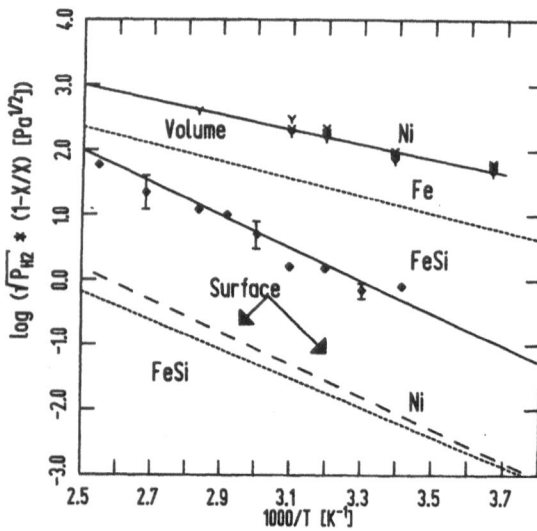

Fig. 6.22. Plot of log $\left[\sqrt{p_{H_2}}\right.$ $\left.(1-x)/x\right]$ vs. $1/T$ (for details see text)

temperature dependence was observed. However, when the data are compared with data for trapping (upper dotted line in Fig. 6.22 [6.64]) it is found that for both materials the apparent binding energies of hydrogen to the crack tip are larger than the binding energies obtained from permeation transients in thin deformed sheets [6.65, 66]. In addition, to fill volume traps higher pressures are needed than to fill crack-tip sites.

The results presented above let us conclude that the filling of special crack-tip sites, which are neither simple surface traps nor simple volume traps controls hydrogen embrittlement. Hydrogen is more strongly bound to the crack tip than to interfaces or dislocation cores. With increasing coverage, the balance between bond breaking and dislocation emission is shifted towards bond breaking; experimentally this resulted in a smaller crack-tip opening angle. From these experiments nothing is learned about the local failure mechanisms. The experiments only suggest that even at low external pressures a hydrogen coverage of nearly one can be obtained in the near-crack-tip region, which would be enough to weaken the atomic bonds locally. They give no information about the size of this region. In the next section, experiments on the kinetics of crack growth will be given from which this size can be estimated.

6.3.3 Kinetic Aspects of Hydrogen Damage

In the next sections the transport processes which control the crack growth rate, and the role of gas purity are discussed in detail.

(A) Kinetic Models of Crack Growth

The various processes, which might control the kinetics of crack growth in hydrogen gas are illustrated schematically in Fig. 6.23 and are as follows [6.67]:

1) Transport of the gas to the crack tip
2) Physical adsorption, activated or non-activated
3) Dissociative chemical adsorption
4) Hydrogen entry (absorption)
5) Diffusion of hydrogen in the stress field to the FPZ
6) Trapping at internal interfaces

The pressure and temperature dependence of the stage II crack growth rate (Fig. 6.4) is controlled by the transport of hydrogen to the crack tip. In the range of stage II crack growth, the crack growth rate is nearly independent of the driving force, described by K_I in fracture mechanics. The crack grows steadily maintaining a constant crack-tip opening angle, α. The opening angle increases when the driving force is increased since the amount of hydrogen within the FPZ decreases. Hence the resistance against crack growth increases as well, which results in a nearly constant crack growth rate. However, the crack grows faster when the pressure or temperature is increased since in the low-temperature region hydrogen gets to the crack tip more rapidly when the temperature or pressure is raised. *Wei* et al. [6.68, 69] identified a number of rate-controlling processes including transport of the deleterious gas to the crack tip, reaction of the gas with the newly created surface, and diffusion of hydrogen to the FPZ. Which of these processes controls the crack growth rate depends on the experimental conditions and the material used. In the following only surface adsorption and diffusion are considered in detail.

Fig. 6.23. Various rate controlling steps for the transport of hydrogen to the fracture process zone (FPZ)

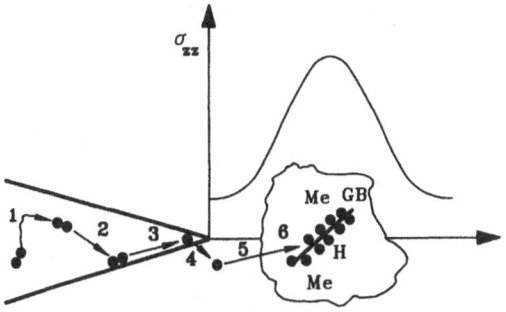

Transport Processes
1. Gas Phase Transport
2. Physical Adsorption
3. Dissociative Chemical Adsorption
4. Hydrogen Entry
5. Diffusion
6. Trapping at Internal Interfaces

In the section above, arguments are given which demonstrate that the hydrogen which adsorbs only on top of the crack-tip surface is not responsible for embrittlement, rather hydrogen must enter the FPZ (Fig. 6.23). However, this does not mean that surface reactions cannot control the rate of stage II crack growth. Assuming dissociative adsorption, the rate laws given in Sect. 6.4.1, can be used to calculate the coverage at the tip of a running crack.

Solving (6.25) for a steadily growing crack $-d\theta/dt = ad\theta/da$ gives

$$\frac{1}{\theta} = 1 + \sqrt{\frac{k_d}{k_a}} \coth(\sqrt{k_a k_d} x_0/\dot{a}) \ . \tag{6.40}$$

For slow crack growth, $\dot{a} \rightarrow 0$, i.e. the range of steady-state crack growth, this yields:

$$\frac{\theta}{1-\theta} = \sqrt{\frac{k_a}{k_d}} = \sqrt{\frac{s_0 A_a}{v_D \sqrt{2\pi m k T}}} \cdot \sqrt{p_{H_2}} \cdot \exp\left(\frac{E_d - E_a}{2kT}\right) \ . \tag{6.41}$$

Relevant data for Fe and Ni single crystals are given in Table 6.3. In cases where hydrogen embrittlement is only controlled by the surface coverage of hydrogen at the crack-tip, (6.41) describes the temperature and pressure dependence of crack growth. However, for most materials definitely higher pressures are needed to cause embrittlement than predicted by (6.41) with the data of Table 6.3. However, equations of the functional type given by (6.41) fit most crack-growth data for high-strength steels remarkably well [6.70].

For the other limiting case, large crack growth rate, $\coth(\sqrt{k_a k_d} x_0 \ \dot{a})$ in (6.40) can be approximated by $1/(\sqrt{k_a k_d} x_0/\dot{a})$ which yields

$$\dot{a} = \frac{1-\theta}{\theta} \ s_0 \ A_a Z \exp(-E_a/kT) \cdot x_0 \ . \tag{6.42}$$

Equation (6.42) predicts a decrease in the crack growth rate with decreasing temperature for thermally activated adsorption. However, adsorption of hydrogen on clean Ni and Fe surfaces was found to be non-activated. But, most materials which do not form hydrides show a strong decrease in embrittlement with decreasing temperature (Fig. 6.6). Therefore either hydrogen adsorbs over a precursor state to the stressed bonds, or the kinetics of embrittlement cannot be explained by surface reactions only.

(B) Single Crystals

Pasco and *Ficalora* [6.71], guided by their adsorption studies on deformed surfaces, proposed that the adsorption behavior of stressed surfaces is different. They assumed that under otherwise identical conditions the coverage at a crack tip is much lower than on an undeformed surface, and that the adsorption into the state which causes embrittlement occurs via a precursor state. Likewise for Ni, it was found, that below 200 K hydrogen adsorbed into a different surface state than at higher temperatures. This surface state was the only state occupied below 200 K [6.72]. It has been hypothesized that the low-temperature form corresponds to the classical picture of chemisorption: the hydrogen atom resides just outside the surface, probably on top of a metal atom, whereas at room temperature the hydrogen has moved into or underneath the first layer of metal atoms [6.73]. If this β_2-state, which is energetically higher, is responsible for embrittlement we have an effective energy of adsorption, E_a, into this state, and a decrease in coverage of this state with decreasing temperature. The arguments given above support the view that embrittlement is caused by hydrogen which segregates to special crack-tip sites.

Klameth [6.74] has measured the crack growth rate in Ni single crystals in a temperature range in which the transport of hydrogen to the crack tip controls the growth rate. His results are given in Fig. 6.24 in which the crack growth rate is plotted vs. $1/T$ for $a_n = $ const. according to (6.42). This plot yields an activation energy for the adsorption to the crack-tip trap of $E_a = 23$ kJ/Mol$_{H2}$. But direct evidence for a thermally activated entry or adsorption of hydrogen is still lacking. According to current knowledge it is not possible to calculate the coverage on a crack-tip from surface data alone.

So far, we have considered only surface reactions as possible rate-controlling steps. However, as is shown in Sect. 6.4.1, the functional dependence on temperature and pressure is similar for segregation of hydrogen to surface states or to deep binding states (traps) within the lattice. Hence, again an

Fig. 6.24. Arrhenius plot of the crack growth rate, da/dt vs. $1/T$ (constant pressure, Ni single crystals)

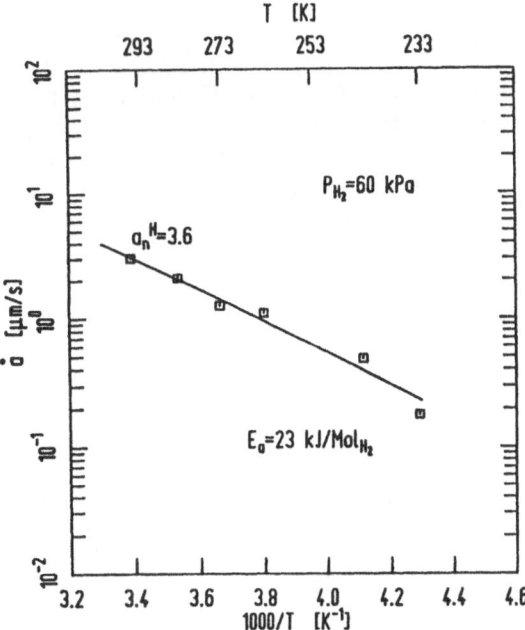

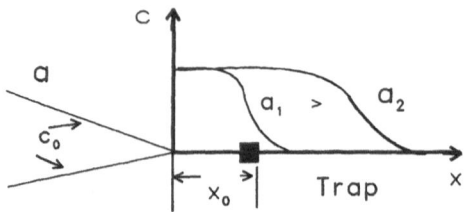

Fig. 6.25. Hydrogen concentration profile seen by a trapping site in front of a stable growing crack

increase in coverage at trapping sites with decreasing temperature is expected. However, provided that for embrittlement hydrogen has to enter the lattice and a finite zone, the FPZ, must be filled with hydrogen, diffusion processes might become rate limiting.

If the hydrogen adsorption at the crack-tip is fast enough to maintain a constant concentration, C_0, at the tip, the concentration profile in front of a steadily growing crack is given by [6.75, 76]:

$$\frac{c}{c_0} = \operatorname{erfc}\left(\sqrt{\frac{(x+r)}{2D}\dot{a}}\right) , \qquad (6.43)$$

where $D = D_0 \exp(-E_D/RT)$ is the diffusion constant.

Now, with the help of (6.31) we are able to estimate the coverage of traps, n, in front of a steadily growing crack. The model is sketched in Fig. 6.25. For the simplest case (hydrogen diffusion is not hampered by other traps) the following estimate for the coverage is obtained:

$$\frac{n}{1-n} = \left[\frac{S_0}{N_0}\sqrt{p}\operatorname{erfc}\left(\sqrt{\frac{x_0\dot{a}}{D}}\right)\right] \cdot \exp\left(\frac{E_b(\sigma) - E_s}{kT}\right) . \qquad (6.44)$$

The parameters S_0, N_0, D, E_b, E_s can be measured independently and are given in Table 6.4 for Ni, Fe and a superalloy. The Figs. 6.26a–c show for Ni single crystals measurements of a_n as a function of the crack growth rate, the hydrogen pressure and the temperature, respectively. By combining (6.44) with (6.36), hence assuming again that a_n is proportional to the occupation of traps with hydrogen, a relationship between a_n, T, P_{H_2} and the crack growth rate is obtained. This relationship is used to fit the data given in Figs. 6.26a–c (solid lines). The mean distance of traps from the crack tip, x_0, is the only

Table 6.4. Diffusivities and trap binding energies of some representative materials

Material	D_0 [m²/s]	E_D [kJ/ mol H]	S_0 [H-Atom/m³ $\sqrt{Pa}$]	E_s [kJ/ mol H]	E_b [kJ/ mol H]	E_s-E_b [kJ/ mol H]	Ref.
Nickel	6.4×10^{-7}	40	2.5×10^{23}	12	33	−21	6.131, 74
Eisen	5.0×10^{-8}	4.1	9.5×10^{23}	30	56	−26	6.132
INCO 903	2.46×10^{-6}	52.7	6.52×10^{22}	4.6	20	−15.4	6.133, 64
CMSX-2						70	6.134

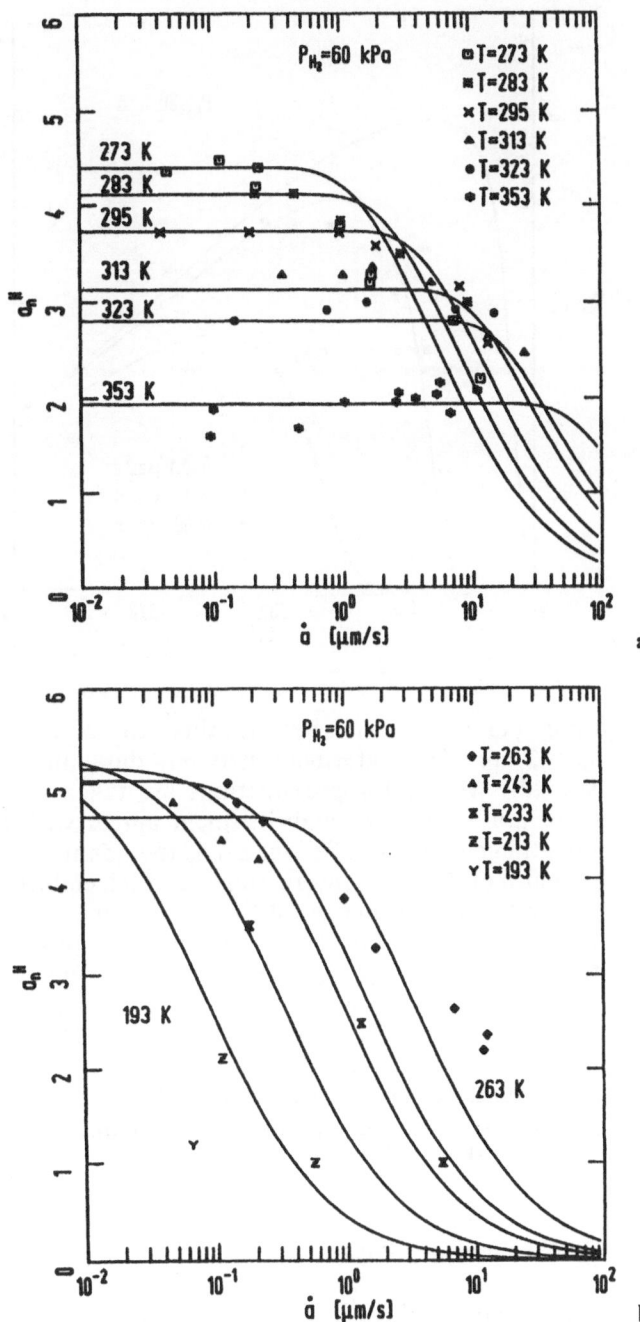

Fig. 6.26. (a,b) Normalized crack growth rates, a_n as a function of da/dt, and **(c)** a_n as a function of T. The curves are calculated according to the model depicted in Fig. 6.25

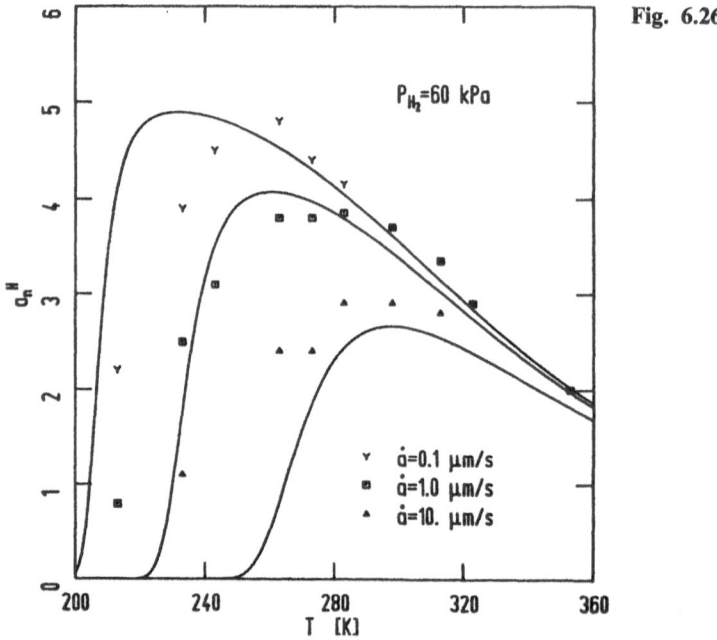

Fig. 6.26c

fitting parameter. The best fit through all the data was obtained with $x_0 = 2$ nm. The agreement between data and prediction is reasonable, especially when all approximations are taken into account. The predicted temperature range for embrittlement agrees with the measured range. The activation energy of diffusion, E_D, is the dominating parameter that controls the width of the temperature range in which embrittlement is predicted at low temperatures. A value of 40 kJ/mol was used in agreement with most literature values for Ni. The slope given by (6.44) for low temperatures is steeper than the slope observed in the experiments. Better agreement can be obtained if the diffusivity in the FPZ is taken as a free parameter or if activated adsorption is assumed, (Fig. 6.24). The curves in Figs. 6.26a,b shift by one decade to the left, if a ten times larger value for x_0 is used. If the value for D is taken as an upper bound for the undisturbed lattice, the assumption is supported that embrittlement is controlled by the filling of deep traps directly at the crack-tip.

(C) Commercial Alloys

In commercial alloys it is often observed that small changes in the alloy composition can drastically change the sensitivity of the alloy against hydrogen embrittlement. For all alloys, the maximum of embrittlement was found near room temperature. With increasing hydrogen pressure the width of the maximum increases towards larger temperatures. Hydrogen effects can

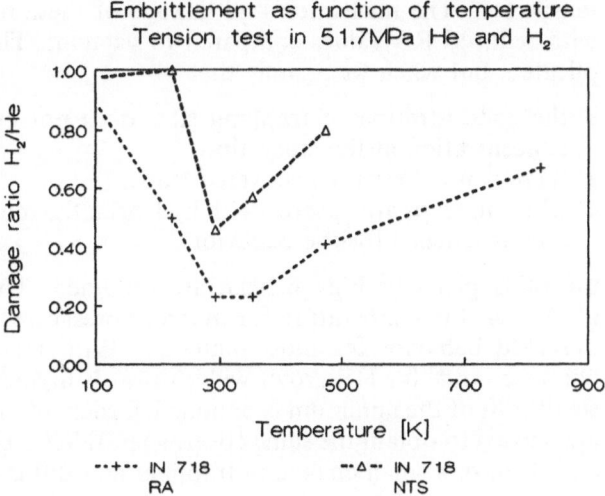

Fig. 6.27. Damage ratio of RA vs. T for tensile tests in He and H_2, IN 718 [6.77]

be observed up to temperatures of 700 K as demonstrated in Fig. 6.27, in which the ratio of the reduction in area, RA, in hydrogen and in helium, RA_{H_2}/RA_{He} is plotted vs. the temperature [6.77, 78]. Compared to pure Ni or FeSi, pressures of more than 10 MPa were needed for HE as shown in Fig. 6.28.

Can the trapping model explain these alloying effects? Taking the data for nickel and INCOLOY 903 from Table 6.4, (6.44) was used to calculate the coverage of traps in front of a micro crack, which nucleates during a

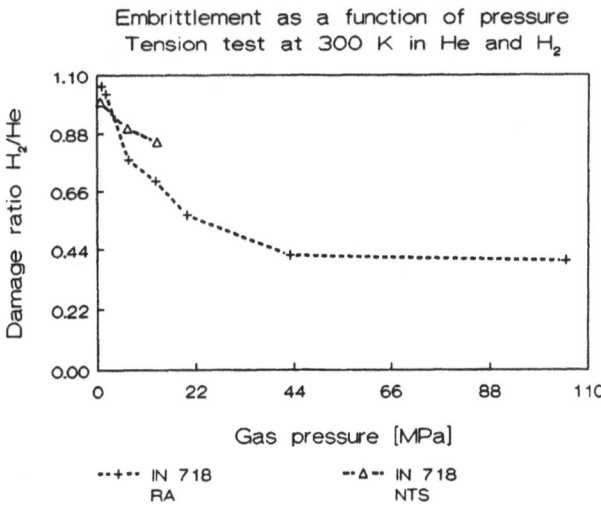

Fig. 6.28. Damage ratio of RA vs. hydrogen pressure at room temperature, IN 718 [6.77, 78]

tensile test. The nucleation and linkage of these micro cracks led to failure with reduced RA-values compared to vacuum. The overall process is complicated, but when we assume that

- the concentration of trapping sites does not change with the hydrogen concentration at the crack tip,
- there is no effect for uncovered traps,
- when all traps are covered with hydrogen the minimum in the reduction of area is reached (brittle behavior),

the plots given in Figs. 6.29a,c are obtained. The plots, in which $1-n$ is plotted vs. the temperature for different pressures, describe qualitatively the observed behavior for most materials. With increasing pressure the temperature range for HE grows wider towards higher temperatures, but only a small shift of the minimum is obtained. Orders of magnitude higher pressures are needed to obtain the same coverage in INCOLOY as in nickel (Figs. 4.29 a,b). Hence, a small change in trapping and diffusion alters the degradation of the material properties due to hydrogen drastically. This explains that even for similar materials large differences in hydrogen embrittlement are observed (Sect. 6.1).

The discussion given above is far from being quantitative. But it has to be kept in mind that the rate processes discussed above have nearly similar activation energies. Hence, the rates are not truly uncoupled. The apparent activation energies measured in crack-growth tests are a combination of all possible reaction rates, and a correspondence with a special rate can even be accidental. However, *Gerberich* et al. [6.79] have shown in a thorough review of most existing data on steels that a functional relationship of the kind given by (6.44) can fit the data.

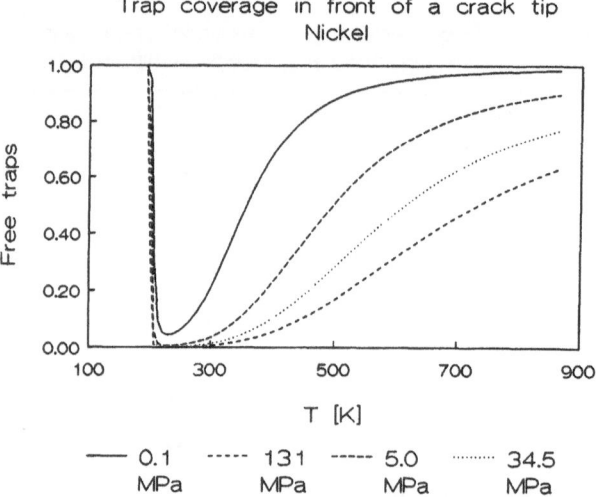

a

Fig. 6.29a–c. Uncovered traps, $1-n$, as a function of T according to (6.44) with data for **(a)** Ni, **(b)** IN 903, **(c)** $1-n$ as a function of pressure for IN 903

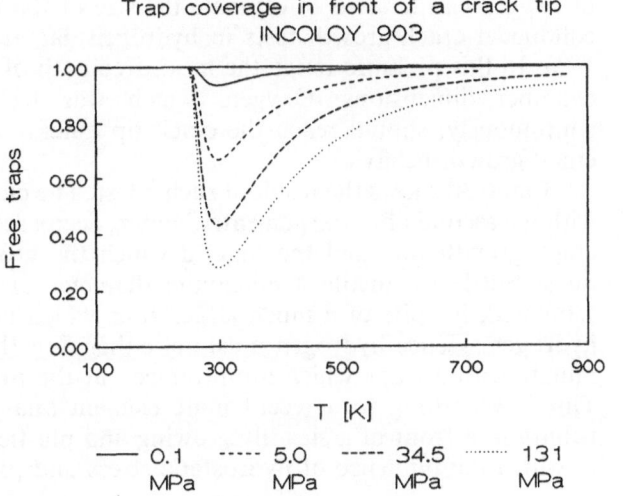

Trap coverage in front of a crack tip
INCOLOY 903

b

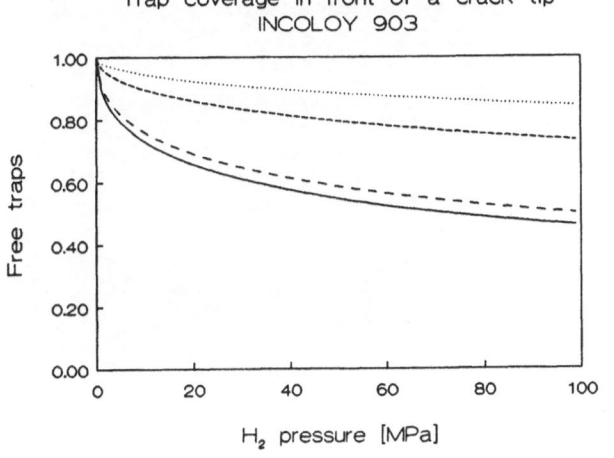

Trap coverage in front of a crack tip
INCOLOY 903

Fig. 6.29b,c

c

(D) Adsorption of Oxygen

So far, we have discussed only the effects of pure hydrogen. How do other gases influence the transport processes? Thermodynamic arguments given in Sect. 6.4 predict that a gas which strongly adsorbs reduces the surface energy and therefore promotes embrittlement. Therefore oxygen, which adsorbs strongly on iron surfaces, should be a strong embrittler in contradiction to observations by *Hancock* and *Johnson* [6.80]. They showed that mixing small amounts of oxygen to the hydrogen environment can suppress HE completely.

At room temperature, oxygen adsorbs preferentially to hydrogen and prevents the entry of hydrogen into the fracture process zone. This behavior of oxygen can be used to estimate the size of the FPZ in FeSi. *Rothe* [6.81] conducted crack-growth tests in hydrogen gas at a partial pressure below 1 Pa. In this pressure range the mean free path of the gas is larger than the chamber dimensions. Oxygen, which was let into the chamber discontinuously, should reach the crack tip quickly and alter the steady-state crack growth behavior.

Fig. 6.30 shows the result of such a test. The curve $P(\delta)$ changes the slope within a second after oxygen entry, and a_n decreases from 10 to 3.3. From the crack growth rate and the time in which the growth mode changed from quasi brittle to ductile a maximum diameter of 50 nm for the FPZ was estimated, in spite of a much larger zone which is definitely saturated with hydrogen. Hence hydrogen must act either directly on the surface or accumulate within traps which are produced at the tip during plastic straining. This is according to a recent finite element analysis of the hydrogen distribution in front of a steadily growing and plastically strained crack [6.82], in which the influence of hydrostatic stress and plastic deformation on dif-

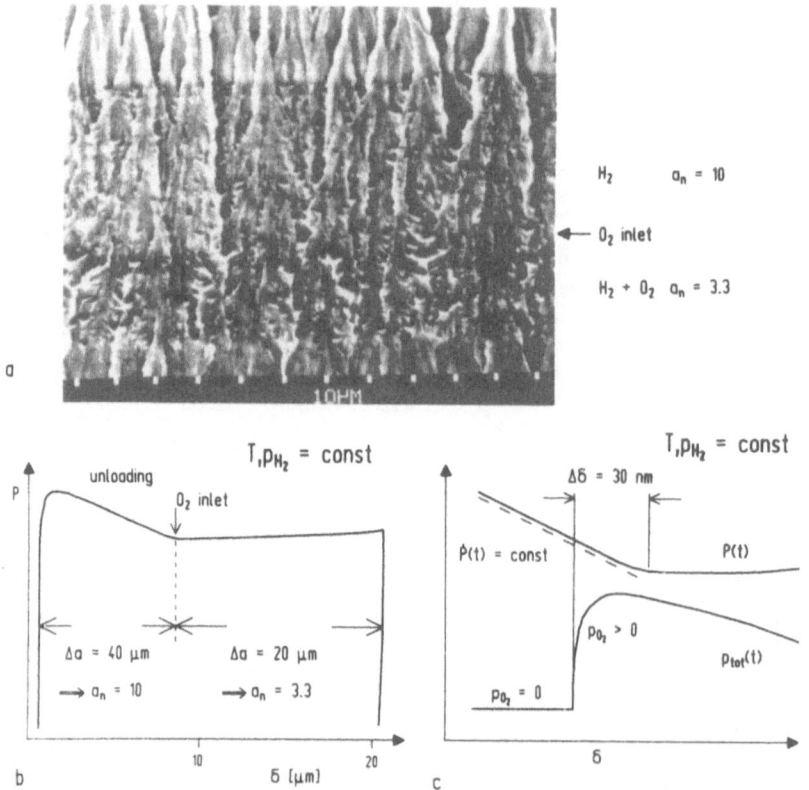

Fig. 6.30. Fracture surface and load elongation curves showing the effect of oxygen on HAC (FeSi single crystal). The $P(\delta)$-curve changes its slope immediately after O_2 inlet

fusion and trapping is simulated. In the calculation it is assumed that the density of trapping sites, which can be considered to be dislocation cores, increases with increasing equivalent strain at the crack tip. The calculations showed that the maximum of the hydrogen concentration lies directly at the crack tip and not in the region of maximum hydrostatic stress in front of the tip, in agreement with the conclusions drawn from the experiments.

Why does oxygen not embrittle? It adsorbs strongly and therefore should reduce the surface energy even more than hydrogen. This argument is an old one but physically correct. So it must be assumed that hydrogen must enter the lattice and segregate at the stressed bonds of the crack-tip core. Oxygen atoms, however, with their larger size cannot enter the lattice and thus do not lead to embrittlement at room temperature.

6.4 Failure Mechanisms

So far, we have become familiar with the many phenomena evoked by hydrogen. Hydrogen embrittles many materials, either by forming new brittle phases (hydrides), or by inducing martensitic transformation, or directly. Detailed experiments have shown, that hydrogen must enrich locally either at crack tips, or interfaces. The transport mechanisms of hydrogen to the locations where cracks or cavities nucleate determine the rate dependence of crack growth. A combination of several rate processes and local equilibria finally results in the observed crack growth rate. But none of the experiments and models given explain how hydrogen embrittles a material. In the following section this question is addressed more closely. Several concepts are presented, which all have in common that a direct experimental evidence is still lacking.

In Sect. 6.4.1, we start with the thermodynamic view of fracture and examine how segregating elements reduce interface and surface energies, then the detailed processes at the crack tip, dislocation emission and shielding will be considered in view of hydrogen effects (Sect. 6.4.2), and in Sect. 6.4.3 some brief remarks on current computer simulations of the hydrogen problem are given

6.4.1 Interfacial Fracture

This section concentrates on the thermodynamic view of HE. The first task is to connect the stress at the crack tip or the energy released during crack growth with an environment sensitive variable. This can be done for an elastic crack neglecting all plastic flow processes, the so-called Griffith crack. For simplicity only Mode III cracks will be considered (Fig. 6.31). The elastic equilibrium equations then reduce to (Laplace's equation)

$$\nabla^2 u_3 = 0 \ , \tag{6.45}$$

where u_3 is the displacement in x_3 direction.

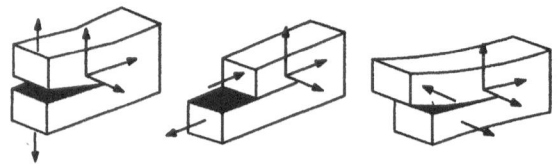

Fig. 6.31. Schematic drawing of the different loading modes in fracture

Mode I	Mode II	Mode III
opening Mode	sliding Mode	shearing Mode

This equation can be solved invoking complex functions. For the stress at the crack tip this yields

$$\sigma = \frac{K}{\sqrt{2\pi z}} \tag{6.46}$$

The constant of proportionality, K, is called the stress intensity factor. The solution for a general crack can be written in similar ways

$$\sigma(r, \theta)_{ij} = \frac{K_M}{\sqrt{2\pi r}} f_{ij}^M(\theta), \tag{6.47}$$

$$K_M = \sigma_M^\infty \sqrt{\pi a} \cdot Y(a/W), \tag{6.48}$$

with M = I,II,III for the different loading modes, (Fig. 6.31), f_{ij}^M are angle dependent geometric functions which do not depend on the specimen geometry, and Y contains the part of the solution which depends on the specimen geometry, namely on the ratio of the crack length, a, to the ligament, W.

Another important parameter, which describes crack growth, is the potential energy release rate, G, during crack growth:

$$G = \alpha \frac{\partial W}{\partial a} . \tag{6.49}$$

α is a parameter, which depends on the geometry of the specimen. For elastic loading, G is correlated to K according to:

$$K_I^2 = E \cdot G, \tag{6.50}$$

where E is the elastic module. This energy release rate is reduced by hydrogen and has to be calculated.

A more microscopic but still a continuum view of interfacial fracture is presented in Fig. 6.32. If the interface is initially unseparated and uniformly stressed, the failure criterion is $\sigma = \sigma_{max}$ (Fig. 6.33). However, if separation occurs during slow crack growth and the bonds attract each other over a finite distance until final separation this gives [6.83]

Fig. 6.32. Crack-tip configuration at cleavage and after the emission of a dislocation

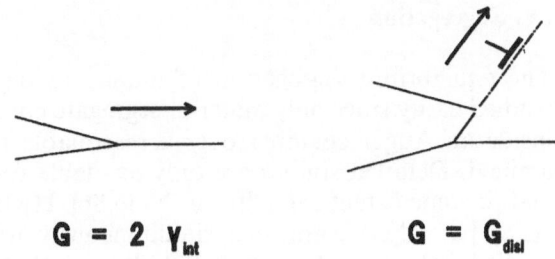

$$G = 2\ \gamma_{int} \qquad\qquad G = G_{dis}$$

$$G = \int\limits_{0}^{\infty} \sigma(\delta)\mathrm{d}\delta = 2\gamma_{int} \ , \tag{6.51}$$

where γ_{int} is the interfacial energy and $2\,\gamma_{int}$ is the work for reversibly separating an interface. If a σ vs. δ relation is assumed which is compatible with the universal bonding correlation of *Rose* et al. [6.84], the schematic curve in Fig. 6.33 is given by

$$\sigma = E_0 \left(\frac{\delta}{a_0}\right) \cdot \exp\left(-\sqrt{\frac{E_0 a_0}{2\gamma}} \frac{\delta}{a_0}\right) \tag{6.52}$$

with a_0 = lattice constant; E_0 = initial elastic module, and a unique relationship between the interfacial energy and the maximum stress is obtained:

$$\sigma_{max} = \sqrt{\frac{2\gamma E_0}{a_0 e^2}} = \frac{k_{IC}}{\sqrt{a_0 e^2}} \tag{6.53}$$

with e = Euler's number. In this view, reducing the interfacial energy by segregation or weakening the bonds by trapping are only different words for the same physical process.

Fig. 6.33. Tensile stress, σ, vs. separation distance, δ, normal to an interface

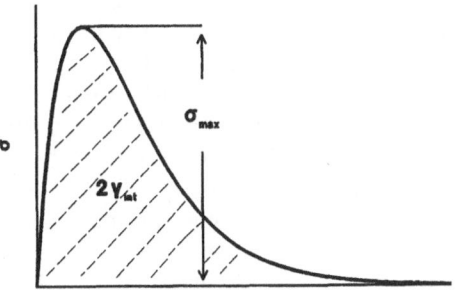

(A) Segregation

The equilibrium segregation of impurities on grain boundaries cannot be studied easily since only material/segregate combinations that can be cleaved inside an Auger chamber over a reasonable range of coverage can be examined. Detailed studies are only available for phosphorus in α-iron [6.85] and to some extent for sulfur in Ni [6.86]. Hydrogen cannot be examined by AES, but when segregated simultaneously with other elements it can be studied indirectly. *Grabke* [6.87] showed that the thermodynamics of the grain boundary segregation of phosphorus in polycrystalline iron can be described by the Langmuir-McLean equation

$$\frac{\theta}{1-\theta} = c_0 \cdot \exp(-\Delta g_b/RT) \ , \tag{6.54}$$

where θ is the grain boundary coverage, c_0 the bulk concentration, and Δg_b the Gibbs free energy of grain boundary segregation. Equation (6.54) can be rewritten to

$$\ln\left(\frac{\theta}{1-\theta}\right) = -\frac{\Delta H^0}{RT} + \frac{\Delta S^{xs}}{R} + \ln c_0 \ , \tag{6.55}$$

where ΔH_0 is the enthalpy of segregation, and ΔS^{xs} is the excess entropy of segregation. The enthalpy of segregation is much less at the grain boundary than on the surface (–180 kJ/mol on the surface compared to –34 kJ/mol at the grain boundary). From these measurements, the reduction of cohesion due to segregation can be estimated directly (see below).

However, segregation enthalpies of hydrogen cannot be measured with current techniques, but the role of hydrogen can be deduced from the simultaneous segregation of hydrogen with other elements. *Bruemmer* et al. [6.88] showed that, after sulfur segregation, polycrystalline nickel fractured along the grain boundaries, whereas they observed only transgranular fracture in pure nickel. The amount of intercrystalline fracture depended on the sulfur coverage of the boundaries. These specimens were tested at different hydrogen overpotentials. It was observed that with decreasing sulfur coverage a higher hydrogen overpotential was needed to obtain similar amounts of intergranular fracture. Conflicting evidence exists if hydrogen can segregate on clean boundaries in nickel. *Kimura* and *Birnbaum* [6.89] were not able to obtain intergranular fracture in pure nickel charged with hydrogen, but they showed convincingly that segregation of C or S respectively enhanced or reduced the cohesion, and influenced the diffusivity of hydrogen. From these measurements, it can be concluded that hydrogen acts like sulfur and reduces the cohesion of nickel bonds across the interface. Again, this conclusion is not fully supported by thin-foil experiments inside a TEM. Tests with nickel showed that the foils ruptured parallel to the boundary and not at the boundary, whereas in more brittle materials fracture along the boundary was observed [6.53, 90]. Symmetrical tilt boundaries in bulk Ni bicrystals, which contained some S on the boundary and which were precharged to cover the boundary with hydrogen, were cleaved inside an AES chamber. The resulting fracture surface showed equal sulfur coverage on both sites of the boundary directly after fracture, indicating, that within the resolution of AES, the crack propagates along the boundary [6.60]. Therefore the different stress states in thin foils (plane stress vs. plane strain in bulk specimens) produce different fracture paths. Detailed crack growth studies on bicrystals with different orientations showed, that depending on orientation the crack path in hydrogen can be either transgranular or intergranular. Therefore the strength of grain boundaries and low index planes is similar, at least in FeSi-bicrystals for which the measurements were done [6.91, 92].

(B) Decohesion of Segregated Interfaces

Many impurities segregate more strongly on free surfaces than on interfaces as shown above. When an interface fractured slowly, according to *Rice* and *Wang* [6.93] two cases must be distinguished:

(i) The segregating element is immobile at room temperature (S,P), hence, during slow separation the coverage does not change (separation at constant θ).
(ii) The element is mobile (hydrogen) at room temperature, during the slow separation process hydrogen can rearrange (separation at constant chemical potential, μ).

The thermodynamic framework now reviewed allows the effects of segregation on $2\gamma_{int}$ to be estimated. The work of separation, given by (6.51), has not a constant value for a segregated boundary if the path in configurational space is not specified. The Helmholtz excess free energy at fixed T is given by

$$df = \sigma d\delta + \sum_i \mu^i d\theta^i , \qquad (6.56)$$

where θ^i is the interfacial coverage of segregant i.

For separation at constant θ (non-mobile segregants) this yields [6.94]

$$(2\gamma_{int})_{\theta\,=\,const} = (2\gamma_{int})_0 - \int_0^\infty [\mu_b(\theta') - \mu_s(\theta'/2)]d\theta' , \qquad (6.57)$$

where $(2\gamma_{int})_0$ is the work to separate a clean interface, $\mu_b(\theta)$ the chemical potential of an unstressed interface with coverage θ, and $\mu_s(\theta/2)$ the chemical potential for a single free surface with coverage $\theta/2$.

For the other limiting case, the segregant is highly mobile, and during quasistatic separation the segregant rearranges keeping the chemical potential constant. It then follows for the work of separation

$$(2\gamma_{int})_{\mu\,=\,const} = (2\gamma_{int})_0 - \int_{-\infty}^{\theta} [2\theta_s(\mu') - \theta_b(\mu')]d\mu' . \qquad (6.58)$$

Two conclusions can be drawn directly from (6.57) and (6.58):

1) It has been proven by *Hirth* and *Rice* that that [6.94]

$$(2\gamma_{\text{int}})_{\mu \,=\, \text{const.}} < (2\gamma_{\text{int}})_{\theta \,=\, \text{const.}} \, , \tag{6.59}$$

i.e., the cohesive energy after separation of the interface at fixed chemical potential is always less than the cohesive energy at fixed coverage. A normal segregant (a segregant that adsorbs more strongly on a free surface than on an interface, like phosphorus in iron [6.87]) always reduces the cohesive energy of the interface more for slow separation than for fast separation.

2) In cases where adsorption can be described by a Langmuir/McLean isotherm (6.54), (6.57) gives directly

$$2\gamma_{\text{int}} = (2\gamma_{\text{int}})_{\text{o}} - (\Delta g_{\text{b}} - \Delta g_{\text{s}})\theta \; . \tag{6.60}$$

For clarity, the slowly varying logarithmic term is neglected. The difference in the free enthalpies of segregation for the boundary and the fracture surface directly gives the reduction of the interfacial energy due to segregation. This yields for the case of phosphorus segregation with the values given above a strong reduction of the grain boundary cohesion since phosphorus adsorbs very strongly on the free surfaces. Comparable measurements for hydrogen are not available, but hydrogen adsorbs very strongly on free surfaces of most metals [6.39, 40], and is therefore a candidate as a grain boundary embrittler.

In the following some experiments are described which, at least indirectly, support the idea that hydrogen weakens interfacial bonds. According to (6.59), immobile segregants should lower the fracture strength to a lower degree than mobile segregants. Hydrogen can be considered to be mobile at room temperature and immobile at the temperature of liquid nitrogen. With this assumption, the reductions of cohesive energy at constant concentration and at constant chemical potential were calculated for Fe with (6.57, 58) to be 0.31 and 1.04 J/m^2 at $T = 343$ K, respectively [6.95]. Compared to the cohesive energy, about 3 J/m^2 for the clean boundary, this reduction is significant. According to (6.50) and (6.51), the reduction of the surface energy can be obtained by measuring the fracture toughness of a covered and uncovered interface. Strictly, (6.51) can only be applied when no dislocation activity occurs during fracture. Most metal single crystals cannot be grown without dislocations. These dislocations alter the stress field at the crack tip, hence, the measured fracture toughness is larger than the fracture toughness expected for ideal cleavage, which complicates the comparison between measured and calculated toughness values. However, theoretical arguments and computer simulations of the brittle to ductile transition show, that even in semi-brittle materials, small changes in γ can drastically reduce K_{IC}. This will be discussed in detail in the next section.

Wang and *Vehoff* [6.95] measured the fracture toughness of pre-cracked FeSi bicrystals for clean boundaries and for boundaries covered with hy-

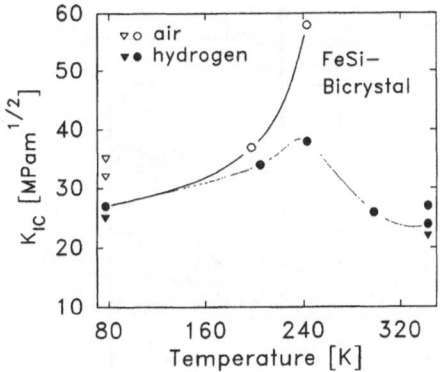

Fig. 6.34. The critical stress intensity factor vs. temperature for hydrogen charged and uncharged FeSi bicrystals

drogen. FeSi is a semi-brittle material in the temperature range of the experiments. The specimens were precharged to obtain a coverage of one on the boundary for the tests at liquid nitrogen temperatures. At medium temperatures, the specimens were continuously charged during fracture. The results are given in Fig. 6.34 in which K_{IC} is plotted as a function of temperature for bicrystals tested in air and hydrogen. The triangles indicate bicrystals grown from alloys of commercial purity, the circles high purity alloys. At 77 K, the fracture toughness of the segregated boundaries was lower than the fracture toughness of clean boundaries (open and filled triangles), but the measured values of 30 MPam$^{1/2}$ were still too high for ideal cleavage fracture, for which K_{IC} values among 1 and 2 MPam$^{1/2}$ are expected. A much larger effect of hydrogen on fracture was obtained at room temperature (filled circles). At room temperature dislocations are definitely mobile, therefore the measured reduction in toughness could be either due to a reduction of the cohesive strength or due to hydrogen effects on dislocation emission and mobility.

In Ni bicrystals, *Vehoff* et al. [6.60] examined the effect of segregation on crack nucleation to search for effects of hydrogen on bond strength. Bicrystals with the geometry shown in Fig. 6.35 were loaded in fatigue. The boundary plane was oriented perpendicularly to the applied load. Two types of boundaries were examined; an asymmetrical tilt boundary that provided a strong obstacle for dislocation glide across the boundary (type I), and a symmetrical tilt boundary which was oriented to allow for an easy passage of dislocations across the boundary (type II). Clean boundaries as well as boundaries which were covered with a known amount of sulfur (the sulfur coverage was checked by cleaving the pre-cracked bicrystals inside an AES chamber) were fatigued in hydrogen gas or cathodically charged in H_2SO_4 during cyclic loading. Cracks did not nucleate at the clean boundaries. Bicrystals and single crystals with comparable orientations showed identical fatigue lives in air and vacuum. However, after an embrittling heat treatment to segregate sulfur on to the boundary only type I boundaries became sensitive to intergranular crack nucleation. In hydrogen gas cracks nucleated earlier at dislocation pile ups and propagated along the boundary. After

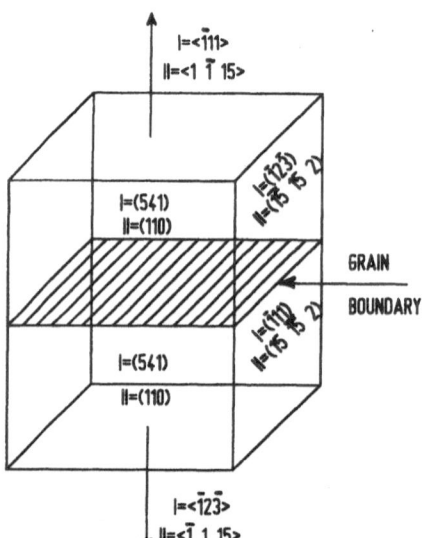

Fig. 6.35. Grain boundary orientations of type I and II Ni bicrystals

hydrogen was pumped away the crack immediately left the boundary. Type II bicrystals still showed transgranular crack nucleation under these conditions. When the hydrogen activity was further increased by cathodic charging during the test, even for type II boundaries the initiation time was drastically reduced, and the bicrystals failed by cyclic cleavage.

Similar results were obtained by *Jones* et al. [6.86]. They showed that in nickel polycrystals, segregated with sulfur, the percentage of intergranular fracture increased when the hydrogen activity was increased. With increasing sulfur coverage (again checked by AES) lower hydrogen activities were needed to obtain the same percentage of intergranular fracture. It can be argued that in polycrystals only boundaries break which are inherently weak and on which segregation is easy. However, detailed examinations on bicrystals with different orientations showed that segregation was relatively insensitive against grain boundary structure. Only special coincidence boundaries which are not very numerous in an untextured polycrystal showed less segregation [6.96, 97].

Intermetallics are better suited to test the effect of hydrogen on cohesion. In many intermetallics, the binding is mainly metallic. But the activation energy for dislocation motion is much higher than for disordered metals, which results in ideal cleavage fracture at low temperatures. This can be directly seen in Fig. 6.36a in which the temperature dependence of K_{IC} for FeSi bicrystals is compared to the behavior of ordered NiAl single crystals. In addition, strongly ordered intermetallics usually have a low grain boundary strength, because the ordering behavior can persist up to the grain boundary forming cavities of atomic size in the boundary [6.98]. If, however, the alloy deviates from the stoichiometric composition, the grain boundary becomes similar to that in fcc metals [6.99]. Ni_3Al is found to be ordered up

Fig. 6.36. (a) The critical stress intensity factor vs. temperature for NiAl single crystals and FeSi bicrystals **(b)** Arrhenius plot of the critical displacement rates and temperatures, T_{BD}, at which the ductile/brittle transitions occurred in different alloys

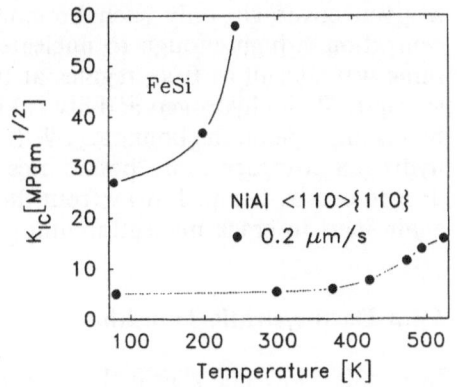

a

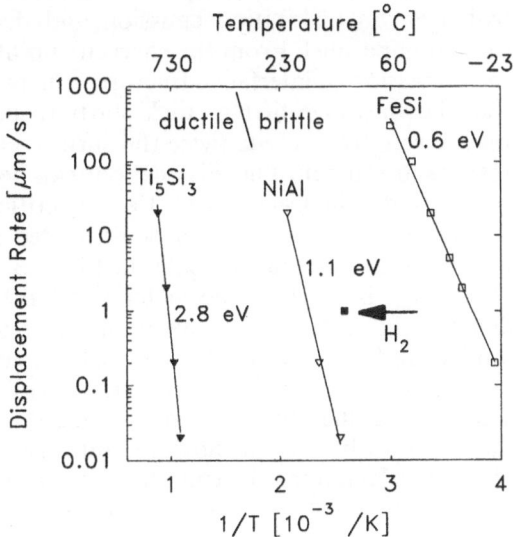

b

to the boundary plane. With a surplus of boron, the boundary strength can be enhanced, which results in transgranular fracture in an alloy which fails intergranularly without boron. Hydrogen, however, causes again a change to true intergranular failure. The observation that cracks in TEM foils propagated in the grain-boundary plane rather than along slip planes even when extensive thinning has occurred suggests that hydrogen decreases the grain boundary cohesion [6.100]. Measurements of the influence of hydrogen on the low-temperature toughness of these alloys are not available since the production of intermetallic alloys with reasonable size for fracture toughness testing is still problematic. Recent results on hydrogen effects on fracture in intermetallics are described in the next section.

Considering the results obtained on polycrystals and bicrystals it is concluded, that segregation alone cannot explain the observed differences in crack nucleation and in the percentage of intergranular fracture. Rather, for a given coverage, only such boundaries cleave on which the stress concentration is high enough to nucleate a brittle crack. Hydrogen acts in the same way as sulfur. It segregates at the boundary and reduces the cohesive strength. If the hydrogen activity is high enough, cracks nucleate where slip bands impinge on the boundary. With increasing stress concentration and/or hydrogen coverage more boundaries can cleave. In contrast to sulfur, hydrogen can be pumped away from the boundary. Clean boundaries are then again inert to crack nucleation and propagation.

6.4.2 Ductile/Brittle Transition

In metals, slow crack growth is always linked to dislocation emission and glide. When the effect of hydrogen on fracture is discussed, the influence of hydrogen on dislocation emission and dislocation mobility must be considered in parallel. From the energetic point of view, which was the basis for the discussion of interfacial fracture, a metal can be either ductile or brittle. It was shown above that a crack starts to propagate when the strain energy release rate, G, exceeds twice the surface energy, (6.51). However, before G reaches its critical value, dislocations can be emitted from the tip, which then results in ductile fracture. At 0 K, the critical strain energy release rate G_{disl} for dislocation emission can be estimated from dislocation theory. The situation is shown schematically in Fig. 6.32. When dislocations emit before $G = G_{cleav}$ is reached, the material behaves ductile. The line $G_{disl} = G_{cleav}$ denotes the boundary between ductile and brittle behavior for an exact model. *Anderson* and *Rice* [6.101] calculated estimates of the critical G_{disl} for several metals using a refined description of the Rice-Thomson model [6.102]. They found that most fcc metals besides Iridium should be intrinsically ductile, most bcc metals, however, intrinsically brittle. In addition, the results of their calculations depend strongly on the loading condition and crystal orientation, which indicates that the ductile versus brittle behavior is affected by the microstructure at the crack-tip.

Hydrogen not only influences the strength of the atomic bonds, but also alters G_{disl}. The emission of a dislocation from the crack tip produces new surface. With the same arguments as given above, the reduction of the surface energy by hydrogen segregation then facilitates dislocation emission as well. The process will be further complicated because the brittle to ductile transition is a dynamic process. The stress at the crack tip not only depends on the critical cleavage or emission stress, but also on the rate with which dislocations can move away from the crack tip. If the mobility of dislocations is reduced due to hydrogen segregation at the dislocation cores, the stress at the crack tip can rise above the cleavage stress for a given loading rate and temperature. For this reason, the distinction if hydrogen reduces the surface energy or decreases the mobility of dislocations is difficult.

That dislocation mobility controls the brittle to ductile transition can be understood when the Figs. 6.36 a and b are considered. At low temperature, the fracture toughness for NiAl single crystals reaches the theoretical limit for true cleavage fracture. In the low-temperature range, the toughness is nearly independent of temperature. The only temperature dependence results from the temperature dependence of the elastic properties of the lattice. At higher temperatures, however, the toughness increases with temperature. The temperature at which the toughness starts to change is called the brittle-to-ductile transition temperature, T_{BD}. This temperature is plotted in Fig. 6.36b vs. $1/T$ for different intermetallic and metallic alloys. Straight lines were obtained with slopes corresponding to apparent activation energies of between 0.6 eV for a disordered metal (FeSi), and 2.8 eV for a covalently bonded intermetallic alloy with large unit cell (Ti_5Si_3). This apparent activation energies point out that either dislocation emission or glide or both are thermally activated.

So far only the competition between dislocation emission and bond breaking was considered. In reality, the response of a crack tip to the external loading rate is a dynamic process. Possible dislocation sources along the crack tip emit dislocations. These dislocations shield the source until the emitted dislocation has propagated far enough to rise the stress at the source above the value for the emission of the next dislocation. This situation is drawn schematically in Fig. 6.37.

The brittle to ductile transition depends on the emission criterion, the dislocation rate and on the number and distribution of dislocation sources. The dynamic three-dimensional problem is not solved, yet. Based on experimental results obtained with Si single crystals, several one- and two-dimensional models were developed by *Hirsch* et al. [6.103] and by *Brede* [6.104] which will be reviewed briefly below and discussed in view of hydrogen effects.

Lin and *Thomson* [6.105] have considered the different solutions of the near-crack-tip stress field for different loading conditions, slip geometries and dislocation distributions. For the simple case of a Mode III crack (Fig. 6.31) they obtained for the force on a screw dislocation in the distance x_j from the crack tip in the force field of other dislocations:

$$f_d(X_j) = \frac{K_{III}b}{\sqrt{2\pi x_j}} - \frac{\alpha\mu b^2}{4\pi x_j} - \sum_{\substack{i=1 \\ i\neq j}}^{N} \frac{\mu b^2}{2\pi} \sqrt{\frac{x_i}{x_j}} \cdot \frac{1}{x_i - x_j} , \qquad (6.61)$$

Fig. 6.37. Crack-tip with dislocation sources, Q, and areas without dislocations, Z

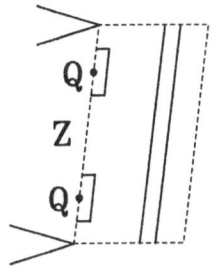

where the first term is the force of the loaded crack tip on the dislocation at x_j, the second is the image term, which attracts the dislocation to the crack tip, and the third is the force due to the other dislocations, which were previously emitted. If this force exceeds a prescribed load (for example to overcome the impurity pinning) the dislocation starts to move by thermally activated glide with a velocity, v_d, according to

$$v_d = A(f_d)^m \cdot \exp(-U/kT) \ . \tag{6.62}$$

When the crack is loaded at a given stress rate, dislocations will be emitted continuously. These dislocations shield the crack from the applied load according to

$$k_{III} = k_{III} - \sum_{i=1}^{N} \frac{\mu b}{\sqrt{2\pi x_i}} = K_{III} - \sum K_D \ , \tag{6.63}$$

where K and k are the applied and local stress intensity factors, respectively, and K_D is the shielding component of an emitted dislocation. In the simple one-dimensional picture of mode III loading, shielding keeps the stress at the crack-tip always below the Griffith stress for sources, which operate below this stress. Therefore *Hirsch* et al. [6.103] have assumed that the dislocation sources are distributed along the crack front. In their still one-dimensional computer simulation, they calculated the stress intensity at a point Z (Fig. 6.37) between the sources. If this stress intensity exceeds the Griffith value, the crack will propagate. The calculations yield $K_{IC}(T)$-curves in which K_{IC} increases smoothly with increasing temperature. The parameters, which are entered into the model, are the distance of the dislocation sources, the operational stress of the dislocation sources and the rate law for the dislocation velocity. *Brede* [6.104] recently simulated the same problem. In his calculations, he used the real rate law obtained from experiments and a more realistic picture of the slip geometry in Si, and he considered all three components of the stress intensity factor in mode I loading. In contrast to the mode III calculations, he found, that for K_I loading the stress at the crack tip increases with the number of emitted dislocations. No assumptions for a discrete distribution of dislocation sources were necessary.

So far, only the case of cracks, which emit dislocations, has been considered. For metals, however, the case of external preexisting dislocation sources, which emit dislocations that might blunt or shield the crack from the external load, can be equally important. The situation is shown schematically in Fig. 6.38 together with a schematic representation of the results of a preliminary calculation by *Thomson* [6.106]. A crack that approaches a dislocation source will be shielded or anti-shielded depending on the sign of the Burgers vector and the relative position between crack and dislocation, when the source starts to emit dislocations. Depending on the relative velocities of the crack and the dislocations, the crack will be either slowed down or stopped. Calculations of this type are currently in progress. These models describe correctly the processes at a crack tip, but are far from being quantitative.

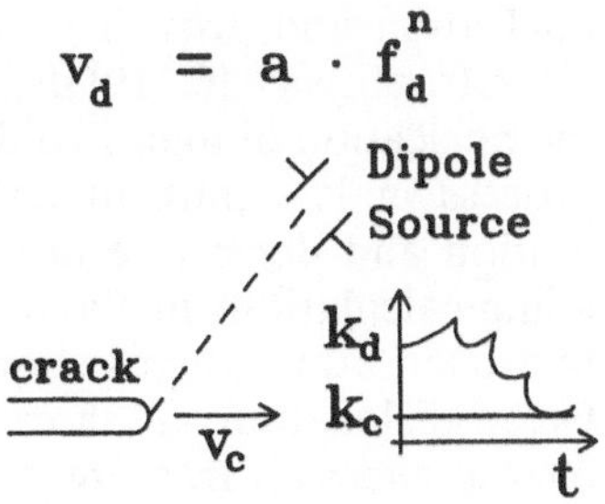

Fig. 6.38. Crack-tip approaching external dislocation sources; v_c is the velocity of the crack, v_d the velocity of the dislocations, k_d the stress intensity at the crack-tip

$$v_d = a \cdot f_d^n$$

Dipole
Source

crack $\xrightarrow{v_c}$ k_d
k_c
t

(A) Effect of Hydrogen

Hydrogen can alter the ductile to brittle transition in two ways: either it reduces the cohesive forces as discussed in Sect. 6.4.1, thus the crack starts to grow at a lower local stress intensity factor, k_{III}, according to (6.63), or hydrogen alters the dislocation velocity, (6.62). Again, it can be seen, that for the brittle to ductile transition reduction of the surface energy or altering the dislocation rate is closely connected. Recent results by *Kimura* et al. [6.100] showed that hydrogen is strongly trapped at the dislocation core in Ni. Serrated yielding was observed indicating that for the experimental conditions used, hydrogen hinders the motion of dislocations. Ongoing work of *Kimura* showed that hydrogen causes softening at low applied stresses and hardening at higher stresses. High stresses predominate at crack tips. In high-purity iron softening due to hydrogen was observed as well. This softening was explained by *Moriya* et al. [6.107] in terms of hydrogen-dislocation interactions. They assumed that hydrogen trapped at the dislocation core increases the mobility of screw dislocations and reduces the mobility of edge dislocations. Direct evidence of increasing dislocation activity by hydrogen was obtained from experiments in thin foils which were kept under stress in an environment cell inside a TEM. After hydrogen was introduced into the chamber, the dislocations started to move [6.108]. In relationship to the brittle to ductile transition this means that hydrogen can either ductilize or embrittle an alloy.

(B) Hydrogen Enhanced Local Plasticity

Birnbaum [6.109] concludes from the experiments, which were done in his group on thin foils and from the effects of hydrogen on dislocation mobility discussed above, that hydrogen embrittles an alloy by locally enhanced plasticity. This idea seems to be at variance with embrittlement, but the distribution of hydrogen can be highly nonuniform and localized. Thus, locally the flow stress can be reduced, and local slip with high dislocation densities might occur. These local slip bands can fail by normal fracture modes, which then results in an overall reduction of the work to fracture. The macroscopic result is similar to the reduction of the fracture stress by cavity nucleation and growth that is also a highly localized process (Sect. 6.2.4). Therefore *Lynch* [6.110] proposed that even low-fugacity hydrogen facilitates the nucleation of micro voids at the crack tip and gave some evidence for this process in less pure materials. The reasons for the enhanced dislocation motion and density resulting from hydrogen are not well established. Ongoing calculations in *Birnbaum*'s group on elastic shielding of dislocation by hydrogen atmospheres showed that hydrogen decreases the interaction between dislocations at short range and has no effect at large distances. The overall effect on fracture is difficult to understand and will not be discussed here. Further details can be found in the reviews of *Lynch* [110] and *Birnbaum* [6.109].

(C) Hydrogen Enhanced Brittle Fracture

Since the stress at a crack tip in a bulk is high, the experiments mentioned above predict that hydrogen reduces the mobility of the dislocations emitted from the crack tip. Equations (6.61) and (6.62) then predict that the brittle to ductile transition temperature, T_{BD}, shifts to higher temperatures. In principle, measurements of the brittle to ductile transition as a function of the hydrogen pressure cannot distinguish between the effect of hydrogen on dislocation motion and on surface energy. But the reduction of the surface energy can be measured directly by fracture toughness tests at low temperatures for which dislocations and hydrogen atoms are immobile. However, with this technique only such interfaces can be tested which can be completely covered with hydrogen. In addition, the alloys must be brittle and nearly dislocation free to avoid dislocation shielding obscuring the fracture toughness values. Measurements on intermetallic alloys showed true cleavage fracture along grain boundaries and crystallographic planes due to hydrogen, but quantitative results on the reduction of surface energy are still not available.

The influence of hydrogen on the brittle to ductile transition cannot be measured easily. This transition is controlled by dislocation kinetics, and as discussed in Sect. 6.3 in detail, the kinetics of hydrogen assisted crack growth, HAC, are controlled by the transport of hydrogen to the fracture process zone. Therefore the brittle to ductile transition must be measured in a temperature and velocity range in which hydrogen is in local equilibrium at the crack tip (Sect. 6.3.2). At higher temperatures (above 400 K) the hydrogen concentration at trapping sites will be too low for hydrogen embrittlement at reasonable pressures. At temperatures below 250 K, the diffusivity of hydrogen in most alloys is too low for crack velocity measurements.

Figure 6.39 shows a typical brittle to ductile transition curve for FeSi single crystals in hydrogen gas. The brittleness in hydrogen (measured by a_n, Sect. 6.3 increases at 373 K with increasing crack tip opening rate until a plateau value is reached, in contrast to the behavior observed when the brittleness is transport controlled (below 273 K). In that case, a_n decreases with increasing crack tip opening rate. The temperature range, which could be examined, was not large enough to obtain an apparent activation energy

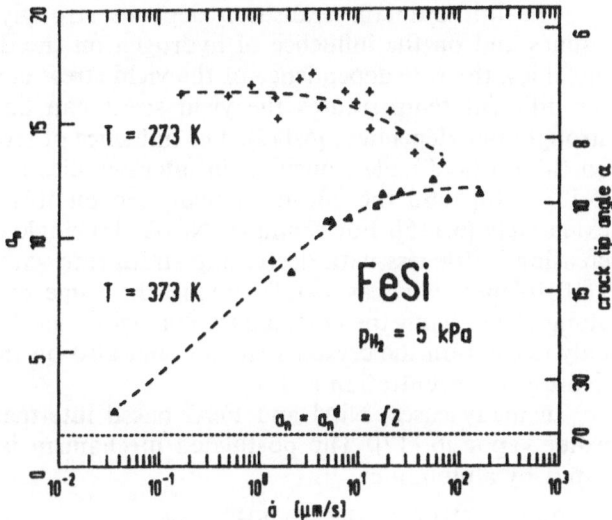

Fig. 6.39. Normalized crack growth rate, a_n, as a function of the crack growth rate

for the brittle to ductile transition in hydrogen gas in the way described above, but the point indicated by an arrow in Fig. 6.36b could be obtained from the lower dashed curve in Fig. 6.39. Hydrogen shifts the ductile to brittle transition by nearly 200 degrees towards higher temperatures. These measurements support the view that hydrogen reduces the mobility of dislocations, which then results in a shift of T_{BD} towards higher temperatures. Direct measurements of the influence of hydrogen on the dislocation velocity are lacking due to experimental difficulties in decorating single dislocations in metals (high dislocation density and rate). Measurements of this kind should be possible in intermetallics since the density and mobility of dislocations can be much lower in special intermetallic alloys.

(D) Intermetallics

Aluminium and Silicon based intermetallics, which are of great interest for future high temperature applications are also strongly sensitive to hydrogen. Compared to simple metals, intermetallic alloys can have complex crystal structures, and depending on the composition, the bonding character can vary from metallic to covalent and/or ionic. Intermetallics with nearly cubic structures are of primary interest, since these alloys can have sufficient slip systems for ductile deformation. The alloys with L_{12}– (for example Ni_3Al, ordered fcc with Ni on each face center) and with B2-structure (for example, NiAl, ordered bcc with Al in the center) are extensively examined. Since some alloys show rising yield stress with rising temperature hence combining high temperature strength with good room temperature toughness [6.111].

The brittle/ductile transition depends strongly on the dislocation dynamics and on the influence of hydrogen on the dislocation rate. In intermetallics, the rate dependence of the yield stress is much more complicated: for different temperatures the yield stress can be independent of rate or strongly rate dependent [6.112]. The influence of strain rate and temperature on the brittle/ductile transition in intermetallics is an actual research topic [6.113, 114]; and the effects of hydrogen on this transition are examined extensively [6.115]. For example, Ni$_3$(Al,Ti) single crystals (L_{12}) showed increasing brittleness with decreasing strain rate with cleavage fracture along {110}-planes, whereas Co$_3$Ti having the same crystal structure fractured along {100}. Thus the crystallographic plane that fails in hydrogen does not only depend on the crystal structure, but also on the alloy composition and hydrogen concentration [6.116].

In many cases, NiAl and FeAl based intermetallics are embrittled by water vapor [6.117]. The postulated mechanism is the reduction of water vapor by aluminium:

$$2Al + 3H_2O \rightarrow Al_2O_3 + 6H^+ + 6e^- \ . \tag{6.64}$$

In particular, the grain boundaries in Ni$_3$Al and NiAl are severely embrittled by hydrogen [6.118]. For example NiAl single crystals are nearly immune against hydrogen, bicrystals, however, are severely embrittled. In Fig. 6.40 the effect of hydrogen charging on the fracture toughness is shown. The observed embrittlement depended on grain boundary orientation and charging conditions, but was always more pronounced than for the single crystals that had the orientations of the adjacent grains [6.3]. The other system of practical interest is Ti-Al. Most data exist for Ti$_3$Al; a review is given by *Eliezer* [6.119]. Ti$_3$Al forms hydrides of the type (Ti$_3$Al)H$_x$ [6.120]. These hydrides reduce the ductility, fracture strength and toughness [6.121]. Whether γ-TiAl (L_{10}) suffers from hydrogen effects cannot be decided from the data which are currently available. From our data, which showed only a transition from unstable to stable crack growth and no effect of hydrogen on the fracture toughness [6.1] it can be concluded that for comparable condi-

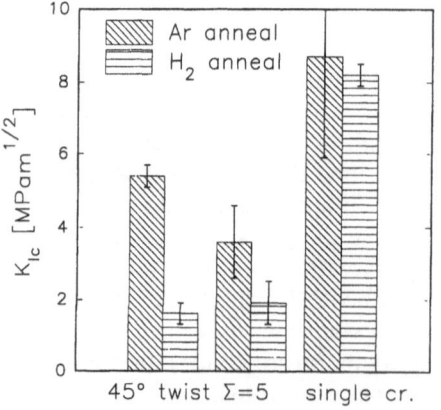

Fig. 6.40. Fracture toughness of NiAl Bi- and single crystals charged in hydrogen gas

tions hydrogen makes no problems. However, only the two phase, γ-α_2-alloy can be used for commercial application since, even without hydrogen, TiAl is too brittle.

Detailed TEM examinations of Co_3Ti single crystals fractured in hydrogen gas showed that in hydrogen a large number of planar stacking faults formed at the crack tip which were not observed in vacuum [6.122]. It is most likely that the stacking faults that formed in hydrogen at the crack tip, are local barriers for the motion of dislocations. This will increase the local stress at the crack tip for a given strain rate as described in detail above. In this case, the macroscopic fracture stress is reduced by making the motion of dislocations more difficult and not by reducing the surface energy. The theoretical picture of the ductile/brittle transition in intermetallics cannot be proven directly, since currently it is not possible to measure the dislocation rate in the near-crack-tip region. Rates obtained by in situ straining of thin foil inside a TEM are not representative, since for this special geometry the stresses at the crack tip are much lower than in bulk crystals. The different crystallographic fracture planes observed in intermetallics with the same crystal structure can only be predicted by atomistic computer simulations [6.123]. That is discussed in the next section.

6.4.3 Atomistic Computer Simulations of Hydrogen Embrittlement

Two techniques are currently the most powerful for describing hydrogen effects on material strength; first principle calculations and methods based on empirical potentials. The first principle calculations yield, for small atomic clusters, information on bond strength, interatomic forces, surface energies, structure, and possible effects of hydrogen on bonding [6.124]. For the more empirical methods potentials must be constructed, which fit crystal properties like the elastic constants, lattice constants, and stacking fault energy or others for the material of interest [6.125]. The most powerful method for the simulation of large numbers of atoms able to address problems like impurities, internal interfaces, and other defects in materials, is the embedded atom method (EAM) [6.61]. The basic assumptions of the EAM model and some results will be summarized below.

In the EAM method each atom is viewed as embedded in a host lattice consisting of all other atoms. The calculations of the EAM are based on the assumption that the energy of an impurity in a host is a function of the electron density of the unperturbed host. A simple local approximation would be to assume that the embedding energy depends only on the local environment around the impurity. The total energy of a system of atoms can then be expressed as the sum of the energies to embed each atom into the local electron density of all other atoms, plus the sum of short-range pair interaction energies as follows [6.61].

$$E_{\text{tot}} = \sum_i F_i(\rho_i) + \frac{1}{2} \sum_{ij, i \neq j} \Phi_{ij}(R_{ij}) \ . \tag{6.65}$$

In this equation, ρ_i is the atomic electron density at the site of atom i caused by the other atoms in the system, $F_i(\rho_i)$ is the energy to embed an atom of type i into the electron density ρ_i, and $\Phi_{ij}(R_{ij})$ is the interaction energy between two atoms separated by a distance R_{ij}. The electron density, ρ_i, is approximated by the sum of atomic electron densities, $\rho_j^{a}(R_{ij})$, seen by each atom i due to each atom j at a distance R_{ij} from the nucleus as follows:

$$\rho_i = \sum_{j \neq i} \rho_j^{a}(R_{ij}) \ . \tag{6.66}$$

The pair potentials are short-ranged and repulsive, and assume the form

$$\Phi_{ij} = \frac{Z_i(R)\, Z_j(R)}{R} \tag{6.67}$$

so that only the two effective charges $Z_{Ni}(R)$ and $Z_{H}(R)$ are needed to specify the three Φ's for Ni-Ni, Ni-H, and H-H interaction in case of hydrogen embrittlement in Ni. The potentials and embedding functions F_i for hydrogen in Ni are best known, the fitting procedures for the embedding functions, effective charge functions, and pair interaction energies are described in detail by *Foiles* [6.126].

The following section describes some results concerning the effects of hydrogen on brittle fracture and dislocation motion. In particular, calculations are presented for the binding energy of hydrogen to a crack tip, the effect of hydrogen on the brittle fracture stress, the binding energy of hydrogen to a dislocation [6.127], and for the segregation of hydrogen to grain boundaries [6.128]. The calculations proceed by setting up a configuration of atoms representing a crack or a dislocation, with or without some hydrogen atoms present. The motion of the atoms are obtained by solving Newton's equation with the internal forces obtained from (6.65) according to the external stresses or strains applied. A typical slab with periodic boundary conditions in two directions consists of 14 planes in the z direction, 62 planes in the x direction and 1 plane in the y direction (thickness direction, plane strain).

The binding energy of hydrogen to a crack tip is found by placing a hydrogen atom at different places near the crack tip and minimizing the energy of the system with respect to the position of hydrogen. Due to the periodic constraints imposed on the small atomic clusters, dislocation emission is inhibited and the results represent the behavior of a brittle nickel. The calculations show that hydrogen segregates preferentially at the crack tip, and that hydrogen weakens the bonds between metal atoms, but its energy is only 0.06 eV below its energy on the free surface [6.127].

For a large-angle grain boundary in Ni, *Moody* and *Foiles* [6.128] showed that the boundary structure changes significantly with increasing hydrogen coverage, and that hydrogen segregates at the boundary with a binding energy up to 0.5 eV in agreement with the experimental data discussed in Sect. 6.3. Figure 6.41 shows the trap binding energy as a function of the distance from the grain boundary plane according to the calculations of *Moody* and *Foiles* [6.128]. Trapping was found to be limited to a distance less than 0.4 nm from the boundary plane. With increasing bulk concentration,

Fig. 6.41. Trap binding energy at a grain boundary in Ni as a function of the distance from the boundary

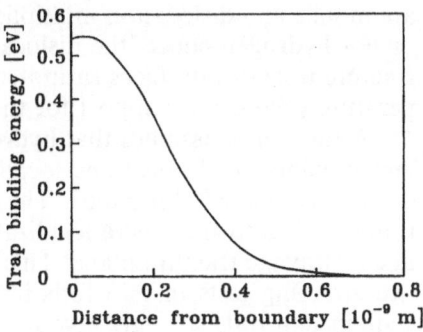

grain boundary sites fill according to their energy until all possible sites are filled. Further increasing the lattice concentration had no influence on grain boundary coverage and structure in accordance to the observation of a lower limit in crack growth susceptibility at high concentrations [6.129].

The examples given above show that the EAM method can describe many effects of hydrogen on dislocations, surfaces, and fracture. But the theory still contains severe approximations, and it is not known which features of the local binding behavior or of the dependence of the electron density on the location must be incorporated in a model to obtain correct predictions in finite computing times. New techniques for computer simulations develop rapidly. Combined with first principle calculations and experiments it might be possible to develop physically based simplifications which still give correct predictions of the mechanical and chemical properties of different alloys.

6.5 Conclusions

Several theories of hydrogen-related fracture have been discussed in the context of critical experiments. The discussion of the voluminous literature was limited to those experiments designed to test theoretical ideas. From the existing evidence it can be concluded that stress-induced hydride formation followed by cleavage, and nucleation and growth of pressurized cavities after hydrogen charging are two mechanisms that are well established theoretically as well as experimentally.

Still conflicting views exist of the effects of low-pressure hydrogen on fracture under non-hydride forming conditions. Hydrogen can enhance or reduce the mobility of dislocations depending on strain rate. A reduction in mobility would increase the stress at the crack-tip and hence shift the ductile to brittle transition temperature to higher temperatures. Computer simulations and measurements of grain boundary segregation suggest that hydrogen reduces the cohesive forces. Experiments that directly prove the reduction of the bond strength as a function of the local hydrogen concentration are not available. Fracture experiments at medium temperatures, where dislocations are mobile and dislocation multiplication is easy, cannot distinguish whether or not hydrogen effects the dislocation velocity or the bond strength. Only fracture tests of interfaces saturated with hydrogen and cleaved at low temperature yield direct support for the decohesion point of view.

A third view assumes that hydrogen enhances the density, multiplication and mobility of dislocations locally. This results in very localized slip on densely populated slip bands. These slip bands should then fail by normal rupture. Due to the severe localization of slip the fracture surface can look like a cleavage fracture plane. This mechanism was directly confirmed by in situ straining tests in thin foils inside a TEM. However, the special stress state in thin foils strongly favours slip band cracking. The mechanisms discussed above compete with each other. Several mechanisms may even operate in the same system, depending on the local conditions near the crack tip or the external environment. In addition the crack growth rate is controlled by the kinetics and thermodynamics of the various hydrogen transport mechanisms. For long-term life-time predictions the dominating transport and failure mechanisms must be known.

The combination of recent developments in dislocation theory and first principle calculations should soon allow the solution of the question of the effect of hydrogen on bond strength, dislocation emission and mobility. Experimentally, intermetallics with their various crystal systems, dislocation structures and with their large sensitivity to hydrogen containing environments offer a new tool to measure the effect of hydrogen on bond strength and dislocation dynamics. Combining the new theoretical tools with new experimental approaches on specially designed alloy systems will help in gaining a deeper understanding of the complex processes discussed above.

Acknowledgments. The author is grateful to G. Bergmann for his extensive literature research, for the compilation of the Figs. 6.1, 6.7 and 6.8, and for proofreading the manuscript. Special thanks are extended to W. Vogt for the recompilation of the figures to obtain a unified nomenclature, and to L. Kalter for writing parts of the manuscript.

References

6.1   G. Bergmann: "Untersuchungen zu Wasserstoffversprödung von NiAl, TiAl und CMSX-6, VDI Verlag, Reihe 5, Nr. 406 (1995)
6.2   Y. Liu, T. Takasugi, O. Izumi, H. Suenaga: J. Mater. Sci. **24**, 4458 (1989)
6.3   G. Bergmann, H. Vehoff: Mat. Sci. & Eng. A **192**, 309–315 (1995)
6.4   J.E. Hack, J.M. Brzeski, R. Darolia: Scripta Metall. Mater. **27**, 1259 (1992)
6.5   R.A. Oriani, P.H. Josephic: Acta Metall. **25**, 979 (1979)
6.6   M.O. Speidel: *Adv. Fracture Research*, **6**, 2685 (Pergamon, Oxford 1982)
6.7   W.W. Gerberich: Fatigue in *Hydrogen Degradation of Ferrous Alloy*, ed. by R.A. Oriani, J.P. Hirth, M. Smialowski (Noyes Publ., Park Ridge, New Jersey 1985) p. 366
6.8   P. Gangloff, R.P. Wei: Met. Trans. **8A**, 1043 (1977)
6.9   D.L. Graver: Corrosion of nickel-based alloys, in Conf. Proce., Cincinnati, Ohio, USA (1984) p. 79
6.10   N.R. Moody, M.W. Perra, S.L. Robinson: Metallography of advanced materials. 20th annual technical meeting of the Int. Metallographic Soc., Monterey, Cal., USA (1987) p. 177

6.11 J.O.M. Bockris, J. McBreen, L. Nanis: J. Electrochem. Soc. **112**, 1025 (1965)
6.12 A.N. Frumkin: In *Adv. in Electrochemistry and Electrochemical Engineering*, ed. by P. Delahay, (Interscience, New York 1963) Vol. 3 p. 375
6.13 A. McNabb, P.K. Foster: Trans. Met. Soc. AIME **227**, 618 (1963)
6.14 E. Riecke: Arch. Eisenhüttenwes. **47**, 247 (1976)
6.15 H.P. Van Leeuwen: *Fugacity of gaseous hydrogen in Hydrogen Degradation of Ferrous Alloys*, ed. by R.A. Oriani, J.P. Hirth, M. Smialowskie (Noyes Publ., Park Ridge, New Jersey 1985) p. 16
6.16 A. Turnbull: *Embrittlement by the Localized Crack Environment*, ed. by R.P. Gangloff (AIME Warrendale 1984) p. 3–33
6.17 A. Turnbull, J.G.N. Thomas: J. Electrochem. Soc. **129**, 1413 (1982)
6.18 H.W. Pickering, A. Valdes: In *Embrittlement by the Localized Crack Environment*, ed. by R.P. Gangloff (AIME Warrendale 1984) pp. 33–49
6.19 R.P. Gangloff, A. Turnbull: Crack electrochemistry modeling and fracture mechanics in *Modeling Environmental Effects on Crack Growth Processes*, ed. by R.H. Jones, W.W. Gerberich (AIME, Warrendale 1986) pp. 55–85
6.20 L.C. Wiener: Corrosion **17**, 137 (1961)
6.21 P.G. Shewmon: Met. Trans. A **7**, 279 (1976)
6.22 D.A. Westphal, F.J. Worzala: In *Hydrogen in Metals*, ed. by I.M. Bernstein, A.W. Thompson (ASM 1974) p. 78
6.23 U. Gramberg: Corrosion cracking. Int. Conf. on Fatigue, Corrosion cracking and Failure Analysis, Salt Lake City (ASM 1986) p. 147
6.24 M.L. Grossbeck, H.K. Birnbaum: Acta Metall. **25**, 135 (1977)
6.25 J. Pielaszek: Nickel hydrides, in *Hydrogen Degradation of Ferrous Alloys*, ed. by R.A. Oriani, J.P. Hirth, M. Smialowski (Noyes Publ., Park Ridge, New Jersey 1985) p. 167
6.26 S. Gahr, H.K. Birnbaum: Acta Metall **26**, 1781 (1978)
6.27 H. Matsui, M. Koiwa: Acta Metall. **38**, 1175 (1990)
6.28 S. Koike, T. Suzuki: Acta Metall. **29**, 553 (1981)
6.29 D.S. Shih, I.M. Robertson, H.K. Birnbaum: Acta Metall. **36**, 111 (1988)
6.30 J.P. Hirth, B. Carnahan: Acta Metall. **26**, 1795 (1978)
6.31 H.J. Maier, H. Kaesche: Aspects of hydrogen effects on fracture processes in low alloy steel, in *Hydrogen Effects on Material Behavior*, ed. by N.R. Moody, A.W. Thompson (The Minerals, Metals & Materials Society, 1990) p. 733
6.32 G.M. Scamans: J. Mater. Sci. **13**, 27 (1978)
6.33 W.R. Wampler, T. Schober, B. Lengeler: Phil. Mag. **34**, 129 (1976)
6.34 S.X. Xie, J.P. Hirth: Mater. Sci. Eng. **60**, 207 (1983)
6.35 O.A. Onyewuenyi, J.P. Hirth: Met. Trans. A **14**, 259 (1983)
6.36 H. Cialone, R.J. Asaro: Met. Trans. A **12**, 1373 (1981)
6.37 H. Matsui, H. Kimura, S. Moriya: Mater. Sci. Eng. **40**, 207 (1979)
6.38 G. Ehrlich: J. Appl. Phys **32**, 4 (1961)
6.39 F. Bozso, G. Ertl, M. Grunze, M. Weiss: Appl. Surf. Sci. **1**, 103 (1977)
6.40 J. Benzinger, R.J. Madix: Surf. Sci. **94**, 119 (1980)
6.41 K. Christmann: Z. Naturforsch. **34a**, 22 (1979)
6.42 K. Christmann, O. Schober, G. Ertl, M. Neumann: J. Chem. Phys. **60**, 4528 (1974)
6.43 P. Marcus, J. Oudar: Gas-iron surface equilibria, in *Hydrogen Degradation of Ferrous Alloys*, ed. by R.A. Oriani, J.P. Hirth, M. Smialowski (Noyes Publ., Park Ridge, New Jersey 1985) p. 36
6.44 M. Kiskinova, D.W. Goodman: Surf. Sci. **108**, 64 (1981)
6.45 J.D. Eshelby: In *Solid State Physics* ed. by F. Seitz, D. Turnbull (Academic, New York 1956) p. 79
6.46 J. Li, C.M. Oriani, L.S. Darken: Z. Phys. Chemie **49**, 271 (1966)
6.47 J.P. Hirth: Met. Trans. **11A**, 861 (1980)
6.48 R.P. Wei, G. Shim, K. Tanaka: Corrosion fatigue and modeling, in *Embrittlement by the Localized Crack Environment*, ed. by R.P. Gangloff (AIME Warrendale 1984) p. 243
6.49 H. Vehoff, W. Rothe: Acta Metall. **31**, 1781 (1983)

6.50 H. Vehoff, K.H. Klameth: Acta Metall. **33**, 955 (1985)

6.51 H. Vehoff, P. Neumann: Acta Metall. **27**, 915 (1979)

6.52 H. Matsui, A. Kimura, H. Kimura: The orientation dependence of the yield and flow stress of highpurity iron singlecrystals doped with hydrogen. Conf. Strength of Metals and Alloys, ICSMA5, (Aachen 1979) p. 977

6.53 I.M. Robertson, T. Tabata, W. Wei, F. Heubaum, H.K. Birnbaum: Scripta Metall. **18**, 841 (1984)

6.54 H. Matsui, H. Kimura, S. Moriya: Mater. Sci. Eng. **40**, 207 (1979)

6.55 Y. Felefli, W. Dahl, K.W. Lange: 3rd Int'l Congr. on Hydrogen in Materials, H3, Paris, 1, E14 (1982) p. 539

6.56 F.A. McClintock: Int. J. Fract. Mech. **4**, 101 (1968)

6.57 H. Vehoff, P. Neumann: Hydrogen cracking of single crystals of Fe-alloys, in *Hydrogen Degradation of Ferrous Alloys* ed. by R.A. Oriani, J.P. Hirth, M. Smialowski (Noyes 1985) p. 686

6.58 W.W. Gerberich, P. Marsh, J. Hoehn, S. Venkataraman, H. Huang: Hydrogen/plasticity interactions in stress corrosion cracking, in *Corrosion-Deformation Interactions*, ed. by T. Magnin, J.M. Gras, les editions de physique les Ulis, France, **325** (1993)

6.59 M.J. Lii, X.F. Chen, Y. Katz, W.W. Gerberich: Acta Metall. **12**, 2435 (1990)

6.60 H. Vehoff, C. Laird, D.J. Duquette: Acta Metall. **35**, 2877 (1987)

6.61 M.S. Daw, M.I. Baskes: Phys. Rev. B **29**, 6443 (1984)

6.62 J.S. Blakemore: Metall. Trans. **1**, 145 (1970)

6.63 E.M. Riecke: 3rd *Int'l Cong on Hydrogen and Materials*, H3, Paris, E7 (1982) p. 497

6.64 N.R. Moody, S.L. Robinson, S.M. Myers, F.A. Greulich: Acta. Metall. **37**, 281 (1989)

6.65 A. Atrens, D. Mezzanotte, N.F. Fiore, M.A. Genshaw: Corros. Sci. **20**, 673 (1980)

6.66 A.J. Kumnick, H.H. Johnson: Acta. Metall. **18**, 33 (1980)

6.67 A.J. McEvily, R.P. Wei: Fracture mechanics and corrosion fatigue in *Corrosion Fatigue*, ed. by D. Deveraux, A.J. McEvily, R.W. Staehle (NACE 1973) p. 381

6.68 G.W. Simmons, P.S. Pao, R.P. Wei: Met. Trans. A **9**, 1147 (1978)

6.69 M. Lu, P.S. Pao, T.W. Weir, G.W. Simmons, R.P. Wei: Met. Trans. A **12**, 805 (1980)

6.70 R.W. Pasco, K. Sieradzki, P.J. Ficalora: Scripta Metall. **16**, 881 (1982)

6.71 R.W. Pasco, P.J. Ficalora: Acta Metall. **31**, 541 (1983)

6.72 T.E. Fisher: In *Atomistics of Fracture*, ed. by R.M. Latanision, J.R. Pickens (1983) p. 39

6.73 W. Eberhardt, F. Greuter, E.W. Plummer: Phys. Rev. Lett. **46**, 1085 (1981)

6.74 K.H. Klameth: *Wasserstoffversprödung an Nickel Einkristallen*; Dissertation, RWTH Aachen (1985)

6.75 C. Atkinson: J. Appl. Phys. **42**, 1994 (1971)

6.76 H.H. Johnson: *Hydrogen in Metals*, ed. by I.M. Bernstein, A.W. Thompson (A.S.M. 1974) p. 35

6.77 L.G. Fritzemeier, W.T. Chandler: Hydrogen embrittlement – rocket engine applications, in *Superalloys, Supercomposites, and Superceramics* (Academic, London 1989) p. 491

6.78 G.C. Smith: Effect of hydrogen on nickel and nickel-base alloys, in *Hydrogen in Metals*, ed. by I.M. Bernstein, A.W. Thompson (ASM 1974) p. 485, 4.16

6.79 W.W. Gerberich, T. Livne, X. Chen: in *Modeling Environmental Effects on Crack Growth Processes*, ed. by R.H. Jones, W.W. Gerberich (AIME, Warrendale, 1986) p. 243

6.80 G.G. Hancock, H.H. Johnson: Trans. Met. Soc. AIME. **236**, 513 (1966)

6.81 W. Rothe: *Experimente zum Mechanismus der Wasserstoffversprödung bei niedrigen Partialdrücken*; Dissertation, RWTH Aachen (1983)

6.82 P. Sofronis, R.M. McMeeking: J. Mech. Phys. Solids **37**, 317 (1989)

6.83 J.R. Rice: In *Chemistry and Physics of Fracture*, ed. by R.M. Latanision, J. Pickens (Nijhoff, Dordrecht 1987) p. 22

6.84 J.H. Rose, J.R. Smith, F. Guinea, J. Ferrante: Phys. Rev. B **29**, 2963 (1984)

6.85 H.J. Grabke, H. Erhart, R. Möller: Microchimica Acta Wien Suppl. **10**, 119 (1983)
6.86 R.H. Jones, M.J. Danielson, D.R. Baer: J. Mater. Energy Syst. **8**, 185 (1986)
6.87 H.J. Grabke: In *Chemistry and Physics of Fracture*, ed. by R.M. Latanision, J. Pickens (Nijhoff, Dordrecht 1987) p. 388
6.88 S.M Bruemmer, R.H. Jones, M.T. Thomas, D.R. Baer: Metall. Trans. A **14**, 233 (1983)
6.89 A. Kimura, H.K. Birnbaum: Acta Metall. **36**, 757 (1988)
6.90 G.M. Bond, I.M. Robertson, H.K. Birnbaum: Acta Metall. **37**, 1407 (1989)
6.91 H. Stenzel, H. Vehoff, P. Neumann: In *Intergranular Stress Corrosion Cracking in Bicrystals*, Proce. of Modeling Environmental Effects on Crack Initiation and Propagation, TMS-AIME Meeting, Toronto, Canada, October 13–17, 1985, ed. by R.H. Jones, W.W. Gerberich Metallurgical Soc., Warrendale, PA (1986) p. 225–243
6.92 H. Stenzel: "Untersuchungen zur interkristallinen Spannungsrißkorrosion an FeSi-2.7%-Bikristallen"; Dissertation, RWTH Aachen (1987)
6.93 J.R. Rice, J.S. Wang: Mater. Sci. Eng. A **107**, 23 (1989)
6.94 J.P. Hirth, J.R. Rice: Metall. Trans. A **11**, 1502 (1980)
6.95 J. Wang, H. Vehoff: Scripta Metall. **25**, 1339 (1991)
6.96 A. Roy, U. Erb, H. Gleiter: Acta Metall. **30**, 1847 (1982)
6.97 H. Vehoff, H. Stenzel, P. Neumann: Z. Metallkde. **78**, 550 (1987)
6.98 G.J. Ackland, V. Vitek: MRS, **133**, 105 (1989)
6.99 J.J. Kruisman, V. Vitek, J.Th.M. De Hosson: Acta Metall. **36**, 2729 (1988)
6.100 A. Kimura, H.K. Birnbaum: Acta Metall. Mater. **38**, 1343 (1990)
6.101 P.M. Anderson, J.R. Rice: Scripta Metall. **20**, 1467 (1986)
6.102 J.R. Rice, R. Thomson: Phil. Mag. **29**, 73 (1974)
6.103 P.B. Hirsch, S.G. Roberts, J. Samuels: Proc. R. Soc. (London) A **421**, 25 (1989)
6.104 M. Brede: Acta. Metall. Mater. **41**, 211 (1993)
6.105 I.H. Lin, R. Thomson: Acta Metall. **34**, 187 (1986)
6.106 R. Thomson: Private communication (1991)
6.107 S. Moriya, H. Matsui, H. Kimura: Mater. Sci. Eng. **40**, 217 (1979)
6.108 G. M. Bond, I.M. Robertson, H.K. Birnbaum: Acta Metall. **36**, 2193 (1988)
6.109 H.K. Birnbaum: Mechanisms of hydrogen-related fracture of metals in Environment-Induced Cracking of Metals (NACE-10, Houston 1990) p. 21
6.110 S.P. Lynch: J. Mater. Sci. **21**, 692 (1986)
6.111 S.M. Copley, B.H. Kear: Trans. Metall. AIME **339**, 977 (1967)
6.112 D.F. Lahrman, R.D. Field, R. Darolia: MRS Proc. **213**, 603 (1991)
6.113 H. Vehoff: Fracture mechanisms in intermetallics, Proce. NATO ASI on "Ordered Intermetallics-Physical Metallurgy and Mechanical Behaviour", ed. by C.T. Lui, R.W. Cahn, G. Sauthoff, (Kluwer, (1992) p. 299
6.114 H. Vehoff: Fracture and thoughness of intermetallics, in *High Temperature Ordered Intermetallic Alloys* V, MRS Proc. *Vol.* **288,** 71 (1993)
6.115 Y.T. Liu, Takasugi, O. Izumi, T. Yamada: Acta Metall. **37**, 507 (1989)
6.116 T. Takasugi: Acta Metall. Mater. **39**, 2157 (1991)
6.117 C.T. Liu, C.G. McKamey: Environmental embrittlement - a major cause for low ductility of oriented intermetallics in *High Temperature Aluminides and Intermetallics* (MRS, Warrendale, PA, 1990) p. 133
6.118 X.J. Wan, J. Zhu, K.L. Jing: Scripta. Metall. Mater. **26**, 473 (1992)
6.119 D. Eliezer, F.H. Froes, C. Suryanarayana: JOM **43**, 59 (1991)
6.120 D.E. Matejczyk, C.G. Rhodes: Scripta. Metall. **24** 1369 (1990)
6.121 W.Y. Chu, A.W. Thompson, J.C. Williams: Acta. Metall. Mater **40**, 455 (1992)
6.122 Y. Liu, T. Takasugi, O. Izumi, T. Yamada: Acta Metall. **37**, 507 (1989)
6.123 M.H. Yoo, C.L. Fu: Scripa Metall. Mater. **25**, 2345 (1991)
6.124 M.W. Finnis: In *Chemistry and Physics of Fracture*, ed. by R.M. Latanision, R.H. Jones (Nijhoff, Dordrecht 1987) p. 177–195
6.125 M.S. Daw, M.I. Baskes: Phys. Rev. Lett. **50**, 1285 (1983)
6.126 S.M. Foiles: Phys. Rev. B **32**, 7685 (1985)

6.127 M.S. Daw, M.I. Baskes, C.L. Bisson, W.G. Wolfer: In *Modeling Environmental Effects on Crack Growth Processes*, ed. by R.H. Jones, W.W. Gerberich (AIME, Warrendale 1986) pp 99–124
6.128 N.R. Moody, S.M. Foiles: Proc. MRS Conf.: Structure and Properties of Interfaces in Materials (Boston, MA 1992) p. 381
6.129 N.R. Moody, S.L. Robinson, W.M. Garrison: Res. Mechanica **30**, 143 (1990)
6.130 M.R. Shanabarger: In *Hydrogen Effects in Metals*, ed. by I.M. Bernstein, A.W. Thompson (The Metallurgical Society of AIME, New York 1981) pp. 135–141
6.131 W.M. Robertson: Z. Metallkde. **64**, 436 (1973)
6.132 A.S. Tetelmann: In *Hydrogen in Metals*, ed. by I.M. Bernstein, A.W. Thompson (ASM 1974) p. 1
6.133 W.M. Robertson: Met. Trans. A **8**, 1709 (1977)
6.134 C.L. Baker, J. Chene, I.M. Bernstein, J.C. Williams: Met. Trans. A **19**, 73 (1988)

7. Metal-Hydride Technology: A Critical Review

P. Dantzer

With 13 Figures and 5 Tables

Historically the first interest in metal hydride technology was initiated after the second world war for nuclear reactor applications, inasmuch as the capability of hydrogen to scatter and dissipate energy from neutrons was satisfied by the high density of hydrogen atoms stored in metal hydrides. The major development of these materials as moderator, reflector or shielding components was carried out for high-temperature mobile nuclear reactors, where the utmost important properties of metal hydrides were mechanical strength and stability, as well as excellent hydrogen retention. Despite numerous and tremendous difficulties encountered, the fabrication of hydrides for a major industrial use has been successful for the first time.

It is obvious that during that period, the contributions coming from materials science, including the search of new appropriate materials, the utilisation of specific processes in powder metallurgy or the need of accurate and extensive thermodynamical characterisations of hydrides were of paramount importance. Thus, the first intermetallic–intermetallic hydride $ZrNiH_3$ was prepared in 1958 by *Libowitz* et al. [7.1], but it took some time before it was realised how many possibilities could be opened up, by varying the composition of the different metallic components of the alloys. Although the potential applications are still based on the high hydrogen density stored in the materials, now the novelty is brought out by the attractive operation of transferring hydrogen through rechargeable metal hydrides.

The development of these applications and the interest in metal-hydrides started in 1970 with the discovery of hydrogen absorption by $LaNi_5$ [7.2] at Philips, and by FeTi [7.3] at the Brookhaven National Laboratory. This came at the right time, when public interest was aroused in the possibilities of diversifying energy resources driven by fear of shortages of petroleum supplies and continuous expansion of the world energy needs. In the present decade attention is being focused upon the serious problem of atmospheric pollution. Even if the 'greenhouse' effect remains an opened debate, the main problems of the environment can no longer be avoided. These problems are correlated to the combustion of fossil energies with emission of 5.4 milliards of tons of carbon per year; more than 100 000 km^2 of land are deforested per year in the under-developed countries contributing for 1.6 milliards of tons of carbon, mainly to satisfy an elementary need: eating. Knowing that (1) sea pollution through hydrocarbons is growing, (2) the long-term storage of

Topics in Applied Physics, Vol. 73
Wipf (Ed.)
© Springer-Verlag Berlin Heidelberg 1997

nuclear materials of reduced radioactivity is far from being solved, and (3) drastic shortage of petroleum is predicted around 2050, then any alternative energy possibility should be seriously taken into consideration.

The fact that hydrogen plays an important role as a nonpolluting fuel in internal combustion engines, as well as a nonpolluting working fluid in chemical heat pump systems, lends strong arguments for supporting research in clean energy systems based on metal hydrides.

Several national research programs were launched around 1975 in the US, Japan, and Germany, in the aftermath of the first petroleum crisis. These programs were mainly oriented toward solving hydrogen storage problems. It should be said that current commercial applications are limited for economic reasons, because the commercially available storage reservoirs are small units used mainly for laboratory purposes. However, progress has been made in materials production to the point where it is now possible to obtain on an industrial scale a large spectrum of metal alloys of sophisticated composition. Beginning in 1980, the energetic aspect associated with energy transfer and storage was proposed in the US [7.4] and in Japan [7.5] with the innovation of new heat management processes, the hydride chemical heat pumps. The largest contribution to this particular field came from the Japanese scientists, but the proposed machines are still at the prototype level. In 1990 the first hydride batteries reached the marketing stage.

The commercial and economical aspects will not be discussed here, but will remain as a shadow for metal-hydrides technology. In general, the situation is as follows: With the exception of isotope separation, where expense is not a limiting factor for development, most other hydride applications are in competition with well established and cheaper solutions, at least at the present time. In view of this economic fact, it should be clear now that metal-hydrides have been proven viable, and that the effort to optimise the various known processes should be pursued in order to increase the competitiveness of the proposed systems.

7.1 Outline

The purpose of this chapter is to evaluate the current status of research and development in the field of hydride applications, and to draw attention to investigations which must be encouraged to improve our understanding of hydrogen behaviour in metal-hydrogen systems with respect to nonequilibrium processes occurring during hydriding and dehydriding reactions. Reviews and references of the latest major developments on applied research are available in books [7.6–9] and in the proceedings of international symposia [7.10–15]. We should also mention the work of experts who have been continuously reporting, for more than 20 years, on a future hydrogen-based economy, including hydrides [7.16–19]. National programs related to hydride applications may be found in several works [7.20–22]. Throughout this chapter, the term 'Metal-Hydrogen', 'InterMetallic Compound', and 'Intermetallic Hydrides' will be abbreviated as 'MH', 'IMC', and 'IMH', respectively.

7.2 Current Problems in Hydride Applications

The two basic properties which make metal hydrides attractive are their high and reversible storage capacity per mole of compound, and the high energy stored per unit volume. The mass of hydrogen that can be stored by unit volume of hydride is almost twice what can be stored in liquid form, and the energy contained may run as high as one MJ per litre of hydride. These figures fully justify the current applied research on hydrides as fuel or for energy transfer. The appeal of the former stems from the fact that the automobile industry is the world's foremost civilian industry. The latter subject arouses greater interest and support in energy dependent countries and in those heavily engaged in environmental politics.

Hydride devices have two operating modes, those operating as 'closed systems' where the same hydrogen is cycled over and over, such as heat pumps and detectors, and those operating as 'open systems' where the storage unit is either fed with new hydrogen once the existing charge is spent, or is fed continuously with hydrogen. Open systems are more sensitive to impurities in the hydrogen gas than closed systems. This implies that the stability of the IMC in the presence of chemical impurities should be well established, as it may be either a detrimental or a quality factor, depending on the type of open system application, storage reservoir or purification device.

The main drawback of hydrides, with the exception of magnesium hydride, is their low hydrogen weight percentage, which is never more than 2%. Of course this weight limitation is less important in hydrogen purification processes, as well as in actuators or miniature detectors and to an even lesser extent, in stationary hydrogen storage. The first step in hydride technology is to find an alloy with appropriate properties offering optimum performance in order for it to produce a marketable unit. The ability to maintain absorption properties over the long term will be of paramount importance. Consider applications in cars with an average lifetime of 200 000 km. A hydride reservoir fuelling a car for 200 km will have to be filled only 1000 times with H_2 gas. For a chemical heat pump, in continuous operation in 20 minutes cycles over an expected 10 year lifetime, the number of required cycles will be 300 000. For batteries, the number of charge/discharge cycles should be of the order of a few hundred.

One thing is certain, amongst the many alloys proposed, none is universal in view of the spectrum of the potentiality of applications which cover a wide temperature domain, ranging from few Kelvin up to 600 K. This is why the prime criterion concerns the thermodynamic characterisation of the IMH systems. If these informations are supplemented with the knowledge of the parameters influencing the long term retention of the absorption properties of a given material, then any attempt to develop a hydride device would be conducted successfully. However, this latter task does not seem possible if we consider the hundreds of alloys available. Finally, as each application exhibits its own quite definite technical and economical constraints, the alloys selected should provide the best compromise between constraints and property changes during the hydriding and dehydriding cycles, for extensive use. This shows that a clear inventory is needed and that the factors influencing the absorption properties must be understood, otherwise the expectancy for developing a hydride device will remain very limited.

7.3 Relevant Properties

The parameters used in selecting an IMC have been subjected to many investigations and are classified according to their metallurgical aspects, including elaboration, structural characterisation and activation, their thermochemical properties, including thermodynamic parameters, their chemical stability with regard to impurities and cycling, their dynamic behaviour, i.e., mass and heat transport properties, and the technological problems of handling the materials.

The metallurgical problems have recently been discussed by *Percheron-Guegan* and *Welter* [7.23], who described the preparation techniques of the most representative families of IMCs. *Manchester* and *Khatamian* [7.24] have recently discussed the activation mechanisms.

7.3.1 Thermo-Chemical Reactivity

From a technological viewpoint, the thermochemical reactivity concerns the IMC in its fully activated state, which usually means a mechanically disrupted finely divided powder, with grain sizes of the order of a few microns. This is the starting material for the investigations devoted to applications.

(A) Thermodynamics

Flanagan and *Oates* [7.25] have recently discussed the thermodynamics of IMH systems. They introduced para equilibrium notation to qualify the low temperature metastable state, compared with the complete equilibrium state where the IMCs disproportionate according to the decomposition reactions associated with the metallic elements [7.26, 27]. In the following, as for hydride applications, the systems are considered as remaining in the para-equilibrium configuration that is as pseudo binary metastable hydrogen systems.

Most experiments devoted to the characterisation of hydrogen gas-solid reactions are investigated by volumetric analysis, the so called Sievet's type apparatus and X-ray diffraction analysis. A modified version of the former, which combines the use of microcalorimeter and gas volume technique is presented in [7.28]. *Ryan* and *Coey* [7.29] described a volumetric instrument with a detection limit of 10^{-7} mole of gas, showing that such a simple equipment may even outperform thermogravimetry in some cases for solid-gas reactions. Nevertheless if these measurements are routinely produced, there have been ample difficulties of gathering reliable thermodynamic data for IMH systems [7.30, 31]. Leaving aside the hypothesis of poor quality material or of contamination by any impurity, a great number of reported results shows that the experimental aspects of the hydride phase growth have been overlooked by many scientists in the field, who restricted their ambition to data production leading solely to enthalpy and entropy of formation of IMHs. It should be strongly emphasised that a deep understanding of the thermodynamic properties of activated IMHs which belong to the family of highly defective materials is unavoidable for their optimal use in technological conditions.

Whatever the application, the most important properties are related to the presence of a 'plateau pressure' which represents the coexistence of two condensed phases when hydrogen is loaded over the solubility limit. Assuming reversible and equilibrium conditions Gibbs phase rule gives a variance of one for two components, H and IMC, and three phases, one gas phase and two solid phases. With reversible behaviour, all the thermodynamic informations are schematically plotted in Fig. 7.1. For the two coexisting solid phases the hydride grows according to the chemical reaction:

$$(b-a)^{-1}\mathrm{MH}_a + \frac{1}{2}\mathrm{H}_2(\mathrm{plat}, g) \rightarrow (b-a)^{-1}\mathrm{MH}_b \ , \tag{7.1}$$

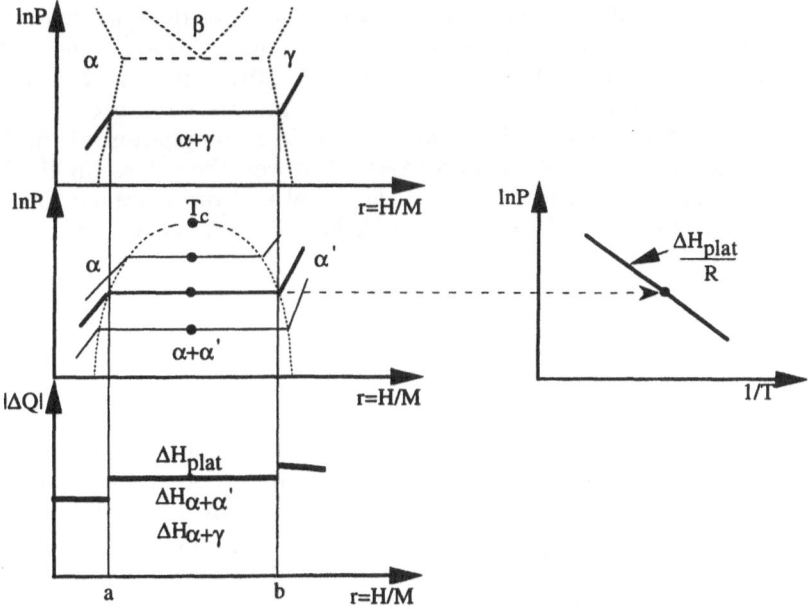

Fig. 7.1. Isotherms for two types of phase diagrams (top and middle). van't Hoff plot derived from the two phases region of the miscibility gap system (right). Heat evolved during hydrogen absorption (bottom)

where a, b correspond to the hydrogen contents of the phase boundaries expressed as H/M, and 'plat' indicates that the system is in a two-phase region. The following wellknown relations are deduced:

$$\Delta\mu_{\text{plat}} = \mu_{\text{H},a} - \frac{1}{2}\mu^{\circ}_{\text{H}_2} = \mu_{\text{H},b} - \frac{1}{2}\mu^{\circ}_{\text{H}_2} = \frac{1}{2}RT\ln(p_{\text{plat}}) \ , \tag{7.2}$$

where the relative plateau chemical potential of hydrogen represents the standard Gibbs energy change for reaction (7.1) when the hydrogen gas is at 1 atm, and MH_a and MH_b are in their standard states. Thus,

$$\Delta H_{\text{plat}} = \frac{\text{d}(\Delta G^{\circ}/T)}{\text{d}(1/T)} \ , \tag{7.3}$$

or

$$\frac{|\Delta H_{\text{plat}}|}{R} = \frac{1}{2}\frac{\text{d}\ln(p_{\text{plat}})}{\text{d}(1/T)} \ . \tag{7.4}$$

The later equation is referred to as the van't Hoff equation and relates the amount of heat evolved during the process of formation of a hydride compound to the plateau pressure. It is illustrated by the $\ln P$ versus $1/T$ in Fig. 7.1.

As a first step, we can take ΔH_{plat} to be a reasonable approximation of the heat of formation of the hydride, ΔH^{f}, which determines the stability of the material. The errors introduced are correlated to the relative partial heat of solution and then to the extension of the solubility of hydrogen in the solid solution or in the nonstoichiometric hydride. Predicting the heat of formation is a very difficult task. Semiempirical models have been reviewed by *Griessen* and *Riesterer* [7.32], who also made a compilation of the heats of formation of various IMHs and MHs. However, the values for the heats of solution and heats of formation of the metal-hydrogen systems have not all been updated, in particular for the V, Nb, Ta, Pd, Ti, La, Y, and Th–H_2 systems [7.33–40].

Equation (7.2) can be written in the form,

$$\frac{1}{2}\ln(p_{\text{plat}}) = \frac{\Delta\text{H}^{\text{f}}}{RT} - \frac{\Delta S^{\text{f}}}{R} \ , \tag{7.5}$$

where ΔS^{f} is the entropy of formation. This expression provides a useful way to estimate the success of a given IMH in an application. Equation (7.5) assumes that the reaction is reversible, at equilibrium, and so its use should be restricted to a first principle energy calculation. Using this relation to describe the dynamics of processes overextends its usefulness and leads to errors, because irreversibilities have to be introduced when modelling the hydride cycles (second principle induced effects).

Related to the thermodynamic of IMHs, we note that plots of the isotherms could have been used to elaborate phase diagrams as shown by the

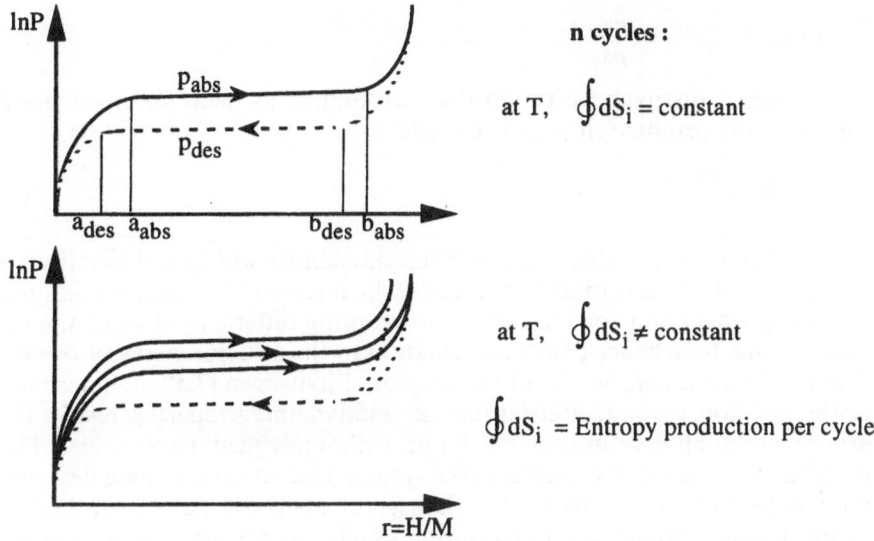

n cycles :

at T, $\oint dS_i$ = constant

at T, $\oint dS_i \neq$ constant

$\oint dS_i$ = Entropy production per cycle

Fig. 7.2. Schematic Plot of reproducable (top) and nonreproducable (bottom) isothermal hysteresis

thin lines in Fig. 7.1 for hypothetical systems with miscibility gap and structural transformation [7.41]. However, up to now, this opportunity has not been found to be attractive for IMH because of the lack of accuracy due to the broadening of the phase boundaries in the reported isotherms. Recently, *Luo* and *Flanagan* [7.42] drew a series of correlation for checking the consistency between the phase diagrams and the thermodynamic data for metal hydride systems.

The behaviour encountered in real systems is schematically shown in Fig. 7.2. The isothermal hysteresis is illustrated by the fact that thermodynamic paths associated with the processes of phase growth, hydride formation or hydride decomposition are different, leading to pressures and phase boundaries changes such that $p_{abs} > p_{des}$, $a_{abs} > a_{des}$, and, $b_{abs} > b_{des}$. Equations (7.1–3) have to be written for each process. Hysteresis and sloping plateau effects are largely responsible for the difficulties in determining reproducible stationary states of the IMH systems, which as a consequence currently limits the validity of predictions of long term thermodynamic behaviour for the IMH in use. The latest academic contributions to our understanding of hysteresis are due to *Flanagan* et al. [7.43], *McKinnon* [7.44], *Quian* and *Northwood* [7.45], *Shilov* and *Kuznetsov* [7.46], *Baranovski* [7.47], and *Balasubramanian* [7.48].

Since the realistic view for an application coincides with the knowledge of reliable state equations, the degree of irreversibility per cycle has to be evaluated. It is given by the free energy loss,

$$\Delta G_{\text{loss}} = \frac{1}{2} RT \ln \frac{p_{\text{abs}}}{p_{\text{des}}} \ ,$$ (7.6)

this energy is dissipated through the surrounding as heat, and provides the total entropy production per cycle with,

$$\oint dS_i = \frac{\Delta G_{\text{loss}}}{T} \ .$$ (7.7)

Stationary metastable states and reproducibility will be satisfied if we are able to find the appropriate parameters which generate a constant minimum entropy production over the cycle. Disregarding different values of the ratio $p_{\text{abs}}/p_{\text{des}}$ due to artefacts, such as deviation in the stoichiometry of the IMC [7.49] or influence of the size of the aliquot of hydrogen [7.50] in determining isotherms, nonreproducibility in the thermodynamic properties results from experimental observations carried out with undefined parameters. These parameters, internal or external to the system, lead to various time dependent thermodynamic states. Recently, *Dantzer* et al. [7.51] have identified and quantified the parameters acting on the LaNi$_5$–H$_2$ system. Accurate control of the mass flow rate of hydrogen gas, drastic control of the temperature maintaining quasi-isothermal conditions, <0.1 °C, makes it possible to satisfy a fine control of the driving force during the hydride phase growth. The strict condition on the variation of the temperature imposes a maximum flow rate, whose limitation is only given by the heat transfer capability of the reactor. Minimising coupling effects with heat transfer and temperature gradients in the activated powder insures the maximum homogeneity for the reaction. Respecting these elementary experimental conditions, reproducible thermodynamic states are obtained during the generation of complete cycles, moreover, the sensitivity of the method allowed to point out a memory effect in LaNi$_5$–H$_2$ system. This work tends to prove that if homogeneous hydride phase growth is insured, then no difference should be noted between stationary or quasi-static and dynamic states, with continuous absorption of hydrogen. Work in the latter domain was initiated by *Goodell* et al. [7.52], *Josephy* and *Ron* [7.53, 54], and *Groll* et al. [7.55].

Different reasons have been offered to explain the sloping plateau effect, such as inhomogeneity within the compound introduced during its preparation and degradation imposed by a highly stressed material correlated to the extensive volume variations during cycling. X-ray analysis of cycled compounds showed broadening of lines [7.56], whereas analyses of the line profiles showed the presence of residual strains after dehydriding, for specific symmetry axes of the compounds.

Because the problems involved in thermodynamics of activated materials are now well identified, one can expect more rigorous thermodynamic characterisation in the future. Noting that for the LaNi$_5$–H$_2$ system, the first calorimetric studies above room temperature were reported in 1989 [7.57], a first tentative phase diagram was proposed in the same year by *Shilov* et al. [7.58]; as a result, scientists should be convinced that solving problems of well known prototype IMCs, is always helpful for a practical use of IMHs.

(B) Chemical Contamination

The large specific area after activation (0.2–$5\ m^2/g$) corresponds to a highly reactive powder that acts as a getter attracting all contaminant molecules present in the hydrogen gas. Keeping in mind the technological aspect, the important points to be investigated here are the surface modifications leading to surface segregation, the deterioration of the catalytic effects, and the slowing down of absorption kinetics.

The time evolution of metal-H_2 impurity interactions has been studied extensively and their effects were quantified [7.59–62] for AB_5 and AB prototype alloys. The alloy impurity effects were classified as poisoning, retarding, reactive, and innocuous. These studies were complemented by an analysis of the alloys capability of recovering their absorption properties. A model was developed correlating the cyclic loss of H_2 storage capacity to a damage function and to a contaminant concentration dependence. This work has been corroborated by other studies in this area [7.63–67].

Interest in metal-H_2 surface interactions grew as batteries were developed and with the idea of hydrogen storage in thin film metal-hydrides. The latter field was initiated in 1983 by *Wenzl* et al. [7.68], who prepared a thin film of NbV solid solution alloy. The greatest advantage to this approach resides in its overall heat transfer capability, as heat transfers much more easily in this configuration. *Jain* et al. [7.69] recently reviewed the hydrogen storage in thin film metal-hydrides. Thin film deposition has also been used to produce a highly reactive surface, thereby removing the barrier to hydrogen dissociative absorption, such as occurs with Pd-coated Mg samples [7.70]. This achieves a catalytic and protective effect. A $LaNi_5$ thin film has been used by *Shirai* et al. [7.71], *Sakaguchi* et al. [7.72] to inject protons into an amorphous WO_3 thin film which changes its colour with the absorption of hydrogen. By using this phenomenon, different metal films can be inserted between $LaNi_5$ and WO_3 films for investigations of their permeability. This technique was recently applied to study the penetration of hydrogen through an amorphous V_2O_5 film deposited on WO_3 and covered with $LaNi_5$. Thin film samples of FeTi, TiNi, TiPd, $TiMn_2$, and $LaNi_5$ were prepared, layer by layer, by controlled metal evaporation [7.73] while the effects of oxygen precoverage on the reduction of the hydrogen absorption rate was investigated. The results show trends similar to those found for metal films and metal/titanium sandwich films.

Even if contamination is unavoidable, there is no longer any major problem in holding the impurities down to the lowest level, thereby increasing considerably the lifetime and high performances of the IMCs, at least for solid-gas applications. For example, according to *Wang* and *Suda* [7.74], certain hydride forming alloys treated with a fluorine containing aqueous solution will exhibit surfaces with high hydrogen affinity, together with protection from impurities. The efficiency of the treatment in long term cycling studies remains to be proved. The degradation encountered in metal hydride electrodes will be discussed in the relevant section.

(C) Stability and Cycling

Disproportionation may occur in cycling, and will result in phase separation and in the formation of more stable hydrides, reducing the H_2 storage capacity of the material. Cycling experiments are time consuming and obviously require great care to avoid coupling with other phenomena that will accelerate the disproportionation process, such as segregation induced by gaseous impurities. Different methods of ageing/cycling are currently being reported, such as temperature or pressure induced cycling, or cycling using two coupled reactors, as in a heat pump device [7.75]. A static technique was recently proposed by *Sandrock* et al. [7.76] for studying disproportionation; the IMC is loaded at high temperature while the hydrogen gas pressure remains above the dissociation plateau pressure of the hydride. To date, no general explanation has been given to a better understanding of the disproportionation process, although the results of the experiments seem to depend upon the alloys used and upon the method adopted. The safest way of determining the long-term stability of an IMH is still by performing experimental tests.

In 1978, the first report [7.77] on temperature cycling concerned the $EuRh_2$–H_2 system, and this was followed by similar studies on $(La_{0.9}Eu_{0.1})Ni_{4.6}Mn_{0.4}$ [7.78–80]. Degradation of this compound was reported as the temperature was increased from room temperature up to 300 °C. After 1500 cycles, the hydrogen absorption capacity was only 26 % of the initial value. It was confirmed that disproportionation was an intrinsic process and was not induced by any oxidation of the sample, as that would have been detected by Mössbauer spectral evidence of Eu_2O_3. As cycling continues, the metal atom diffusion processes are enhanced by the large lattice strains induced by the volume changes during hydriding. Moreover, the heat generated during absorption can lead to local temperature gradients, which in turn, promote diffusion and rearrangement of the metal sublattice [7.77].

Buschow [7.81] developed a model to describe the tendency of an IMC to disproportionate. Although the model is based on an oversimplified description, it does show that decomposition does not depend on the stability of the ternary hydride, nor on the stability of the corresponding IMC, but it rather suggests that it is correlated with metal atom diffusion activation energy in ternary hydrides. Following this reasoning, low temperature cycling and good heat transfer capability of the reactors are needed in order to prevent degradation. These conditions were satisfied in ageing experiments performed with a dual bed configuration for the $LaNi_5$ and $LaNi_{4.7}Al_{0.3}$ compounds between 25 and 80 °C, experiments which avoided any over pressure on the samples [7.75, 82]. It was shown that the compounds retained their hydrogen absorption capacity up to 1250 cycles, at which point the experiments were stopped.

The same compounds studied by *Park* and *Lee* [7.83], *Han* and *Lee* [7.84], by thermal cycling, between 30 and 185 °C, at a cycling frequency of 10 minutes, showed reduced capacity of the order of 50 and 12% after 2500 cycles for LaNi$_5$ and LaNi$_{4.7}$Al$_{0.3}$, respectively. Similar results were obtained by *Gamo* et al. [7.85], who reported degradation of the LaNi$_5$ during pressure cycling, where fresh hydrogen gas was introduced at each cycle. The stability of multicomponent mischmetal-nickel compounds were also investigated by cycling with impurities [7.86, 87]. The latest degradation studies on AB$_5$ were reported by *Chandra* et al. [7.88]. LaNi$_{5.2}$ was thermally cycled between room temperature and 400 K and showed loss of hydrogen storage capacities after the production of 10 000 cycles. X ray diffraction analysis reveals the precipitation of Ni microphases indicating disproportionation as the possible cause of degradation of the IMC. The degradation could have been accelerated by the fact that the system was studied above 343 K, temperature at which the intermediate hydride phase LaNi$_5$–H$_3$ is reported [7.89], this phase is considered as a strain induced metastable phase. To overcome degradation *Bowman* et al. [7.90] substituted tin for nickel in LaNi$_5$. Up to 1300 cycles were produced between room temperature and 500 K; the results indicate a noticeable reduction of the rate of degradation by a factor of 20 for LaNi$_{4.8}$Sn$_{0.2}$ compared with LaNi$_5$.

Long term cycling ability experiments have also been carried out on TiFe. *Reilly* [7.91] subjected TiFe to 13 000 absorption cycles without noticeable deterioration, but thermally induced cycling between 30 and 200 °C produces a drastic reduction in storage capacity [7.92, 93]. The cycling stability of TiFe$_{0.8}$Ni$_{0.2}$H$_x$ was tested between 20 and 185 °C [7.94, 95] using up to 65 000 cycles where a full cycle lasted for about 8 minutes. Periodically X-ray, reaction rate, and PCT measurements were taken. After 65 000 cycles, the system exhibits a slight increase in absorption pressure, 0.1 atm at $H/M = 0.3$ and 55 °C, and a reduction of about 16% in the hydrogen uptake.

As far as a few hundreds and up to thousands of absorption-desorption cycles are required, degradation induced by cycling does not seem insurmountable. Remember that the losses of hydrogen storage capacities severely impedes any development of devices that require over 10^5 cycles, such as compressors, heat pumps... If, as it is suggested degradation is accelerated by thermally activated processes such as metallic diffusion, then IMCs should currently be limited in use to low temperature domains less than 100 °C. It has been shown that additional metallic elements contribute in some cases to reduce the decomposition tendencies, this will be an improvement for the development of multicomponent alloys. Noting that once degradation has been established, it is usually confirmed by structural analysis which give quantitative information on the consequences but do not provide any hint on the origin of the phenomenon. More work should be attempted to clarify the origin of degradation. It is rather surprising that the influence of strong thermal gradients induced in the powder by thermal cycling has been currently neglected in the analysis. This problem will be discussed in the paragraph on thermal properties.

Notten et al. [7.96, 97] contributed greatly towards a better understanding of mechanical degradation phenomena in AB$_5$ system. The authors were able to establish that the mechanical stability of ternary nonstoechiometric compounds is correlated to the discrete $\alpha \rightarrow \beta$ lattice expansion, rather than the total lattice expansion. They found that materials with small discrete lattice expansion are mechanically much more stable, emphasising the important role of Cu in the mechanical powder stability and in the electrochemical long term cycling stability. More research in that direction should be investigated. The path to follow is once again opened by fundamental work performed at Philips laboratories.

7.3.2 Transport Properties

(A) Kinetics Survey

It has been argued that cycling rates will not be limited by the chemical kinetics of the reactions, but rather by the capacity to add or remove heat from the hydride bed. This argument has often been used to support the assumption that quasi-thermodynamic equilibrium is reached instantaneously, greatly simplifying the modelling of hydride reactor dynamics. Though this is true for most IMC's, this assumption should be reconsidered in some cases, especially for low temperature uses of hydrides. Note that for any hydride storage unit, once we know the kinetic law and the temperature dependence, we can determine the theoretical limits for the exchange rate of the device.

It is beyond the scope of this section to review such a broad field as heterogeneous IMH kinetics in detail. However, this does deserve comment in the context of hydride device optimisation. IMH kinetics have been extensively studied. Concerning the earlier reports, before 1983–85, the attempts to interpret the data in terms of a microscopic model are questionable since no proof was provided on the isothermal conditions of the experiments. This aspect was ignored for a long time, although evidence of temperature increase of the sample on hydriding has been given by *Goodell* and *Rudman* [7.98]. Note that the technique of using thermal ballast to insure quasi-thermal conditions, first proposed in [7.98], solves only apparently the problem of coupling between hydriding reaction and heat transfer in the powder because the temperature increase at the particle level remains driven by the over pressure on the bed and of course by the cooling power to expel heat from the mixed powder bed. *Dantzer* and *Orgaz* [7.99] analysed with the help of a simple heat transfer model, the modifications of a postulated rate law which results from the assumption of a nonisothermal regime. The quality of the heat transfer was characterised by an overall parameter, the thermal time constant, which evaluates the thermal performance of a specific experimental set up. Some results of the simulation are shown in Fig. 7.3 with the effect of the dissipation of heat on the shape of the transformed fraction curves, calculated for temperatures of 25, 50 and 75 °C. The hydrogen absorption controlled by heat transfer shows that the rate can be reduced at the higher temperature as compared with the rate at the lower temperature. The time corresponding to the same transformed fraction usually decreases with increasing temperature. As indicated in Fig. 7.3, this behaviour is displayed

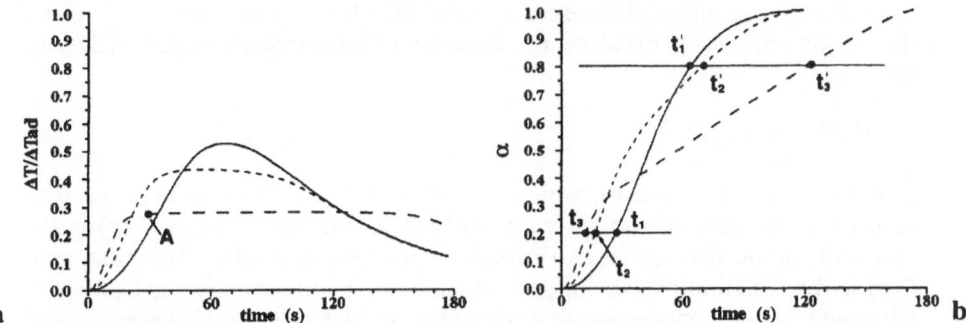

Fig. 7.3. Heat dissipation effect in temperature dependent hydriding kinetics: (**a**) normalized temperature profile $\Delta T(t)/\Delta T_{adiabatic}$ vs. time; (**b**) effect on the transformed fraction vs. time. Thermal time constant = 1min; $T = 25$ °C(——); 50 °C(- - -); 75 °C(– –)

initially, $t_1 > t_2 > t_3$, but when the reaction proceeds, the simulation demonstrates that the rate obtained at 75 °C decreases and the initial behaviour is reversed ($t_1 < t_2 < t_3$). Apparent anomalous behaviour is reported, when the same amount of transformed product is attained with increasing time, as the initial temperature increases due to the thermal resistance of the bed.

In later years, considerable effort has been made to improve the experimental observations, chiefly to satisfy quasi-isothermal conditions for which only intrinsic kinetic parameters are involved. The excursion of several decades in rate constant for the LaNi$_5$–H$_2$ system, as compiled by *Goodell* and *Rudman* [7.98], has now been reduced to one decade, but there is still poor agreement in the reaction mechanisms and in the overall kinetic laws suggested.

Here again, the difficulties result from the presence of activated powder of undefined grain shape, and from the presence of hydrides which are unstable at room temperature. So there is some difficulty in associating the kinetics experiments with metallographic studies to identify the mechanisms controlling the reactions, as was done by *Mintz* and *Bloch*, [7.100] who studied the hydriding reaction mechanisms of rare-earth hydrogen systems for a given sample geometry.

The fact that the rate equation has to account for more than one process seriously complicates the interpretation of kinetics experiments, to the point that modelling is the most serious tool of progress. *Rudman* [7.101] proposed extending the nucleation and growth formulation to solve the kinetics of LaNi$_5$. The model was developed on the basis of three intrinsic processes operating in series, which were finally reduced to a surface and a bulk process, leading to the identification of four parameters, two rate constants and two activation energy parameters. It is interesting to note the similarity of this model with the unreacted core model [7.102] widely used in chemical reaction engineering, a model which involves three consecutive steps, surface limitation, penetration, and diffusion through the reacted layer to the surface of the unreacted core, and transformation reaction.

Others have analysed the experimental data by a yet simpler model where the various processes involved are described by an overall empirical law of the form

$$\text{Rate} = k f(p) f(x) \ , \tag{7.8}$$

where k is the rate constant, which should be only temperature dependent to reflect the mobility of the reaction through an Arrhenius term, $f(p)$ is the thermodynamic driving force through a pressure dependent function, and $f(x)$ is the function expressing the dependency of the transformed product. Although such an expression is apparently simpler to solve numerically and certainly easier for modelling hydride reactors, the recent literature does not provide any such unambiguous law. So here we will make no attempt to discuss the different mechanisms proposed, but we will limit the discussion to the laws proposed for AB_5 and AB_2 compounds. The literature concerning earlier studies will be found in the various papers listed.

The $LaNi_5 - H_2$ system was investigated by *Josephy* and *Ron* [7.103], *Koh* et al. [7.104], and *Han* and *Lee* [7.105] who also studied the $LaNi_{4.7}Al_{0.3} - H_2$ system, as did *Wang* and *Suda* [7.106–108]. All authors claimed to work with high-performance reactors that maintained isothermal conditions, while the experiments were performed under isobaric conditions for [7.103–105] and isochoric and variable pressure conditions for [7.106–108], and over a temperature range extending from room temperature to 90 °C.

Josephy and *Ron* [7.103] found that the decomposition of the hydride satisfied a first-order type reaction with a linear pressure dependence. The interesting point is that they were able to identify an intrinsic rate constant from which an activation energy of 29 kJ mol^{-1} could be determined from the temperature dependence. The experiments of *Selvam* et al. [7.66] are very similar, but their data did not satisfy a first-order type plot, while a logarithmic dependence of pressure is suggested and activation energies of 27 and 37 kJ mol^{-1} are obtained for the absorption and desorption processes, respectively. These values cannot be compared with the previous ones, because the rate constant includes a pressure dependence. Although the thermal ballast technique was used to insure good heat transport in the bed, the data indicate temperature variations for both processes, variations which definitively influence the interpretation of the experiments [7.99], [7.109]. *Han* and *Lee* [7.105] did not propose overall rates, but analyse the various possible mechanisms in depth.

Wang and *Suda* [7.106–108] derived empirical equations where $f(p)$ and $f(x)$ in (7.8) are transformed into an analytical function $f(x^a, p^b)$ where a and b are assimilated to a reaction order with respect to hydrogen concentration and hydrogen pressure. This technique allowed them to discern different kinetic behaviours in the single and two phases region. For the two phase region, they derived an intrinsic rate constant with activation energies of 36.8 and 54 kJ mol^{-1} for the absorption and desorption processes, respectively. Recently, *Wang* and *Suda* [7.110] reported a new experimental technique using a highly heat-sensitive reactor, with controlled cooling by injection of liquid carbon dioxide. The results for the hydriding reaction in the two-phase region are reasonably close to the earlier results, where the difference amounts to 6% for the rate constant and for the activation energy (39.2 kJ mol^{-1}). The only common feature for all the reported works is that the absorption process is faster than the desorption.

  Bernauer et al. [7.111] investigated the reaction kinetics of the multi-component system Ti$_{0.98}$Zr$_{0.02}$V$_{0.43}$Fe$_{0.09}$Cr$_{0.05}$Mn$_{1.5}$ at –90 to –100 °C and found that the data could be explained by first-order kinetics with two different rate constants for low and high hydrogen concentrations.

  Iron, cobalt and copper have been partially substituted for nickel in the LaNi$_5$ system to study the effect this has on the reaction rates [7.112, 113]. The effect of surface contamination on the hydriding behaviour of LaNi$_5$ has also been thoroughly analysed by *Uchida* and *Ozawa* [7.114], *Ohtani* et al. [7.115]. These authors were able to correlate the changes in hydriding rates with the alterations of the highly reactive surface due to contamination, and they made a critical analysis of the different rate equations which are currently used.

  Reilly et al. [7.116], *Gavra* et al. [7.117] proposed to use metal hydride slurries, hydride particles suspended in n-undecane, to study the isothermal kinetics of fast chemical processes. LaNi$_5$–H$_2$ and Pd$_{0.85}$Ni$_{0.15}$–H$_2$ systems were investigated. It was shown that for both systems the kinetics were found to be in agreement with a phase boundary controlled model. The rate limiting process is the phase transformation that takes place at the interface between the unreacted hydride and the hydrogen-saturated layer of the metal product.

  In view of the experimental progress over recent years, we look forward to future studies that will provide the appropriate intrinsic kinetic parameters which, when combined with the other physical properties, will make it possible to model the dynamics of hydride processes with greater precision.

(B) Thermal Properties and Confinement

The low thermal conductivity of powdered hydride beds, (packed beds or porous media, as it is termed in thermal physics) was recognised early on as a key parameter in reactor design. There is no doubt that the greater the thermal conductivity of the sample, the shorter the cycling rates that will be tolerated, thereby increasing the performance of the reactors.

  Steady state and transient methods have been used to determine the thermal properties of packed hydride beds, and the thermal conductivity has been correlated with hydrogen concentration and pressure. *Suda* et al. [7.118, 119] pointed out that in nonsteady state experiments, the uncertainties in the thermal conductivity are closely related to the determination of the reaction rate use in solving the heat equations. The effective (overall) thermal conductivity, K_e, of powder hydride beds will rarely exceed 1–2 W m^{-1} K^{-1}, while all measurements reported in a plot of K_e versus the log p_{H_2} lie on an 'S' shaped curve. This reflects gaseous transport properties, Knudsen effect, where different heat flow regimes exist in the powder, depending on the mean free path of the gaseous molecules relative to the size of the pores of the beds.

Effective thermal conductivity is essentially interpreted using theoretical models, which should be able to predict this property. In view of the numerous parameters affecting K_e, such as the individual thermal conductivities of the solid and gas phases, the bed porosity, and the size distribution and shape of the particles, each of which are considered to affect the functional dependence of K_e differently. Care should be taken in selecting the model. Heat flow analyses have, to date, essentially been conducted using models formulated in the sixties, such as those developed by *Kunii* and *Smith* [7.120] and *Yagi* and *Kunii* [7.121, 122] (referred to as K.S. and Y.K. in the following). A recent survey of the thermal conductivity of packed beds was published by *Tsotsas* and *Martin* [7.123].

Suda et al. [7.118, 119] selected the K.S. model to identify the K_e of the reference material, but retained an empirical polynomial form for the expression of K_e for the hydride beds. A similar representation was maintained in succeeding reports, where a three-dimensional structure [7.124] and then an inclusion of a copper wire matrix [7.125] were used to enhance heat transmission. *Suissa* et al. [7.126] based their interpretation on the Y.K. model and came to the conclusion that K_e remains within a very narrow interval even if the thermal conductivity of the solid changes over a very broad range. Finally, the Batchelor model used by *Kempf* and *Martin* [7.127] seem to be inappropriate as the value it produces for K_e is too low.

Detailed heat flow analysis is not entirely without interest as it does give some idea of the order of magnitude of K_e, but the various techniques for improving thermal conductivity should now be approached in a more rigorous way than by trial and error. Most of the earlier works on the thermal properties of powdered hydride beds were performed through the measurements of stationary temperature gradients in a monodimentional volume, usually of cylindrical symmetry. The effective thermal conductivity K_e is known by solving the Fourier heat equation,

$$\frac{\partial T}{\partial t} = \nabla(a\nabla T) \tag{7.9}$$

with $a = K_e/\rho C_p$ where a is the thermal diffusivity, ρ is the packed bed apparent specific weight and C_p is the specific heat for the solid in the packed bed.

Unfortunately a steady-state method means that the coupling of heat transfer with a reaction cannot be investigated. The thermal studies of reactive hydrides require transient techniques, where now (7.9) includes the source term, dQ/dt, due to the reaction,

$$\frac{\partial T}{\partial t} = \nabla(a\nabla T) + \frac{Q}{\rho C_p} \ . \tag{7.10}$$

Equation (7.10) has to be solved numerically in the whole reactor, and the two-dimensional effects can no longer be neglected. An advantage of removing the monodimentional constraint is found in the design of the reactor, since sample volume can be drastically reduced, from a few liters in steady state measurements down to a few dozen of cm^3 in transient method. This approach was recently proposed by *Pons* et al. [7.128] who developed a new technique for studying K_e, the wall heat transfer coefficient, α_w, and the interactions between temperature and concentration gradients in the LaNi$_5$– H$_2$ system. α_w involves the heat transfer resistances at the wall-material interface, and is an important parameter in optimising hydride reactors. It was unknown before this work. The method is a combination of a measurement reactor, described in Fig. 7.4, with a numerical model solving the Fourier equation in the whole reactor. K_e and α_w are obtained by identification, i.e., adjusting the calculated temperature and pressure to the measurements. Obviously, the model requires an accurate state equation of the system being investigated to calculate the mass balance at each computation step. An analytical expression for the H/M, P, T correlation within the hysteresis loop was also developed. More details about the numerical phenomenological expressions, the calibration of the experimental set up, the reactor model as well as the identification procedure can be found in [7.128, 129].

For the pressure range studied, $p_{H_2} < 2.5$ MPa, K_e values extend over a range of 0.13 up to 2.3 W m^{-1} K^{-1}, including the published values. An interpretation of the conductivity was obtained with the support of the model developed by *Pons* and *Dantzer* [7.130], whose main advantage is to introduce a theoretical particle to describe the packed bed. The thermal conductivity of the bulk LaNi$_5$ was also determined, $K_{e,bulk} = 30 \pm 10$ W m^{-1} K^{-1}. A rather interesting result is reported for α_w which amounts to 2000

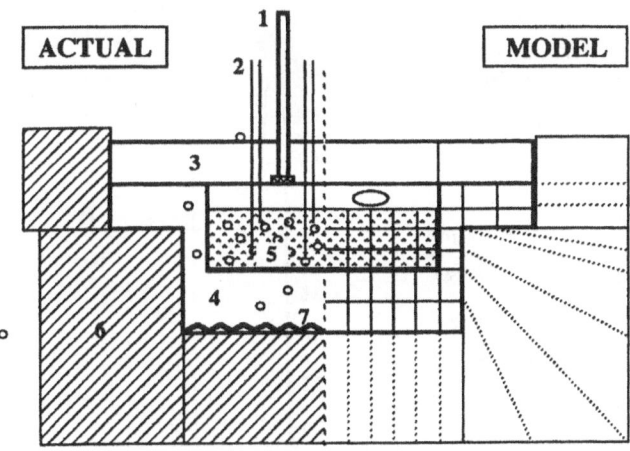

Fig. 7.4. Measurement reactor for thermal properties studies (Actual). Discretization for modelling (Model). 1: gas transfers, pressure and measurement. 2: Internal thermocouples. 3: Reactor lid. 4: Reactor body. 5: LaNi$_5$ packed bed. 6: Thermal insulation. 7: Electrical heating o: Different thermocouples location. Modelled flux lines in insulation are also presented

W m^{-2} K^{-1} for hydrogen pressure larger than 100 kPa. Such large values are due to the small size of the particles, according to the interpretative model. However, α_w decreases by more than an order of magnitude when the pressure is reduced to 3 kPa.

The thermodynamic states and the thermal properties of the system being well characterised, the coupling between heat transfer and reaction is demonstrated experimentally, when characteristic times of both phenomena are of the same order of magnitude. Experiments are reported, either with temperature induced reaction, temperature scans in the hysteresis loop, and with largely slowered reactions, several hours long [7.131]. Comparison of experimental results with results of simulations permits an interpretation. The analysis shows that the temperature gradients induce composition gradients within the hydride bed, making the local temperatures, composition or reactions rates very different from the average values, as it is illustrated in Fig. 7.5, for the LaNi$_5$–H$_2$ system submitted to a heating-cooling cycle.

The initial conditions, $T = 297$ K, $p_{H_2} = 200$ kPa correspond to the isobar on the left in Fig. 7.5, where the initial homogeneous composition of the hydride, $H/M = 3.1$ in an absorption state, is indicated by A/B. A and B are located in the powdered hydride bed, respectively, the closest and the farthest from the heat exchanger, i.e., at the bottom and top of the bed described in Fig. 7.4. During the heating period, the local temperature change at point A is faster than the average temperature and thus faster than the pressure evolution. This means that the system desorbs at A, while it absorbs at B. The final compositions reached at the end of the heating period are imposed by the thermodynamic paths imposed by the hysteresis effect, as shown by A and B on the isobar at $p_{H_2} = 450$ kPa. As the reactor cools, the situation is reversed, A reabsorbs, B desorbs. At the end, when the reactor is

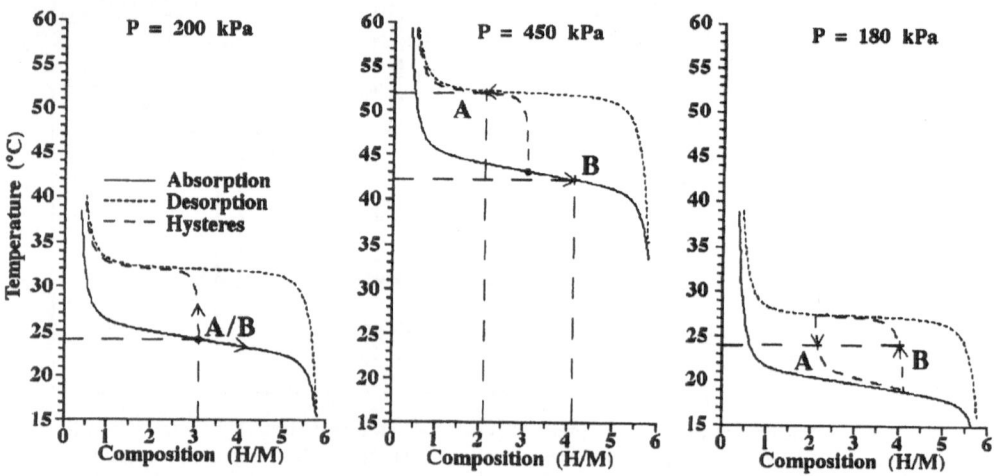

Fig. 7.5. Isobaric hysteresis diagram: (1) Initial state, hydrogen absorption (T_i $P = 200$ kPa); (2) End of the heating period ($P = 450$ kPa); (3) Back to T_i ($P = 180$ kPa)

back to the initial temperature, the whole system is found in the hysteresis domain, a lower pressure is measured and confirmed by the simulation. Despite uniform pressure and temperature, macroscopic inhomogeneities are always generated in the bed. Due to the temperature gradients, redistribution of hydrogen with heat of reaction has taken place inducing a heat transfer parallel to conduction, thus now the final thermodynamic state of the system is largely weighted by the hysteresis effect. ZrCo–H_2 was studied in detail by *Devillers* et al. [7.132] who attributed a pressure decrease after a first heating cooling cycle to hydrogen trapping in the ZrCo. This seems to be an erroneous interpretation. Recently, *Kisi* and *Gray* [7.133], *Gray* et al. [7.134] confirmed the above prediction by performing in-situ X-ray and in-situ neutron diffraction to study the growth of $LaNi_5$ hydride at the free surface and in the bulk of powdered samples. The results demonstrated that macroscopic compositional inhomogeneity does occur.

Various techniques have been tested to improve the thermal characteristics of the beds. The methods use different ways of binding a metallic element to the powder to form a composite mixture, such as aluminium foam added to the powder by *Supper* et al. [7.135], preparation of composite compacts (PMH) by *Ron* et al. [7.136], *Tuscher* et al. [7.137] and *Wang* et al. [7.138], a cold pressing and sintering technique developed by *Töpler* et al. [7.139], and micro encapsulation within a copper layer a few microns thick by *Ishikawa* et al. [7.140–142]. So the problems of confinement of small particles in materials can be eliminated as long as cracks do not form in the pressurised hydride pellets. A gain of a factor of 10 to 50 in K_e is reported for the PMH of *Ron* et al. [7.136] while the technique of *Töpler* et al. [7.139] provides a material with K_e of the order of 7 to 9 W m^{-1} K^{-1}, a value that is also of the same order of magnitude as that reported in [7.125].

In their careful and extensive study, *Josephy* et al. [7.143], *Bershadsky* et al. [7.144, 145], showed that while K_e can be greatly increased by compacting the IMC with a metallic matrix, unfortunately the permeability of the bed decreases drastically above a critical amount of the binding metal. This leads to mass transfer limitations which, in turn, affect the heat rate output of the reactor. They were able to describe K_e for the composite material with an empirical law and then concluded that K_e and the permeability factor would have to be optimised to obtain high hydrogen and thermal yields. The conclusion should rather have been that the coupling between heat and gaseous mass transfers are affected by numerous parameters, such as different heat flow mechanisms, bed porosity and permeability, and particle size and shape, all of which should be taken into account in analysing and optimising the performance of packed beds.

Other methods have been employed to increase the heat transfer capacity of the IMH, such as the obvious one of modifying the reactor design. Different reactor configurations have been proposed [7.146–149]. A metallurgical approach was attempted by *Ogawa* et al. [7.150] who increased the alloy's resistance to disintegration by unidirectional solidification on the $LaNi_5$–Ni eutectic alloy, with the idea of achieving a 'fibrous' composite material. *Uchida* et al. [7.151] mixed the IMC powder with silicone rubber, making a sheet that could be given various shapes for battery applications. Finally, metal-hydride slurries have been proposed by *Johnson* and *Reilly* [7.152] to overcome the main difficulties with powder beds. Thermal conductivity is increased, bed expansion eliminated and reactor designs simplified. There are other disadvantages due to the solvent vapour pressure, solvent degradation, agitation of the bed, and larger reactor size. The LaNi$_5$–H$_2$–n–undecane system has retained greater attention.

(C) Dynamic Properties

From the properties discussed so far, we may now advance the basic criteria for further development of hydride technology. Considering the dynamic behaviour means that the real performance of a hydride has to be evaluated in terms of the mass of hydrogen stored or transferred, and the power produced or required. The discussion now shifts to the present situation in modelling of hydride reactors.

It is well agreed that metal hydride reaction beds have to progress in two directions, increasing the available hydrogen content or the hydrogen mass really transferred, and increasing the thermal conductivity of the beds. The former goal may be reached by searching for new alloys or improving existing ones, while the latter calls for more experimental investigations and a better understanding of the thermal transport phenomena as explained above. The best way to clarify the state of research and development of hydride reactors consists of adopting an engineering viewpoint using two hypothetical ideal coupled reactors (R_1, R_2), which are associated with heat sources (HS$_1$, HS$_2$) and heat exchangers (HE$_1$, HE$_2$), as shown in Fig. 7.6. The equations governing the overall process of hydrogen transfer are written below for the three levels:

$$\text{Heat Source 1} \quad M_{s1} C_{P_{s1}} \frac{dT_{i1}}{dt} = \dot{m}_1 C_{Ps1}(T_{o1} - T_{i1}) + W_1; \tag{7.11}$$

$$\text{Heat Source 2} \quad M_{s2} C_{P_{s2}} \frac{dT_{i2}}{dt} = \dot{m}_2 C_{Ps2}(T_{02} - T_{i2}) + W_2; \tag{7.12}$$

$$\text{Heat Exchanger 1} \quad \dot{m}_1 C_{P_{s1}}(T_{o1} - T_{i1}) = \dot{m}_1 C_{Ps1}(T_{i1} - T_{r1})\eta_1; \tag{7.13}$$

$$\text{Heat Exchanger 2} \quad \dot{m}_2 C_{Ps2}(T_{o2} - T_{i2}) = \dot{m}_1 C_{Ps2}(T_{i2} - T_{r2})\eta_1; \tag{7.14}$$

$$\text{Reactor 1} \quad \left[\sum m C_P\right]_1 \frac{dT_{r1}}{dt} = |\Delta H_1|\frac{dMH_1}{dt} + \dot{m}_1 C_{Ps1}(T_{i1} - T_{r1})\eta_1; \tag{7.15}$$

$$\text{Reactor 2} \quad \left[\sum m C_P\right]_2 \frac{dT_{r2}}{dt} = |\Delta H_2|\frac{dMH_2}{dt} + \dot{m}_2 C_{Ps2}(T_{i2} - T_{r2})\eta_2. \tag{7.16}$$

The energy balance equations (7.11–14) assume that the operational technical parameters are known, such as the power of the heat source mass flow

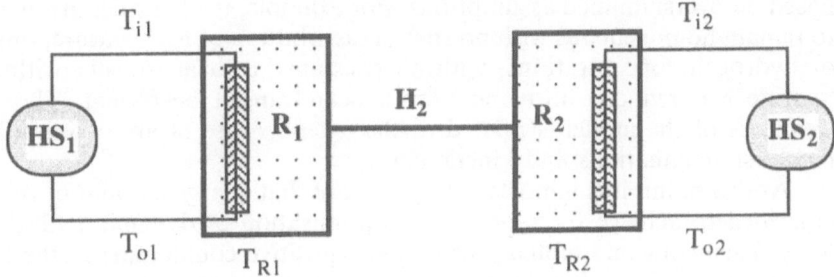

minimum physical parameters needed :

Heat Source, HS$_i$:

M_{si} mass of fluid i

m_{si} mass flow rate of fluid i

Cp_{si} specific heat of fluid i

W_i heating, cooling rate of the source i

Heat Exchanger, HE$_i$:

T_{ij}, T_{oj} inlet, outlet temp. of fluid j

η_j efficiency of the exchanger j

Reactors, R$_i$

T_{ri} temperature of hydride

$\left[\sum m\,Cp\right]_i$ total heat capacity of reactor

MH_i mass of hydride i

$|\Delta H_i|$ heat of formation of hydride i

Fig. 7.6. Representation of ideal coupled hydride reactors

rate of the fluid exchanger, the efficiency of the heat exchanger. Equations (7.15–16) represent the thermal balance for the chemical reactions evolved in each reactor where the first term of the second member corresponds to the chemical heat source, part of which is transferred through the heat exchanger and part of which is accumulated. The energy equations are complemented by the mass balance equations, which include the state equation for each IMH and the corresponding kinetics laws. This set of ordinary differential equations describes the full system and the solution is found numerically. Depending on the accuracy with which the physico-chemical and thermal properties of the materials are expressed in the model, the numerical methods can be used for elementary dimensioning or simulation or, at best of cases, for process optimisation.

The last goal has not been reached to date. The proposed models, for coupled or single reactors, can be used for dimensioning and for reasonably good numerical simulations, but no more, because the set of equations are based on oversimplified assumptions. For example, the hydrides are assumed to remain homogeneous with no spatial distribution of temperature, pressure or hydrogen concentrations, with all resistance to heat transfer within the hydride bed reduced into one overall heat transfer coefficient. The other weakness of the model is related to the extended use of approximated empirical state equations and kinetic laws.

At this point, it is important to point out that a large amount of research is needed to achieve thermodynamic optimisation of dynamic hydride systems. This work undoubtedly will have a positive counterpart in the future development of reactor design and the development and optimisation of adaptive hydride systems.

7.4 Applications

7.4.1 Hydrogen Storage

A common feature of all the various means of hydrogen production is the need for storage. This need can be satisfied by 'conventional' methods, such as pressurised gas and liquid storage, or by 'recent' technologies, such as metal hydrides systems and cryoadsorber storage, where hydrogen is stored at 70 K in specially manufactured activated carbon [7.153].

The notion of size has to be introduced when speaking of storage systems, because, depending upon what use is considered for the hydrogen, different constraints have to be obeyed. For example, large stationary storage units should, in principle, be available at the production site, and have a storage capability of the order of 2000–3000 m^3 S.T.P. Smaller stationary storage units will take over for the local distribution of hydrogen, while mobile storage systems will be reserved for transport and distribution to different sites, or for hydrogen fuelling vehicles.

Hydride storage capacity is evaluated in terms of the mass of hydrogen stored either per kilogram of hydride, or per liter. This is summarised in Table 7.1 for the most familiar IMHs, and is compared with other storage methods. Compared with 'conventional' storage, metal hydrides offer ob-

Table 7.1. Storage properties of various materials

Material	H_2 Weight density [wt%]	H_2 volume density [g/dm^3]	Energy [MJ/kg^{-1}]	Density [MJ/dm^3]
MgH_2	7	101	9.9	14
$FeTiH_{1.95}$	1.75	96	2.5	13.5
$LaNi_5H_{6.7}$	1.37	89	2	12.7
Liq. H_2 (20 K)	100	70	141	10
Gas H_2 (100 atm)	100	7	14	1
Cryoadsorber (77 K)	3.8 → 5.2	15 → 30		

vious advantages of compactness in reactor design, moderate pressure and safe uses, and energy savings relative to the energy costs of liquefaction. Other parameters also favour hydride storage, such as the range of storage unit sizes possible, and the quality of the gas.

Since the first studies on metal hydride storage around 1973, storage has been proven viable, but its economical feasibility has not been achieved. This is particularly true for large storage units, for which development has practically ceased. The high cost is blamed on the cost of the alloys, but also on the investments in heat exchangers to remove heat from or supply it to the reservoirs. Smaller storage units (2–50 m^3 S.T.P.) are now commercially available, and whether or not they will come into extensive use depends on whether or not applications can be identified where the advantages mentioned above will raise hydride storage over cryogenic and compressed gas methods of storage. After a short survey of earlier work on hydrogen storage, the commercial units will be discussed.

(A) Large and Medium Sized Stationary Storage

The first research into hydrogen storage began in 1973 at Brookhaven National Laboratory (BNL), where the possibility of storing electric energy produced in off-peak hours by producing hydrogen from an electrolyser followed by hydrogen storage was investigated [7.154]. A demonstration prototype was built by Public Service Electric and Gas Co. of New Jersey, with the cooperation of BNL. The system underwent 60 cycles without difficulty. The hydride storage received hydrogen from an electrolyser via a compressor. Then hydrogen was released from the reservoir to a fuel cell to produce electricity. The storage unit contained 400 kg of FeTi and an effective charge of 6.4 kg of hydrogen (72 m^3 S.T.P.). Heat was removed by water circulation through an internal heat exchanger operating between 15 and 45 °C. It is clear that this earlier prototype did not produce the optimum performance but it did have the merit of showing that when combined with other hydrogen related technologies, large scale hydride applications were possible.

This idea was soon taken up by *Suzuki* et al. [7.155], who loaded a storage reservoir with $MmNi_{4.5}Mn_{0.5}$, with a storage capacity of 16 m^3 S.T.P. In a demonstration plant at Kogakuin University, the hydride storage units insure purification and compression up to 7 MPa [7.6] with a capacity of 240 m^3 S.T.P. *Ono* et al. [7.156] reported a semi-commercial scale reservoir based on $LaMmNi_{4.8}Al_{0.2}$ and a hydrogen capacity of 175 m^3 S.T.P. Though other storage units have been developed since, they will be mentioned in other types of applications, such as commercial size hydride heat-pump reservoirs.

The largest storage system for which information is now available was developed with the support of the European Economic Community by HWT, a division of Daimler-Benz and Mannesmann [7.157]. It has a capacity of 2000 m^3 S.T.P stored in 10000 Kg of the AB_2 Laves phase compound, $Ti_{0.98}Zr_{0.02}V_{0.49}Fe_{0.09}Cr_{0.05}Mn_{1.5}H_3$. The hydrogen is loaded and unloaded within one hour, provided the appropriate energy is supplied for cooling and heating during the absorption and desorption processes. The plant has an assembly of 32 reservoirs containing the active material. Each storage unit is an expansion of smaller reservoirs, which are described below. This storage facility was constructed mainly to investigate the generation of ultra pure hydrogen on an industrial scale, but it also provides the opportunity to examine the full potential of large scale hydride storage, as well as other possibilities, such as compression, and recovering hydrogen from industrial processes.

(B) Small Stationary Storage

These units can be purchased from companies such as Ergenics Inc., Ringwood, N.J., USA, Hydrogen Consultants Inc., HCI, Littleton, Co., USA, Hydrid Wasserstoff Technich, HWT, Mühleim, Germany and Japan Metals and Chemicals Co., Ltd., Tokyo, Japan. Their main applications are in laboratories, where a strong argument in their favour is in their purifying effect. As shown in technical notes [7.158–159] a hydride storage reservoir fed with industrial gas will release hydrogen with a total contamination of the order of 1 ppm. The storage characteristics of some of the proposed reservoirs are listed in Table 7.2.

Of course, manufacture and characterisation of the IMCs is associated with hydride technology. For low temperature storage, < 100 °C, and moderate pressure, < MPa, Ergenics [7.160] produces a family of AB_5 and AB, HY-STOR alloys, and HWT has developed a series of AB_2 multicomponent compounds. The basic research on the synthesis, structural characterisation, and thermo-physico-chemical properties of the latter compounds are summarised by *Bernauer* et al. [7.111].

The heat transfer media has an influence on the hydrogenation rate, so the storage units are proposed with still-air, forced-air, or water regulation for operation. The heat exchanger is external to the reactor containing the alloys, and consists of an external jacket with circulating water in the Ergenic

Table 7.2. Commercial H_2 storage reservoirs

	Type	Storage capacity [m³]	Total weight [kg]	Working pressure [10⁶ Pa]	Flow rate [dm³/min]	Working temp. [°C]
Ergenics	ST-1	0.04	0.5	< 2.5		depends
	ST-45	1.275	24			on alloy
	ST-90	2.500	36		0–85	selected
HWT	KL 114-1	1	12	< 4	0–20	20–40
	KL 114-2	2	22	< 4		or
	KL 114-3	3	33	< 4		20–100
	KL 114-5	5	51	< 4		

models, and a longitudinally finned core for the HWT units [7.161]. The difficult problem of pulverisation and expansion during hydriding has been solved by including a specially designed micro filter and confining the alloy in Al capsules which, if well adjusted, act as fins that increase the heat flow from the hydride beds to the external walls and within the hydride (HWT). These hydride devices represent the state of the art in the technology of hydride storage. No restriction can be made on the quality of the hydrogen gas supplied and the variety of alloys makes it possible to adapt the storage units to each particular requirement, as proposed by Ergenics.

We will conclude this section with a question. What laboratory or company could afford to invest 3000 DM for 1 m^3 storage reservoir from HWT, where 20 m^3 of compressed gas costs roughly 300 DM?

(C) Mobile Transportation

Magnesium-based hydrides were proposed to be used for vehicular hydrogen fuel storage in 1969 by *Hofman* et al. [7.162]. The exhaust heat of the engine produced the energy required to decompose the hydrides. Unfortunately, the stored hydrogen could not be fully discharged at a high enough rate because the heat had to be supplied at a high temperature (300 °C) and the amount of energy needed was too high to be supplied by the rate of heat generated by the engine. As a consequence, the hydrogen rate was insufficient to follow the engine demand. This problem was partly solved in 1973 with the discovery of the less stable FeTiH$_x$ hydride. Although FeTiH$_x$ presented a lower specific mass and lower energy density than the Mg alloy, it does have the advantage of using low grade heat, low temperature, for discharging hydrogen, while high-grade heat, high temperature, could be restricted to the Mg alloys. A compromise had to be made when the dual bed configuration was employed. The perspective of using metal-hydride for fuel storage generated great interest up to the beginning of the 1980s. During that period, some gasoline vehicles were converted to hydrides and efforts in that field were largely sponsored by public funds. The leader and pioneers were Billings Corporation, now Hydrogen Energy Corporation of Kansas City, and Daimler-Benz. A list of different hydride vehicles is given in Table 7.3.

In spite of their weight penalty, hydride vehicles compare favourably both with the other alternative power sources mentioned at the beginning of the section, and with electric propulsion. The most serious competitor is liquid hydrogen storage, which has been the subject of much continuing research in Germany and Japan over the past ten years. Even though boil off losses and safety problems have not yet been solved, progress has already been made in engine operation [7.163].

During the same period the feasibility of hydrides for automobiles had to be proved. This included the development of new alloys, tests on long-term behaviour, stability with regards to gaseous contaminants, container development, optimisation and safety, and real road tests.

Table 7.3. Hydride vehicles

Type of vehicles converted to hydrides: Billings Corp. (BC), now Hydrogen Energy Corp. of Kansas City: Buick Century, Peugeot sedan, AMC Jeep, Winnebago buses, pick up truck, tractor. Hydrogen Consultants Inc. (DRI): Pick-up truck, caterpillar, forklift. Characteristics: see [7.10, 18, 110]

Mercedes–Benz: (test fleet)	MB 280 TE sedan H_2-gasoline	MB 310 van H_2
Engine	2.8 dm^3	2.3 dm^3
Total weight [kg]	2350	3500
Payload [kg]	400	700
Range [km]	150	120
H_2 storage	280 kg, Ti(VMn) 11 dm^3 gasoline	560 kg, Ti(VMn) 22 dm^3 gasoline
hydride containers	2	4
Toyota - Japan Steel	forklift (for sale since 1990)	

The problem of high mass flow rate of hydrogen was not fully solved with the FeTi alloy which also has the disadvantage of having two different plateau pressures, which increases the difficulties in controlling the hydrogen gas mass flow rate. Daimler-Benz, after trying this possibility, investigated the AB$_2$ compounds mentioned above, which were used on their second generation of vehicles. The tests were conducted using five hydrogen powered vans and five sedans with mixed gasoline-hydrogen reservoirs. The results of the test fleet were reported by *Töpler* and *Feucht* [7.164] after the vehicles had travelled a total of over 250 000 km. The report includes an investigation of various parameters of importance for technical purposes, such as the stability over extensive cycling, sensitivity to gaseous impurities, and their effects on the kinetics of hydrogen exchange. As for most IMC's, the most harmful contaminants are O$_2$, CO, CO$_2$, H$_2$O. They reduce the capacity and cycling stability, whereas N$_2$ and CH$_4$ affect only the kinetics of absorption. The other technical problems associated with refuelling the cars gave satisfactory results over a total of 3900 cycles. In general the tanks were charged to a level 80% in 10 minutes.

No data concerning the environmental impact of hydrogen have been furnished, although it would have been of importance to compare this with gasoline cars equipped with catalyzers. Hydride vehicles should contribute only to NO$_x$ pollutants, due to the combustion of hydrogen with the air mixture, while hydrocarbons would essentially be reduced to traces, owing to lubricant coatings, and that is assuming that hydrogen is not produced from coal.

In the present economic situation, experimental hydride vehicles remain at this stage of development. If there is a future for hydride vehicles, it will be for fleet vehicles, like bus transportation, taxis, and delivery vans. However, in some cases, where ballast is required to improve traction or stability, such as for farm tractors, forklift, caterpillar engines, the weight 'penalty' of the hydride reservoir transforms into a weight 'advantage' [7.165, 166]. For these systems the transition from prototype to commercial unit, only depends on the industrial choice. The improvement of hydride vehicle performances can be expected only by increasing the hydrogen storage capacity of the IMCs. *Orimo* et al. [7.167, 168], in collaboration with Mazda Motor Corporation seem to be engaged in that field. They have synthesised a new composite material $Ti_{0.8}Zr_{0.2}Mn_{0.8}CrCu_{0.2}/Mg$ which, after hydriding, presents the particularity of forming MgH_2 at temperatures below 373 K.

It is easily understandable that further development of hydride vehicles is still a long term affair that can be supported only by public funds, and weighted by environmental problems. Unfortunately, most of the industrial countries have launched new programs on batteries development. At present, the range limit covered by battery powered commercial vehicles does not exceed 90 km. The weight penalty exists also for batteries.

7.4.2 Chemical Selectivity

(A) Getters

Since gettering was first established by Langmuir, getter materials are widely used in the area of vacuum-sealed electronic devices, in vacuum pumping, as well as in other practical applications, such as keeping vacuum insulation in dewars or in thermostated bottles. Getters are classified in two groups, the first comprises evaporable getters such as Ti evaporated film and the second, nonevaporable or bulk getters, which have the advantages of compactness and of being manufacturable in almost any desired shape. Research and development on bulk getters has received greater attention in the last ten years because of the potential applications in thermonuclear fusion reactors where fuel quality and impurity controls are absolutely necessary. The development of these materials benefited directly from the possibilities offered by IMC reactivity with hydrogen and other chemical elements.

Different commercial materials are available from SAES [7.169], Ergenics, and HWT. Depending on temperature and pressure they are capable of sorbing irreversibly reactive species like O_2, N_2, CO, and CO_2. Other molecules, such as H_2O, NH_3, and CH_4 can be decomposed at given temperatures where the basic elements, O, N, and C are then absorbed. As H_2 and its isotopes, D_2 and T_2, are pumped reversibly at low temperature, getters act as storage materials from which the isotopes can be released at higher temperatures at a later time. The characteristics of getters, their selectivity and recommended temperatures, can all be found in technical notes provided by the companies mentioned above.

SAES markets multicomponent alloys based on Zr metals, under the trade names ST 101, Zr-Al alloys, STI 98, Zr-Fe alloys, and ST 707, Zr-Fe-V alloys. *Mendelsohn* and *Gruen* [7.170] investigated alloys of $Zr(V_{1-x}Fe_x)_2$ composition and demonstrated their usefulness as bulk getters. The products proposed by Energics are referenced under the trade names HY-Stor 140 and 105 for H_2 and H_2O gettering, while HY-Stor 402 covers a broad range of active species such as H_2, N_2, O_2, H_2O, CO, and CO_2. These series of alloys are developed with Zr as base material. Gettering with rare earth based alloys is proposed with HY-Stor 501.

The properties of the multicomponent materials developed so far for fusion applications are not fully satisfactory if we consider the overall process of storing, supplying and recovering tritium gas. The absorption pressure is too high at room temperature; the alloys should be able to satisfy an absorption pressure of 10^{-3} Pa at room temperature and one atmosphere at 500 °C [7.171]. According to *Devillers* et al. [7.132] ZrCo should be one of the most suitable IMC for tritium storage.

The problem of gettering hydrogen isotopes has been investigated by *Cann* et al. [7.172], for preventing hydride cracking in Zr-2.5% Nb pressure tubes in nuclear power reactors. During operation of the reactor, the tubes absorb deuterium from the heavy water coolant, through corrosion. Yttrium, selected as getter, was encapsulated in Zr alloy and fixed to the tubes. The effect of yttrium hydride formation on gettering rates and hydrogen isotope concentration in the Zr alloy near the Zr alloy/yttrium interface were determined at 586 K for different hydrogen fluxes. The conditions required to prevent delayed hydride cracking were thereby established.

(B) Purification

The boundary between purification and separation may be set by convention according to the levels of impurities to be removed. 'Purification' will be used if the impurities do not exceed 0.1%, otherwise 'separation' is a more appropriate term to define the process.

The commercial multicomponent alloys mentioned above for gettering processes are the materials used for gas purifiers. The commercial units proposed today offer wide possibilities in many areas ranging from laboratory to industrial uses. They are well suited to the manufacture of semiconductor components, optical fibers, sensors, etc. Several hydride purifiers are marketed by Ergenics [7.173] and HWT [7.174], from simple reactor/canisters to the fully controlled device. These systems are able to produce an output with a level of impurities as low as one part per billion. Such performance of course requires that the parts of the system attached to the output gas must be consistent with the handling of ultra high purity gases. Improvements in electropolished surfaces, high quality welding, high quality bellow valves, purifier elements assembled in dust-free rooms make ultra high purification systems very competitive with conventional purification procedures.

The models proposed by Ergenics remove H_2 and its isotopes at room temperature, while other reactive impurities are removed by applying the appropriate temperatures and gas flow rates to the purifier. The temperature ranges from 250 °C for O_2 to 450–500 °C for CO and CH_4. Hydrogen purification is achieved while the gettering alloys operate in a hydrided state. HWT has developed a flexible system which can be used with industrial gases, with a relatively high level of contamination. The principle of the units

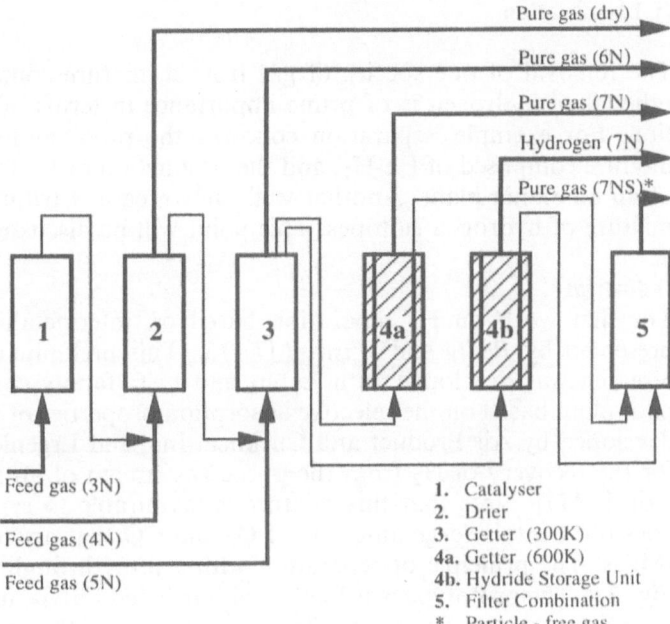

Fig. 7.7. HWT purification concept

is represented schematically in Fig. 7.7. Depending on the quality of the gas input (output), the gas is introduced (withdrawn) at different levels of the purification unit. Then the lowest quality input gas is pre-treated by conventional procedures, including catalysts (1) and dryers (2), before starting the purification with the getter alloy (3), which eliminates all reacting elements at room temperature. At that point, the residual of O_2, CO, and H_2O are already below 0.1 ppm, where a 6N quality is achieved. The high temperature treatment starts at steps 4a, 4b. So for ultra high purity hydrogen production, metal hydrides with high absorption pressures are needed that can trap all the other reactive constituents, while for inert gas production, a stable hydride with low absorption pressure is needed to trap H_2 and the other impurities. The systems operate on the trough flow principle. Finally, the dust-free gas is produced by passing it through specially designed filters. The unit can process gas continuously at flow rates of 15 m^3 S.T.P. per minute.

A small hydrogen purification process has been developed at the University of Unicamp, Brasil [7.175], which insures high quality hydrogen distribution over their laboratories.

(C) Separation

The removal of one species of gas from a mixture containing constituents other than hydrogen is of prime importance in terms of economic production. For example, separation concerns the production of helium from a mixture composed of $He-H_2$, and the production of H_2 from the bleed gases of an ammonia plant. Another well known case is tritium recovery from a mixture of hydrogen isotopes. This point will be discussed separately.

Industrial

The first work on H_2 separation based on intermetallic compounds was presented by *Reilly* and *Wiswall* [7.176]. This preliminary laboratory scale research was developed with LaNi$_5$ and FeTi family compounds. Later, a pilot plant based on the selective absorption properties of metal hydrides was developed by Air Product and Chemical Inc. and Ergenics. It was designed for the recovery of H_2 from the purge gas stream of an ammonia synthesis loop [7.177]. Note that this mixture is favourable to separation because it does not contain large amounts of O_2 and CO that are highly damaging to IMC's. The principle of separation with a flow through reactor is given in Fig. 7.8. The prototype was loaded with pelleted LaNi$_5$, using Ni as a binder to enhance the heat capacity. The composition of the composite pellets are adjusted in such a way that the process can operate almost adiabaticaly i.e. with no heat flow over macroscopic distances but only within the pellets. The energy stored in the material as a result of hydrogen absorption serves as the internal energy source for the desorption step [7.178]. The process was tested with a mixture of 60% H_2-x% NH_3-$(40-x)$% N_2 and pilot tests were run continuously for 6 months. During that time hydrogen recoveries range from 90 to 93%, with a purity of 99%. The advantage of such treatment is that it

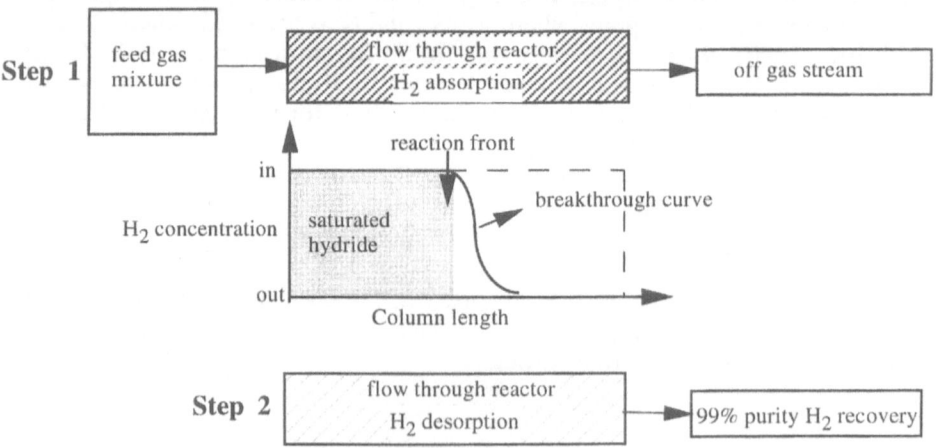

Fig. 7.8. Principle of a flow through reactor for separation

tolerates ammonia, whereas other processes, such as cryogenic or zeolite methods require that the feed gas be free of ammonia. The idea of hydrogen recovery from the bleed gases of an ammonia plant based on IMC was taken up by *Wang* et al. [7.179] who also proposed to recover helium. However, the authors reported experiments carried out with several IMCs of the AB_5 family for which the recovery of H_2 was tested only with Ar-H_2 mixtures of different compositions.

Japan Steel Works, LTD., Kansai Electric Power Co., INC., and Misubishi Electric Corporation have developed a device to improve the purity of hydrogen gas in generator containing as main impurities N_2 and O_2 [7.180]. Hydrogen contained in the generator insures a cooling function. Thus, the loss of hydrogen purity increases friction between the rotor and H_2 leading to a decrease in generator efficiency. The hydrogen separation device, 120 kg of CaNiMmAl alloy, was manufactured and installed on a 600MW generator at a commercial power station. A 96 to 98% purity H_2 gas, coming from the generator is fed to the separation unit providing a 99.9% purity H_2 gas which then is fed back to the generator. This system already operated continuously for more than 2000 cycles. The tests are being pursued.

An industrial separation approach with a metal hydride slurry was reported by *Swart* et al. [7.181], *Snijder* et al. [7.182]. The program resulted in a collaboration between the University of Twente and DSM. Research BV. of the Netherlands. The IMC selected was $MmNi_{4.5}Al_{0.5}$. The method was tested with different processes to insure continuous recovery of hydrogen. Evaluations were carried out for two gases in an ammonia plant, synthesis gas (75 vol % H_2) and gas from cryogenic recovery system (92 vol % H_2). When compared with other methods, such as pressure swing absorption and membrane separation, a metal hydride slurry offers the advantage that high purity levels of 99.99 vol % H_2 can be achieved with lower purge-stream losses resulting in higher chemical efficiencies.

Recently, *Albrecht* et al. [7.183] described an experimental program in the framework of tritium technology research at the Next European Torus (NET). The objective was to investigate the separation of gas impurities from a helium-hydrogen or helium-tritium gas mixture for possible applications of commercial gas purifiers. The gases treated had the composition 85–95% Q_2 (Q = H, D, T), 5% He and up to 10% impurities, such as CO, CO_2, N_2, NQ_3, and Q_2O. The ultimate aim was to recover chemically bound hydrogen isotopes from hydrocarbons, ammonia, and water. The separation/purification works as follows: the first step involves separating a high percentage of molecular hydrogen isotopes. This uses metal getters, including a Pd/Ag diffuser, where NQ_3 is decomposed catalytically at the surface of the membrane. The second step decomposes CQ_4 and Q_2O and removes the N, C, and O elements through the getter beds. In the last step the gases are passed through a Pd/Ag diffuser to separate out the remaining Q_2 from He. The facilities were operated to test the performance of several IMC hydride getters, and to obtain the major process parameters.

Isotopic

The first important application of metal-hydrogen systems was realised for the nuclear industry, due to the potential of hydrides for use as moderators and shielding in nuclear reactors. In 1968, *Huffine* [7.184] described the hydride manufacturing techniques that had been developed for these purposes. The discovery of IMH had opened up new opportunities for this industry where the future fusion reactors will require a careful and safe way of handling tritium. The quantities of tritium that will have to be handled in fusion reactors will be of the order of several kilograms.

All the advantages so far claimed for hydride applications were stressed even more for tritium confinement, where safety must be paramount. Thus purification and separation of isotope mixtures and storage of high purity tritium could all be achieved with a suitable metal or IMC, avoiding at the same time all the problems inherent in large pressure vessels, pumps, compressors, contamination with the walls of the vessel, and tritium decay which produces a mixture of $T_2 - {}^3He$. The latter problem can be solved by bulk storage with temperature scans of the tritide reservoir followed by bleeding-off of the residual ${}^3He$ gas. A review of storage material for tritium has recently been written by *Lässer* and *Schober* [7.185]. In the following, the recent progress in the separation of hydrogen isotopes through the use of bulk materials will be presented after a short survey of the isotopic properties. The notation protium (H) is used in the present section, while Q represents H, D or T.

The isotopic dependence of a metal or of an alloy is evidenced in various properties, such as its thermodynamic properties, diffusion coefficients, and kinetics. Thermodynamics shows that, for the same temperature and composition, the solubility limits and the pressure vary with the isotope and with the host material investigated. Note that the observed effects may be in opposite directions as found with the well-known f.c.c. Pd and b.c.c. Nb, V, and Ta metals. For example, in the Pd–Q_2 system, as the mass of the isotope increases, the solvus line between the α and $\alpha + \beta$ regions is translated to higher concentrations, while for the Nb–Q_2 system it is shifted to lower concentrations. The desorption pressure within the two phase region increases in the order H, D, T for Pd, whereas for Nb the heavier isotopes have lower equilibrium pressures. Concerning the diffusion coefficients of hydrogen isotopes, it is well established that tritium diffuses faster than deuterium and protium in Pd above 100 K, while tritium diffuses slower than protium in Nb. The kinetics will obviously be influenced by the bulk diffusion processes, but also by other parameters such as surface exchange rates of the different isotopes for which particle sizes, surface area and catalytic properties of the material are important factors governing the processes.

In view of the complex behaviour of hydrogen isotopes reported for the most extensively investigated metals, it is not surprising that isotopic dependence in IMCs does not make the determination of the effects any simpler. An example is the CaNi$_5$Q$_2$ (Q=H or D) system that, at room temperature, can form three distinct hydrides phases [7.186]. Deuteride phases of the same compositions are reported, and different isotopic pressure effects for each phase are observed, a desorption isotherm may show inverse $(P_{H_2} > P_{D_2})$, null or normal effects $(P_{H_2} > P_{D_2})$. Recently, *Wei* [7.187] reported opposite equilibrium isotope effects for the solid solution and the hydride phase of LaNi$_5$.

The difficulties in determining accurate thermodynamic data for IMHs apply as well to the isotopic effect determinations. If standard procedures are not established, it will be almost impossible to compare results obtained from different groups. At present the IMCs investigated do not give full satisfaction for tritium storage, separation, or long-term behaviour of the tritide. To satisfy this particular demand, new materials must still be discovered, and efforts are still necessary to contribute to a better understanding of the complex isotope surface exchange phenomena which result in mixtures of HD, HT, and DT [7.188–190]. Westinghouse has investigated new materials for separation. According to *Fischer* [7.191], NdCo$_3$ is a promising candidate for the separation of hydrogen isotopes.

Among the techniques proposed, chromatographic processes have been undergoing tests for hydride isotope separation for many years. *Gluekhauf* and *Kitti* [7.192], in 1957, studied the separation of a H$_2$–D$_2$ mixture through a column of Pd. In 1981, *Altridge* [7.193] obtained a patent for a chromatographic procedure based on an AB$_5$ compound. Several different chromatographic techniques have been brought to bear on the problem of isotopic separation, including conventional and displacement chromatography, pressure and temperature swing methods using multistage reservoirs, mainly for gas-solid separation, and parametric pumping for gas-liquid-solid separation.

Displacement chromatography is the most common technique and is widely used in the laboratory. The Joint European Torus (JET) has selected it for their separation requiring a maximum daily throughput of 5 moles of tritium and 15 moles of deuterium with a minor presence of protium. For large quantities of protium in the mixture, the separation is insured by a cryodistillation system. The column contains Pd on a porous alumina substrate operating at 20 °C and 1 atm. in the β phase of PdQ. The efficiency of the separation is evaluated by the separation factor,

$$\alpha_H^Q = \frac{(Q/H)^G}{(Q/H)^S} , \qquad (7.17)$$

where Q and H represent the concentration of the isotopes in the gas, G, and solid, S, phases, respectively. The column operates in four steps. The column is fed at room temperature first with He, then with the gas mixture, and finally with the protium eluant, and then is regenerated at high temperature. Details of the procedure are found in reference [7.194]. Recently, the flow of hydrogen isotopes through a chromatographic Pd column coupled to a reflux tank was computed by *Nichols* [7.195]. This process operates semicontinuously by thermal cycling while the desorbed gas is collected in the reflux tank. It is then recycled to the cold column for another purification. This problem is analogous to a reflux stream in a distillation column.

The pressure and temperature swing absorption techniques were extensively studied by *Hill* and *Grzetic* [7.196, 197], with a vanadium metal separator. They also contributed greatly to modelling the processes. At the operating temperatures selected, an equilibrium isotope effect exists such that tritium is absorbed by the solid from the H–T mixture, but a kinetic isotope effect is also present. The latter effect shows that T is absorbed and released more slowly than protium during an absorption/desorption cycle. The two effects were then opposed and were used to define different operating conditions. For example, for short two minutes cycles, the kinetic isotope effect controlled the separation and the process produced a high-pressure product enriched in T_2 and a low-pressure product depleted in T_2, with the system operating at 373 K. For long four hour cycles, the system was operated at lower temperature and the process was controlled by the equilibrium isotope effect. The systems investigated required multistage separation, where the stages are assembled in series to form a continuous countercurrent cascade. The enriched stream from each stage flows in one direction and the depleted stream flows in the opposite direction. This was modelled to determine the number of stages required to accomplish a specified separation.

The possibility of using a ZrV_2 bed as a means of recovering hydrogen isotopes from an inert gas has been investigated by *Mitsuishi* et al. [7.198]. The column was studied experimentally from room temperature up to 300 °C for absorption, and regeneration at temperatures between 400–800 °C. Parametric analysis of the concentration of the species during the flow, the so-called 'breakthrough curves', showed that they were greatly affected by heat generation during absorption, by the mass transfer coefficient of the fluid (solid) phase and by the interface area per unit volume of the bed. Hydrogen isotope separation was performed using $LaNi_5$ film deposited on polymer membrane by sputtering technique [7.199]. The separation factor amounts to 1.9; it is of comparable magnitude to the values expected for Pd alloys films.

Isotope separation using metal-hydrides or IMH is certainly feasible on an industrial scale. As far as safety is concerned for the handling of tritium, the search for new materials is still an open field. More investigations are also needed to understand the dynamics of the exchange reactions taking place in order to maintain efficient separation as proposed by *Fukada* et al. [7.200].

7.4.3 Hydrides and Thermodynamic Devices

The potential of hydrides now extends to those applications where a secondary effect, such as compression, heat generation, or heat pumping is important. The hydride reservoirs are the basic tools, and the hydrogen gas acts as the working medium. System performance is evaluated in terms of thermodynamic efficiency which has to take into account the many parameters influencing the process.

The technology described for hydrogen storage and purification is still appropriate for validating new concepts. But higher performance reactors will be needed if hydride engines are to be developed on a larger scale, because here again, they are at a heavy disadvantage in the face of widespread and well established devices. Thus, while improvements are still being made in compressor and hydride heat management devices, efforts should also be brought to focus on hydride systems for specific domains where only hydrides have the required performances, as this will bolster interest in this technology. A good example is the production of water vapour from low grade energy, as well as the refrigeration of frozen food. Advantages, such as compactness and high energy density, should be kept in mind when hydride heat pumps are compared with other heat pump devices. For compression purposes, hydride compressors are noiseless, clean, and vibration free systems.

(A) Thermochemical Hydride Compressors

Principle

The principle of the thermochemical hydride compressor is shown in Fig. 7.9. The temperature dependence of the plateau pressure, according to (7.4), is used to compress the hydrogen between the low temperature, T_1, low pressure plateau, P_1, and the high temperature, T_h, high pressure plateau, P_h. In operation, the reactor is repeatedly heated and cooled between T_1 and T_h,

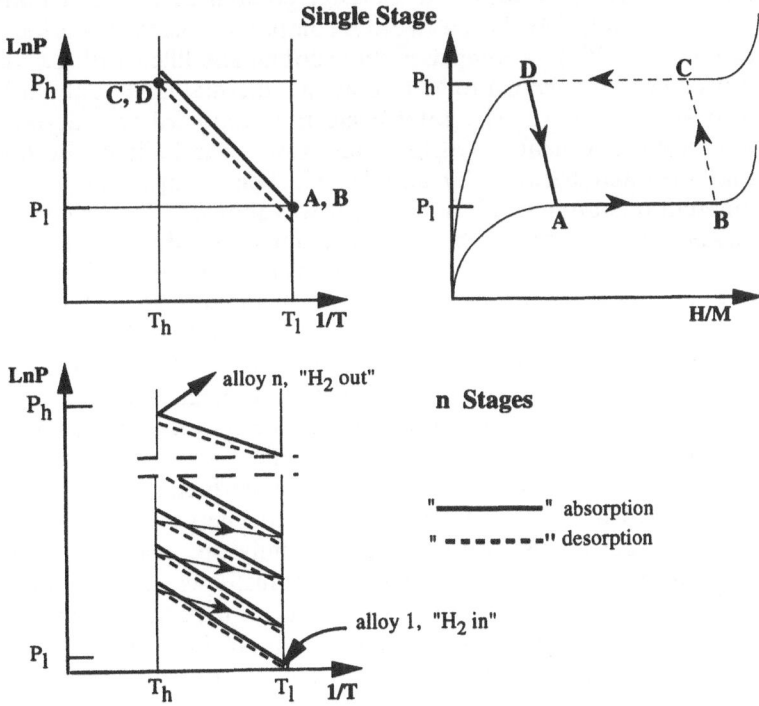

Fig. 7.9. Principle of the hydride compressor (see text for explanations)

resulting in a cycle that follows the path ABCD. Except for the control valves, such a compressor has no moving parts and is free of vibrations. The compression ratio depends on the alloy selected, on the temperature range, and on the overall system dynamics. A 10 to 1 compression ratio is of the order of what can be expected from present hydride compressors. The thermal energy needed to do the work of compression can be produced from waste heat or solar energy.

The maximum efficiency, the Carnot efficiency, of the thermal compressor is given by the relation,

$$\eta_c = \frac{W}{Q_h} = \frac{T_h - T_l}{T_h} \ . \tag{7.18}$$

Of course this theoretical limit will not be reached and the efficiency of the compressor is closely dependent on the design of the heat exchanger. *Solovey* [7.201] published an inventory of the limiting factors in thermal hydride compressors, and quantified these factors to arrive at an evaluation of the true efficiencies that can be expected from such systems.

Development

Around 1971, *Reilly* et al. [7.202] described one of the first applications of hydrides to compressors. The compressor used vanadium hydride and was built to drive periodically a mercury column acting as a pump for other gases. The first IMC based hydride compressor using this principle was built at Philips [7.203]. It comprises three containers filled with LaNi$_5$. Hydrogen was compressed from 4 to 45 atm. with a thermal load at about 160 °C and a heat sink at 15 °C. The total cycle time reported was about 10 minutes. Another experimental hydride compressor was built by Bertin Co. [7.204] that operated between 20 and 80 °C. This compressor was tested under different operating conditions, and it was shown, as expected, that the useful power did not increase when the cycle time was reduced. The mass of hydride required for one mechanical Watt was of the order of 40g, ten times more than expected. As for early prototypes at Philips, this one demonstrated the feasibility of hydride compressors and proved that the efficiencies could be greatly increased if thermal inertia is minimised. One possible solution to this problem remains in the design of new heat exchangers.

A great improvement in the overall design of hydride compressors was produced by *Golben* [7.205] from Ergenics, who developed the first commercial systems. The device is a four stage metal hydride/hydrogen compressor that uses hot water at 75 °C as its energy source. Four stages filled with different alloys from the LaNi$_5$ family are assembled in series in two different beds [7.206]. The hydrogen transfer is insured automatically by unidirectional valves. The author claims an expected compressor life of 20 years, assuming near continuous operation, if the device is set at a 2 minutes cycle time. Based on this work, two models are commercially proposed by Ergenics, a single-stage electrically powered unit that is able to compress H$_2$ between 0.5 and 3.5 MPa at an average rate of 3 STL/min, and a four-stage hot water (75 °C) heated unit that compresses H_2 from 0.4 to 4.1 MPa at a rate of 40 STL/min for a weight of the system of 64 kg [7.140].

Uses of Hydride Compressors

Hydride compressors can be coupled to low-pressure hydrogen sources, such as hydrogen electrolysers. They can be used both to recover the boil off losses from liquid hydrogen and to repressurize a hydrogen tank if necessary. One of the promising possibilities that was first demonstrated by *van Mal* [7.203], concerned the coupling of their prototype compressor to a Joule-Thomson expansion valve with the high pressure hydrogen gas precooled to 78 K with liquid nitrogen. Precooling with liquid nitrogen insures that the high pressure hydrogen gas is below its inversion temperature. Then high pressure hydrogen is expanded through the Joule Thomson valve to produce a gas liquid mixture of hydrogen around 20 K. The cold production is supplied by evaporating the liquid H_2. The principle is shown in Fig. 7.10. Though the refrigeration capacity was reduced by heat exchanger losses, the prototype demonstrated a cold production of about 0.6–0.8 W at 26 K. The low reported efficiency of 9.5% resulted from different parameters that were not optimised. According to the authors, 20% of Carnot efficiency may be achieved by more careful design. Following this idea, *Kumano* et al. [7.207] worked on the hydrogen liquification problem, developing a new closed hydrogen liquifying system, where the hydride compressor is a three stage

Fig. 7.10. Principle of a liquid hydrogen sorption cryocooler

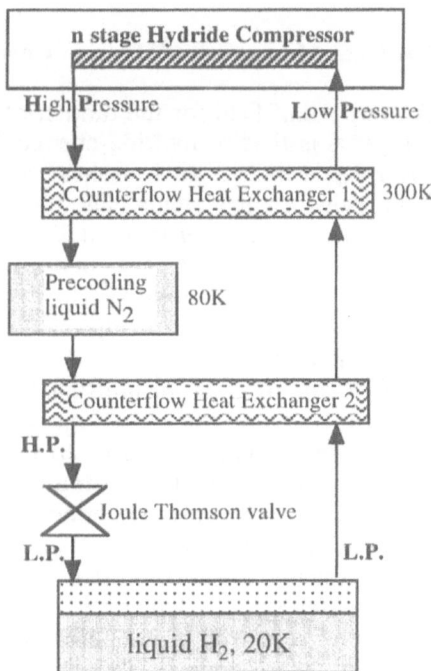

compressor, based on the same principle as the one developed by *Golben* [7.205]. The system was able to reach a minimum temperature of 16.5 K, while tests show that a cold production of 1.2 W at 20 K is possible in continuous operation. Further development of the unit has been announced by the authors.

Sorption refrigerators based on this concept have also aroused interest at Aerojet Electrosystems Co., Azusa, CA, USA for aerospace applications [7.208]. Thus, *Wade* et al. [7.09], *Bowman* et al. [7.210] have designed a new compressor prototype for 25 K liquid and 10 K solid hydrogen. VHx and LaNi$_{0.8}$Sn$_{0.2}$ were tested as the sorbent for the compressor for producing liquid H$_2$. The combination of a low pressure ZrNiH$_x$ compressor in series with a high pressure compressor was investigated to produce refrigeration at 10 K, through the sublimation of solid H$_2$. It was demonstrated that the hydrogen absorption was fast enough to meet the cool down goal of 10 K within 2 minutes [7.211].

One very useful application of hydride compressors was proposed by *Golben* [7.212] to solve the problem of water pumping to meet the requirements, defined by the World Bank, for small-scale water pumps. Thermal energy is supplied from flat-plate solar collectors, while the water being pumped is used as a heat sink. The compressed hydrogen applies pressure to a pair of piston cylinders that operate a reciprocating water pump cylinder. The hydride compressor-water pump was sized to meet the 150 W required, but in addition gives an extra 70 W of electrical power to run the operating valves. This system is now commercially available.

(B) Heat Management: Hydride Chemical Engines

An important field for the industrial use of hydrides has arisen over the last ten years is that of hydride chemical engines, HCE, which can be used for several possibilities such as heat pumps for cooling, heating, or both, and heat transformer systems. The pioneers in this particular area are *van Mal* [7.203] in Delft and *Gruen* et al. [7.213] at Argonne, who developed a machine for cooling purposes, the famous HYCSOS program [7.214]. This exploratory work was followed by extensive investigations in Japan and to a lesser extent by other groups. Efforts have continued in this direction in Germany and Israel.

Overall, the factors limiting the development of chemical engines are both the cost, and the limited number of materials proposed. The last decade has witnessed many attempts to find a new chemical system, new thermodynamic cycles, and also optimum conditions for a specific use of an engine. However, the absorption-chemical heat pumps that are commercially available, such as lithium bromide/water, activated carbon/methanol, and zeolite/water systems, all operate at restricted temperatures, which range from 0–5 °C to 100–120 °C. This range could theoretically be greatly extended with the use of metal hydrides. Other major parameters were also advanced in favour of these engines, the excellent energy density and the concomitant compactness, the high chemical reaction rate and the resulting low cycle time and high power production, and finally the lack of moving parts in a closed system which is only driven by heat.

These HCE are being developed for home, heating and cooling purposes, but they could also satisfy a wide range of applications in industry, where they could make profitable use of thermal effluents. Chemical industries require heat at 150–400 °C, but agrofood industries require heat of no more than 150 °C. Refrigeration for frozen foods is still a growing market, with the important problem raised by the imposed decrease in chlorofluorocarbon production. For the particular case of food transportation by truck or rail, the machines should be able to produce 'deep cold' during all phases of distribution and storage, which means compact refrigeration systems must be found.

Thermodynamic Principles
The principle of an HCE in its most general sense is to transfer energy from one temperature level to another. Hydride machines work cyclically with energy being transferred via the hydrogen gas from the high pressure level, P_h, to the low pressure level, P_l. Depending upon the direction of the hydrogen gas transfer and the temperature levels at which the transfers occur, the type of engine is different and will satisfy different demands, such as cooling, heating, or both, or heat upgrading. The different possibilities are illustrated schematically in Fig. 7.11 with the help of van't Hoff plots.

Any hydride cycle then requires four steps, which are the following for the heat pump mode: hydrogen at high pressure and high temperature is first transferred from the saturated hydride, AH_x, at T_h, to the hydrogen-free alloy, B, at medium temperature, T_m. Next, alloy A is cooled from T_h to T_m, hydride BH_x from T_m to T_l, and then hydrogen at low pressure and low temperature is transferred from hydride BH_x, at T_l, to alloy A, at T_m. Finally, the total system is heated to return to the initial state.

Disregarding the dynamic aspect of hydride engines for the moment and following *Dantzer* and *Orgaz* [7.215, 216], *Orgaz* and *Dantzer* [7.217], if one assumes that the mass of hydrogen transferred at the pressures P_l and P_h occurs with no driving force, $\Delta P = 0$ between the reactors, and that the hydriding/dehydriding reactions are ideal then we obtain relations which will allow us to develop simple thermodynamic models that will be of major help to classify the materials.

Figure 7.11 shows that for a fixed temperature T_m the hydride cycle is ideally defined. The T_h, T_m, T_l with the corresponding pressures P_h, P_l provide the optimum alloy pair for the optimum operating conditions. In such a case there is no other hydride pair offering higher efficiency than this single solution. Thus, in accordance with our description, the ideal efficiencies are,

$$\text{COP}_C = \frac{\Delta H_B}{\Delta H_A}, \quad \text{COA}_C = \frac{\Delta H_A + \Delta H_B}{\Delta H_A}, \quad \text{and} \quad \text{CHT}_C = \frac{\Delta H_A}{\Delta H_A + \Delta H_B},$$

$$(7.14)$$

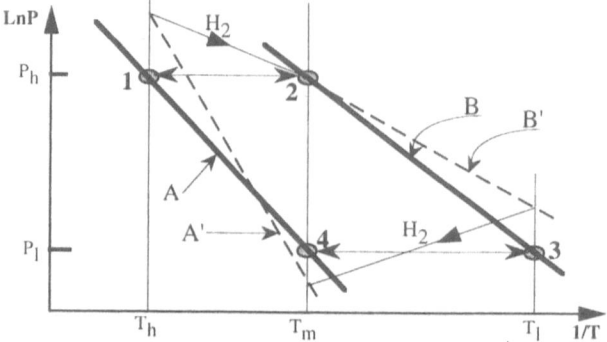

1, 2, 3, and 4 define the ideal cycle for pair A/B, pressure drop =0 between reactors, COP (A/B pair) > COP (A'/B' pair)

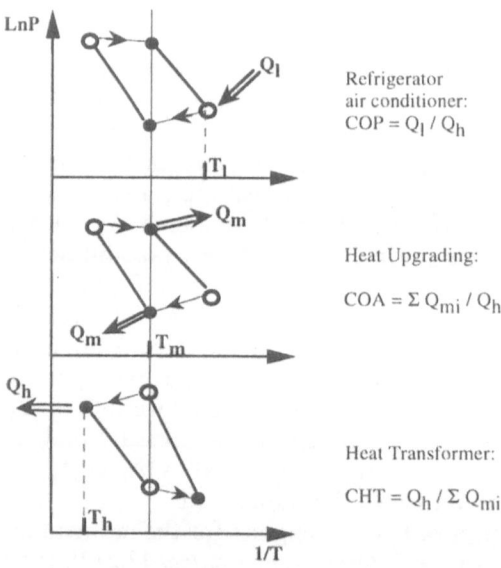

Refrigerator air conditioner:
$COP = Q_l / Q_h$

Heat Upgrading:
$COA = \Sigma\, Q_{mi} / Q_h$

Heat Transformer:
$CHT = Q_h / \Sigma\, Q_{mi}$

Fig. 7.11. Principle of the hydride chemical heat pump (top). Different modes of functioning (bottom), the Q_i indicate the thermal energy of interest (pumped or produced) for the selected mode

where the subscript c is for Carnot, and ΔH_i refers to the heat of formation of hydride i.

It can easily be shown that these ideal efficiencies correspond to the Carnot efficiencies of the engines represented in Fig. 7.11, for cooling, the coefficient of performance is COP_c, for heating, the coefficient of heat amplification is COA_c, and for heat upgrading, the coefficient of heat transformation is CHT_c.

Alloy Selection and Expected Temperature Ranges
The future of HCEs depends very much on what targets are selected. It is hopeless to try to compete with a zeolite-water heat pump that operates in

the temperature range of 5–80 °C, for which the cost of zeolite, as well as of the technology, are already much lower [7.218]. Of course the earlier hydride prototypes had to prove the viability of hydride systems, but this is now well established for single stage systems.

The problem of matching optimum hydride properties with a use in a specific engine is important, because there is no 'universal' pair of hydrides which satisfy all the many applications. As a consequence, in view of the large number of compounds being investigated, modelling was needed to tentatively define the possibilities offered by the hydride heat pumps and heat transformers in order to determine their best mode of operation, i.e., single-stage or multi-stages, and finally to provide the optimised hydrides producing the maximum efficiency for the required temperatures. These problems were analysed by *Dantzer* and *Meunier* [7.219] who succeeded in classifying the performances of pairs of hydrides according to the temperature ranges required to satisfy European Union guidelines.

The method proposed does not suffer from any restriction concerning the number of alloys submitted to the selection, but is limited only by the lack of accurate data concerning the physical and chemical properties of certain alloys. Details of the method, as well as the proposed targets for selected pairs of alloys, are given in reference [7.219]. There are two levels of simulation. First, a coarse screening is needed because of the larger number of combinations possible. Here an oversimplified description of the thermodynamic cycle is sufficient. This elementary approach may be used to estimate roughly the performance that can be expected from the HCE. This was the approach also adopted by *Balakunar* et al. [7.220], *Murthy* et al. [7.21], who reported a comparative study of nine preselected pairs of alloys for heat pump and heat transformer applications. Then a second screening is made in which the chemical constraints are now introduced into the model. These parameters are deduced from a precise and thorough thermochemical analysis of the $LaNi_5/LaNi_{4.77}Al_{0.3}$ pair [7.215]. The final selection insures that the pairs produce machines that will operate within an assigned temperature range δT_m, which, according to the uniqueness of the solution for optimum operation determines the boundaries T_h and T_l. The alloy selection is then reduced to a problem of comparing set temperature thresholds and ordering the proposed pairs according to the computed efficiencies. This is easily solved by microcomputers. This is at present the most appropriate simulation/selection based on the results of quasi-static experiments. It has been applied for five requests, to the selection of hydride pairs out of 67 IMCs representative of AB_5, AB_2 and AB families. The five requests and results are presented in form of diagram in Fig. 7.12. It is clear that the next step will have to be an experimental test of the proposed solutions.

Results show that cogeneration of hot water (70 °C) and chilled water (2 °C), case 1, cogeneration of steam (120 °C) and chilled water (2 °C), case 2, production of hot water (70 °C) and refrigeration (–10 °C), case 3 are improbable with single stage cycles. A temperature increase of 70 °C for the cold side of the heat pump is the upper bound that can be hoped for, with acceptable efficiencies. Two targets are possible with hydride heat pumps.

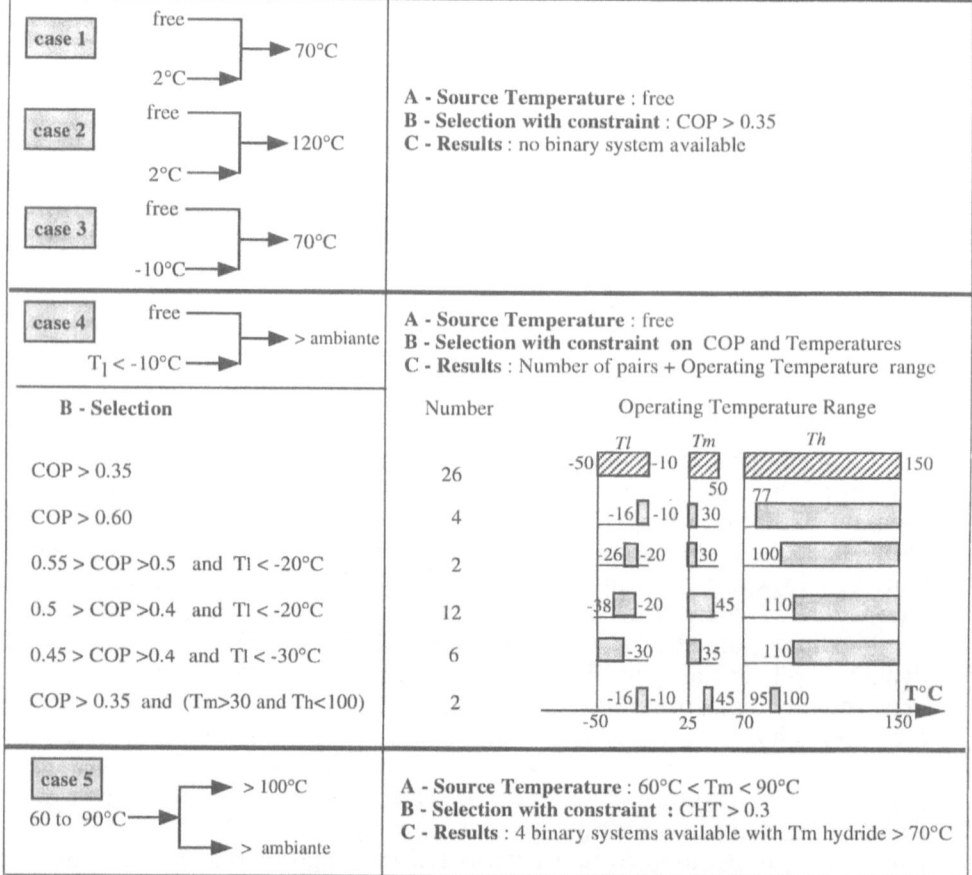

Fig. 7.12. Overall presentation of the results of a screening concerning IMHs (from [7. 219])

They are found with case 4 and case 5. One is cooling in a range running from air conditioning to refrigeration, down to –30 °C. The advantage of HCEs compared with other systems is that they operate at very low eva-porating temperatures which is not the case for solid adsorption heat pumps. But kinetics and heat transfer may then become limiting effects. If these problems are overcome, refrigeration at –30 °C may be a promising field of application. Another application is as a heat transformer for the water va-pour production. However, the results suggest that the materials for use do not allow reaching the 60 °C required at the T_m level, input of the low grade heat source. The high temperature of 300–400 °C may be reached with magnesium alloy hydrides. But we will disregard these possibilities for the present, knowing that work is going on in several directions concerning the new synthesis of the alloys. Their potential for high temperature use is pre-sented in Sect. 7.4.4.

Prototype Development

The simplicity of the hydride chemical heat pump is only in the concept. Unfortunately, the real processes that go on between two coupled hydride reactors involve a relatively large number of parameters which, up to now, have delayed the production of high performance systems. In spite of the lingering gap between predicted and real performances, there is constant progress in the prototypes, which have now reached the size of large, potentially commercial units. Recent investigations and improvements can be seen in patent applications submitted in the last few years [7.222–226].

Suda [7.6] summarises the activities and achievements in Japan and describes the major contribution made by his group. Examples of prototypes operating in Japan are given in Table 7.4, based on information provided by the Japan External Trade Organisation, JETRO. The prototypes studied include all the possibilities of hydride heat pumps from the laboratory scale up to commercial size units. The operating temperatures of the prototypes listed in Table 7.4 call for few remarks. Thus, systems 8, 10, 13 have been developed for a cold production around 5 °C. Such systems have not been proposed in the above selection, because it is the typical domain covered by cheaper devices, such as zeolite/water vapour heat pump. Related to systems 5 and 12, we can see that heat injected at 70 and 80 °C is upgraded at 90 and 100 °C, respectively. In these cases vapour production does not seem to be a priority. Moreover, the interest of the system is weakened by the rejection temperature levels at 15 and 20 °C, meaning that energy withdrawn at this level is costly, unless the system is designed to work during a cold winter season.

Ron and *Josephy* [7.227] have quantified the parameters for hydride chemical heat pumps feasibility, and described the main parameters affecting the specific thermal power output. The net power output at the useful load per unit weight of hydride is the relevant parameter that has to be determined if we want to compare the performance between prototypes. Though predictions are currently feasible, real evaluation is not a trivial problem and requires a properly instrumented system.

Reports on the operational characteristics of hydride chemical heat pumps are found in a limited number of publications. *Nagel* et al. [7.228] investigated the performances of a hydride chemical heat pump in a refrigeration cycle. They obtained a cooling output of 1.28 kW, or 120 W/kg of desorbing alloy, for an optimum cycle time of 13 minutes. Elsewhere, *Argabright* [7.229] reported on a joint project involving the South California. Gas Co., Solar Turbine Inc., and BNL, for the design of high performance heat exchangers for hydride heat pumps. The relatively low cooling power was attributed to the unoptimized designs of certain parts of the system, which increased the heat capacity and affected the cooling performance of the heat pump. *Ron* et al. [7.230–232] designed an hydride chemical heat pump for air conditioner use in buses. The high temperature source was the waste heat of the exhaust gases of the motor. Laboratory tests of this system showed that the lowest possible temperature was in the range $-2 \ °C < T <$

Table 7.4. Examples of metal hydride heat pumps installed in Japan

Output (kcal/h)	Cycle, T_h, T_m, T_1 [°C]	Alloy [kg]	COP
1. 9600 cooling + hot water supply / 15000 heating + hot water supply	Heat Pump/Refrigeration: 92–110, 30, 15	R.E. base: 90	0.9 cool. 1.4 heat.
2. 2000 chilled water generation	Refrigeration: 164, 20, 13	R.E. base: 40	
3. 1500 cooling	Refrigeration: 140, 40, 10	R.E. base: 38	0.4
4. 4000 chilled water generation	Double effect/Refrigeration: 150, 45, 10	R.E. base: 48	1.7
5. 3000 heat source	Heat transformer: 90, 70, 15	R.E. base: 46	0.38
6. 1000 refrigeration	Refrigeration:	R.E. base: 18 / Ti base: 17	
7. 6000 cooling + heating	Heat Pump/Refrigeration: 280, 144, 5	Ti base: 180	0.8 cool. 1.7 heat.
8. 13000 cooling + heating	Heat Pump/Refrigeration: 72, 50, 2–5	Ca base: 800	
9. 150000 cooling + heating	Compression type: heat source Temperature during heating : 60 during cooling : 30	Ca base: 4000	6–7
10. 350000 cooling + heating*	Heat Pump: 80–85, 65, city water Cooling : 80–85, 25, 5	Ca base: 3600	
11. 66000 steam generation	Double effect, Temperature raise: 150, 80, city water	R.E. base: 2600	
12. 1000 heat source	Heat transformer: 100, 80, 20	R.E. base: 12**	
13. ***	Refrigeration: 140–170, 45, 5	R.E. base: 10000	
14. 18000 cooling / 42000 heating	Heat Pump/Refrigeration:	Ti base/Zr base: 600	0.45 cool. 1.1 heat.

*Single cylinder heat exchanger.
**Copper clad hydrogen absorbing alloy utilized.
***Mainly intended to store heat (35000 cal) from nighttime power.

7 °C, with a temperature increase of 27 °C. The cooling power is the highest reported up to date at 250 W/kg.

   Yanoma et al. [7.233] described the design and the operation of a commercial size heat transformer with heat upgrading from 70 to 90 °C. Although, as noted above, the advantage of producing hot water at 90 °C from a heat source at 70 °C does not seem to open a wide field of application, the prototype is of technical interest for the future. Note that, once the system has been designed and assembled, the only experimental factors that can be varied are the temperature levels of the heat sources, the mass flow rate of the fluid through the heat exchangers, the fluid itself, and the cycling time. Given these parameters, the experimental studies will determine the best operating conditions for the machine, conditions which do not necessarily correspond to the optimum performance of the alloys. In this particular case the engine produced 174 kW output, or 49 W/kg of hydride for a cycle time of 10 minutes.

   The last reported heat pump prototype is due to *Nasako* et al. [7.234]. For the first time a device was manufactured to reach the temperature below –20 °C, using a 130 to 150 °C heat source. The heat transfer rate has been greatly improved by careful investigations of the total thermal resistance of a fin type heat exchanger.

   A two-stage laboratory scale hydride heat transformer lab model was developed by *Werner* and *Groll* [7.235, 236] to study the dynamic behaviour of the reaction beds. *Suda* and *Komazaki*, [7.237] published informations related to multistage prototypes.

Dynamics of Hydride Beds, Experiments, Models and Problems

From the literature, the total cycle time that has been determined to be best for system operation is of the order of 10 to 15 min. The estimation of useful power, will depend on the output load temperature, on the alloy selected, and on the cycle time. The results of the simulation of *Dantzer* and *Meunier* [7.219] show that 50 to 150 W/kg could be reached for refrigeration from –30 to 0 °C on a 12 minutes cycle time. This can surely be done with today's technology. Because hydride heat pumps will certainly have to operate with shorter cycle times to increase power and competitiveness with respect to other proposed systems, this will require accurate control of the reaction dynamics and of the operating parameters. The present stage of development of hydride reactors, and the associated problems, were analysed in Sect. 7.3.2C, so the following discussion is limited to the presentation of the work already produced in this field.

   Tusher et al. [7.238] reported the dynamic characteristics of single and dual hydride beds for a cylindrical configuration of the reactors filled with a compact porous metal hydride. The experiments conducted in 1983 demonstrated how difficult it was to achieve isothermal conditions. The authors report that the dynamic isotherms are deeply influenced by the ability to transfer heat between the hydride bed and the heat exchanger. This type of observation has been corroborated by others. Thus *Nagel* et al. [7.239], by changing the operational conditions for several operational modes, showed that the amount of hydrogen transferred is correlated with the real dynamic behaviour of each hydride. A detailed analysis of the data is found in [7.240]. The problem of dynamic pressure was also investigated by *Josephy* and *Ron* [7.53] and *Groll* et al. [7.55].

In 1986, *Bjurstrom* et al. [7.242–244] investigated in detail the dynamic behaviour of coupled hydride reactors with the support of experimental and numerical studies. The experimental work was done in Suda's group, and led to the conclusion that the hydriding kinetics, and the corresponding dehydriding kinetics in the second reactor are both controlled by heat transfer. A mathematical model was constructed using averaged parameters to describe the heat transfer and reaction processes. The model is based on five equations, two per reactor, including the kinetic and energy balance equations and a mass balance equation. Satisfactory agreement was reported between experiment and the simulated process in spite of the oversimplified assumptions. This is the most advanced model to date, and it may be of very great use in predicting the dynamic behaviour of coupled hydride beds if the heat transfer and kinetic parameters are known. Although it is oversimplified for purposes of optimisation, the model could be used in the first step of the design of new reactors. This type of modelling lead *Gambini* [7.244] to make a parametric analysis of a hydride chemical heat pump, an analysis which confirmed the previous results, whereas *Sun* and *Deng* [7.245, 246] tried a purely numerical analysis of the dynamics of hydride beds.

7.4.4 High-Temperature Applications

Magnesium hydride is the prototype material for high-temperature applications, but to date its low dehydriding kinetics has restricted its use to purely academic applications. However, since *Bogdanovic* and *Spliethoff* [7.247] developed new methods of preparation, the magnesium hydride properties have been greatly improved. They succeeded in developing a nickel-doped, catalytically activated, nonpyrophiric magnesium hydride with negligible hysteresis, in which the reaction kinetics are much faster. Following the Bogdanovic method, *Raisi* et al. [7.248] also obtained doped magnesium hydride with comparable behaviour. Heat storage applications have been considered for magnesium hydride prepared in this new way where it is integrated into a small-scale solar thermal power station [7.249]. The development of this prototype is being pursued in a joint project by the Max Planck Institute, Mühleim, Bomin Solar Gmbh Co, Lörrach and IKE, Stuttgart. Detailed information on the solar power station has been given by *Wierse* et al. [7.250]. A specially designed solar collector is connected to Stirling engines to generate electricity. The use of magnesium hydride for heat storage allows the engine to maintain constant power output during cloudy weather and after sunset. The heat is transferred between the different parts of the system, the solar absorber, Stirling engine, and thermal energy storage reactors, by heat pipes. Solar energy produced during daylight hours is transferred to the Stirling engine and to the high temperature source, 200–

300 °C, of magesium hydride, which decomposes to a secondary, low temperature, hydride. During this step, useful heat is produced at 40 to 80 °C. The reverse operation produces refrigeration for the low temperature alloy and heat for the Stirling engine via absorption of hydrogen on magnesium, if necessary. The system has been designed to satisfy a maximum electricity power production of about 1 kW. The amount of nickel-doped magnesium will be of the order of 20 kg. Successfull tests of the prototype led to a second generation engine with a maximum electrical power output of 4 kW.

7.4.5 Sensors and Detectors

Welter [7.251] proposed the use of metal hydrides as active materials for sensors and regulators. A first example is given by a thermostatic expansion valve, in which the temperature sensing element is filled with a metal hydride. The hydrogen pressure variation in response to temperature changes is used to drive a membrane or bellows. When compared with conventional sensors filled with a volatile organic liquid, the metal hydride shows an advantage in avoiding recondensation phenomena in the cooler section. The second case is an electrothermal resistor that is used as a level and flow monitor for liquids.

It is appropriate to mention in this section the most successful application of a metal-H_2 system, which is as a fire detector. It is currently used in most airplanes throughout the world [7.252]. The sensor is based on the Ti–H_2 system enclosed in a long stainless steel tube which controls a switch. Any heat generation along the tube will cause a pressure increase that will activate the switch.

7.4.6 Batteries

Research in favour of the practical uses of hydride storage materials for electrochemical applications began about 30 years ago. Around 1967, *Lewis* [7.253] proposed a Pd–H electrode, and the Battelle Institute in Geneva began studying the use of TiNi electrodes for hydrogen storage purposes [7.254].

In 1970, the need for power sources for space applications led to the development of metal hydrogen batteries to supply energy to satellites. These batteries, which are a combination of a regular storage battery and a fuel cell battery, offered the advantages of high reliability and stability. The disadvantage, due to the presence of relatively large hydrogen gas pressures of 30 bars could be avoided by storing the hydrogen in an intermetallic compound, as proposed and tested by *Earl* and *Dunlop* [7.255]. This was demonstrated by *Justi* et al. [7.256,], *Ewe* et al. [7.257], by using TiNi electrodes, and later was tested with LaNi$_5$. In 1975, a US patent was filed by *Will* [7.258] for a battery in which LaNi$_5$ was the negative electrode and nickel oxide the positive electrode. A French patent was obtained by *Percheron* et al. [7.259] for electrode materials based on LaNi$_5$ substituted compounds, and electrochemical use of such materials. This patent was extended in 1977 to new compounds and in 1978 led to a US patent [7.260, 261]. The search for new compounds up to 1978 is discussed in [7.262–264]. In 1984, *Willems* [7.265, 266] reported on the work performed at Philips on metal-hydride electrodes developed with LaNi₅ related compounds.

For technical reasons, and because of physical and chemical problems [7.267] such as corrosion with oxygen, the development of hydride batteries has been delayed. It seems that environmental considerations, or the possible shortage of Cd, due to the continued and extensive demand of small power supplies, for electronic equipment, computers, or for consumer purposes, have pushed industrial concerns to accelerate their development. According to *Sakai* et al. [7.268] the nickel-metal hydride cell was ready for the market in 1991.

(A) Principle

The concept of sealed rechargeable nickel/metal hydride battery has been reported by *Notten* and *Einerhand* [7.269]. A detailed electrochemical description of this type of battery may be found in [7.270]. This section will be limited to the basic reactions occurring during a charge/discharge cycle, disregarding the side reactions which condition the good functioning of the sealed Ni-IMH battery. A schematic representation is shown in Fig. 7.13. The positive electrode corresponds to the conventional NiOOH electrode, the negative electrode is supporting the IMH. In practice, the technology used for Ni-Cd battery has been adapted for Ni-IMH battery, i.e., rolling of the electrodes which are electrically insulated by a separator impregnated with a KOH solution. During the discharging process, the trivalent Ni state is reduced in divalent Ni state while hydrogen is oxidised, involving the decomposition of the hydride.

According to *Bronoël* et al. [7.264], the measured effective capacity for LaNi₅H₆ is of the order of 320 mAh/g, whereas *Willems* [7.265] calculated a storage capacity of 370 mAh/g. The corresponding value for Cd is only 270 mAh/g.

(B) Development

The discussion refers only to the properties of the IMH electrode. The high energy density, and the possibility of high rates of charging and discharging, were soon recognised, together with the low operating temperature of the hydride electrode. The major difficulties concerned the poor long term behaviour, due to surface damage of the hydride electrode, leading to a drastic drop in capacity after a few cycles. In the case of LaNi₅ electrode the degradation process was attributed to oxidation of the compound. To overcome these problems, a great deal of research has gone into developing appropriate multicomponent alloys to test the large scale process of pulverising and coating the material [7.271, 272], and to study the effect of binding materials in shaping the alloy electrodes [7.273–275].

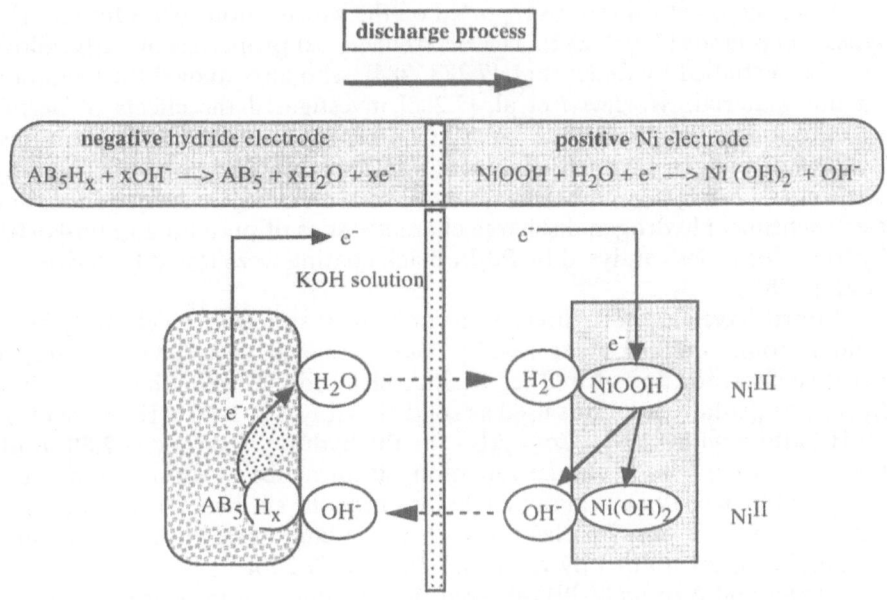

overall reaction : $xNiOOH + LaNi_5H_x \longrightarrow LaNi_5 + xNi(OH)_2$

Fig. 7.13. Schematic representation of Ni–IMH battery

The major improvements were achieved at Philips lab. Thus, *Willems* et al. [7.265, 266] found that the durability of the LaNi$_5$ electrode was drastically increased by substituting the A and B atoms by other elements. It was demonstrated that the new alloy, $La_{0.8}Nd_{0.2}Ni_{2.5}Co_{2.4}Si_{0.1}$ had an excellent cycle lifetime. The cycling stability decay was about 30% of the initial storage capacity after 1000 cycles. They identified the volume expansion as the main cause for material degradation during electrochemical cycling. *Boonstra* et al. [7.276–278] concentrated their studies on an LaNi$_5$–Cu pellet electrode in which the degradation processes, the effect of the electrolyte, and the effect of the pretreatment of powder were successively quantified. The overall mechanism proposed to explain the reduced decay closely follows the model proposed by *Schlapbach* [7.279] for gas-solid degradation in the presence of water vapour and contaminant oxygen, when a newly cleaned, catalytically active, coated Ni surface was regenerated after formation of lanthanum hydroxide and oxide. The authors showed that the oxidised layers at the surface of the particles are formed by reaction with the electrolyte and that encapsulation of LaNi$_5$ particles by copper diminished the oxide growth without affecting the storage capacity of the electrodes. Fundamental research was pursued in that domain at Philips and the latest results related to mechanical stability of the material electrodes during cycling was already presented in Sect. 7.3.1(C).

Since then, several groups reported on the substitution effects for the AB_5 types compounds [7.280–282]. The electrochemical properties of Zr-Ni alloys have been studied by *Sawa* et al. [7.283, 284], who also studied the oxidation on this material. *Moriwaki* et al. [7.285] investigated the effects of partial substitution of V and Mo with Ni in ZrV_2, $ZrMo_2$ for electrode applications. Amorphous Ni-Zr, taken as-cast from a meltspinning machine, was used by *Ryan* et al. [7.286] in an alkaline solution, with no catalytic agent added. The electrochemical hydrogen discharge characteristics of pure Pd and Pd-Ni-Rh hydride electrodes catalysed by Pd/Pt black coating were tested by *Sakamoto* et al. [7.287].

Efforts have also been directed towards the search for new electrolytes. A solid protonic conductor was used by *Poinsignon* et al. [7.288] with $TiNiH_x$ as negative electrode and γMnO_2 as positive electrode. Tetramethylammonium hydroxide pentahydrate was used as solid electrolyte in Ni–IMH and MnO_2–IMH battery with $LaNi_{2.5}Co_{2.4}Al_{0.1}$ for the hydride electrode [7.289]. Both batteries showed poor charge retention performance resulting from a decrease in the density of oxidised metal ions near the surface of active material particles on their positive electrodes. A novel composite electrolyte configuration has been studied by *Liaw* and *Huggins* [7.290].

Kumar and *Saxena* [7.291] showed the usefulness of measuring electrode potential as a function of cathodic charging for the study of the hydrogen absorption characteristics of an IMC electrode, while *Jordy* et al. [7.292] proposed an evaluation of the metal hydride electrode potential based on analysis of the isotherm curves.

The degradation of IMH electrode storage capacity with the number of charge/discharge cycles is still extensively studied by various groups. *Sakai* et al. [7.281] analysed the cycling lives of AB_5 based alloy electrodes, and concluded that $LaNi_{2.5}Co_{2.5}$ offered the highest durability, confirming earlier results of Willem. On the basis of experimental studies, they gave the following qualitative interpretation. The pulverisation rate of the grains may be one of the dominant factors influencing the capacity decay because pulverisation increases the area of the surface, which then oxidises continuously. The influence of the particle size of IMH [7.293], or the selection of a proper binder [7.294] are important factors which have been studied since they affect the discharge performances of the batteries. Mechanical alloying has also been attempted to prepare new alloys materials for electrodes [7.295].

Up to now, the greatest number of contributions has been devoted to the bulk properties of the materials since the prime interest was the hydrogen storage capacity and thus the electrical storage capacity. Moreover, for solid IMC/hydrogen gas application or solid IMC/electrolyte application, the bulk properties remain identical. Differences in properties of the alloys come from the interfaces between solid and gas and solid and electrolyte. In particular for electrochemical cells, current density, corrosion, self discharge are dependent on the composition of the surface layer. In recent years, the interest devoted to surface studies has been growing [7.296–298].

Modelisation contributed also to the development of electrochemical cells. A phenomenological model for the cycle life of AB_2 and AB_5 IMH electrode was proposed [7.299]. The model allows a numerical fit of the measured discharge capacities as a function of the number of cycle with five independent parameters. *Zhuravleva* et al. [7.300] developed a model to compute the charge and discharge curves which characterise the electrochemical process. The authors solved the coupled equations which govern the mass transport at the metal/electrolyte interface and the hydrogen diffusion through the bulk and established the limiting amount of charge that could be extracted at a given grain. The mathematical model of *Vitanen* [7.301] was formulated to describe and to analyse the polarisation behaviour of the electrode under different conditions, in terms of physically measurable properties, i.e., exchange current density, diffusion coefficient of hydrogen, particle size, void fraction and electrode size. This model has been validated by experiment. A close agreement is obtained between the calculated and measured polarisation curves.

Small size batteries are now produced on a large scale in Japan and in the USA. The future cells concern already the development of prototypes of large capacities, up to 30 Ah.

7.4.7 Fuel Cells

Development of hydrogen fuel cells is driven by space exploration, with manned-missions. The highly efficient conversion of chemical energy to electricity, in practice, of the order of 60%, and the water production make the H_2 fuel cell very attractive for such use. A by product of this research should occur for vehicular applications. But, for extensive applications, the cells must overcome their weight handicap and they must be capable of air operation. According to *Staschevski* [7.302], no system exhibits a weight to power ratio of less than 10 kg/kW. Current problems and studies of fuel cells are found in [7.303].

The H_2 fuel cell can be considered as an energy converter working with an invariant electrode/electrolyte system, fuelled by O_2 and H_2 stored in reservoirs. Ergenics [7.304–306] developed an ion exchange membrane fuel cell combined with hydride storage vessels. Although the system was developed for a space station shuttle bus, Ergenics extended their studies to the design of high energy density, portable power equipment. Four models are now proposed with from 200 W, 12 V d.c. to 1000 W, 24 V d.c. power, while the cell reactants are stored as compressed gases, cryogenic liquids or, for H_2 as a rechargeable metal hydride.

7.5 Conclusions

After 20 years of research and development in the field of hydride applications the state of the art may be discussed according to the presentation shown in Table 7.5. Firstly, the successful cases are considered. Successful means that the private companies have taken the control of the development

Table 7.5. Present state of hydride applications

H$_2$ → Hydride Applications for a Clean Environment		
Successful:	alloys elaboration, Getter, Fire detectors, Batteries	
Continuously supported:	Isotope separation, tritium storage	
Stand by:	Purification (small units), Static Storage	
	Hydrides machines:	Compressor
		Heat Pumps Refrigeration
		Heat Transformer
Out at present:	mobile storage for hydride vehicles	

over the national institutions. The industrials insure the fabrication of the products and obviously they are making profits from the sales. Metallurgical industries have benefited from the hydride technology either for the fabrication of multicomponents compounds or for the fabrication of permanent magnets, through powder metallurgy. Thus, NdFeB alloys are pulverised by hydriding before particles alignment in strong magnetic fields and sintering [7.307]. Getters are widely used in any area requiring a high purity control, from the fabrication of electronic components up to nuclear applications. Hydride fire detectors have been mounted extensively in most aircraft, civil or military, for more than 10 years. The hydride batteries for cordless devices are now available on the market since 1992. In this area, the research and development has been reinforced in the USA by new programs to support the fabrication of cells with much larger capacities, taking advantage of the "zero emission vehicle" law passed in California.

It should be pointed out that each of these applications is correlated to a well defined property of the material allowing to reach the specific goal, i.e., chemical reactivity for getters, temperature/pressure relationship for sensors, hydrogen storage capacity converted into electrical current for the electrochemical cells. Thus, the used materials are multicomponents alloys which have been optimised for a proper functioning of the application. Beyond all questions, the fact that hydride batteries are distributed world wide at the consumer level is a tremendous opportunity for continuing the development in other fields. In this regard, most of the research related to the nuclear area such as isotope separation, tritium storage have always been supported by governmental agencies, and there is no doubt that the same behaviour will be pursued in the future. Unfortunately, the success of hydride batteries has expelled the mobile hydrogen storage for vehicles although the advantage of the electrical powered vehicle is not proven yet. The choice is the result of a political decision, based on lobbying rather than on experimental facts.

The disappointment could come from the devices designed for storage and to a larger extend from the hydride machines. In 1992, *Suda* and *Sandrock* [7.308] tried to explain why the commercialisation of these devices have failed. Cost appears as a major argument for many reasons. Thus, the systems are expensive because of the price of the alloys, but the elaboration is also costly in terms of energy. This could be compared with light titanium superalloys elaboration for aeroplanes. Reactors are costly because activation is done in situ, increasing the thickness of the walls and contributing to a loss in efficiencies. This demonstrates that the development should be thought differently. Preparation, activation of the alloy, and device functioning are different steps of the development. Metallurgists can take in charge preparation and activation of the alloys. Special reactors could be reserved only for the activation process prior to loading the alloys in any machine. Finally, high cost contribution is attributed to the heat exchangers. These reasons are complemented by considering the weakness of some material properties such as pulverisation, impurities, poor heat transmission and others. Of course, all the arguments are necessary to justify a "partial failure", but they are not sufficient.

Such analysis call for other remarks. Most of the difficulties encountered in hydride applications were identified since the first report in the field [7.203]. The most serious parameter which now qualifies an IMH is the capacity to retain the absorption properties for an imposed number of loading/unloading cycles. This has been achieved for the hydride batteries, not for solid gas hydride machines. However, few hundred cycles are satisfactory for the former, while several 10^5 are required for the latter. There is no doubt that small prototypes were justified to prove the viability of the engines, but the extension to large scale prototypes, several tons of materials, is rather questionable, knowing that the long term behaviour was not demonstrated. At present, the hydride machines offer too many possibilities to be solved at once. Scientists should restrict their ambition to satisfy in priority one goal, with defined temperature levels, as suggested in Sect. 7.4.3. Moreover, it is well established now that an engineering approach alone is not sufficient when complex chemical reactions with metastable materials are involved in a process. More overlap is needed between fundamental and applied research, where studies of material properties are of paramount importance. Only at these conditions, hydride chemical engines will have a chance of success.

All the applications which have been proposed for metal-hydrides have proved to be viable. Considering that 15 to 20 years is a very short time for developing a new technology, and in view of the progress already accomplished, we may reasonably expect that the use of hydrides at the consumer level will continue to expand.

References

7.1   G.G. Libowitz, H.F. Hayes, T.R.P. Gibb: J. Phys. Chem. **62**, 76 (1958)
7.2   J. Van Vucht, F.A. Kuijpers, H. Bruning: Philips Res. Rept. **25**, 133 (1970)
7.3   J.J. Reilly, R.H. Wiswall: Inorg. Chem. **13**, 218 (1974)
7.4   D.M. Gruen, F. Schreiner, I. Sheft: Int. J. Hydrogen Energy **3**, 303 (1978)
7.5   S. Suda: Int. J. Hydrogen Energy **12**, 323 (1987)
7.6   W.M. Mueller, J.P. Blackledge, G.G. Libowitz: *Metal Hydrides* (Academic, New York 1968)

7.7   G. Alefeld, J. Völkl (eds.): *Hydrogen in Metals II*, Application - Oriented Properties, Topics Appl. Phys., Vol **29** (Springer, Berlin, Heidelberg 1978)

7.8   R. Barnes (ed.): *Hydrogen Storage Materials*, Mater. Sci. Forum, Vol. **31** (Trans Tech Switzerland 1988)

7.9   G. Sandrock, S. Suda, L. Schlapbach: Applications, in *Hydrogen in Intermetallic Compounds II*, ed. by L. Schlapbach, Topics Appl. Phys., Vol. **67** (Springer, Berlin Heidelberg 1992)

7.10   *Hydrogen in Metals*, Proc. Int'l Symp., Belfast, Ireland (1985) ed. by F.A. Lewis, E. Wicke: Z. Phys. Chemie N.F. **143–147** (1986) with references to earlier conferences in this series

7.11   *Int'l Symp. on the Properties and Applications of Metal Hydrides V*, Maubuisson, France (1986) ed. by A. Percheron-Guegan, M. Gupta: J. Less-Common Met. **129–131** (1987), with references to earlier conferences in this series

7.12   *Metal-Hydrogen Systems: Fundamentals and Applications*, Proc. 1st Int'l Symp. combining "Hydrogen in Metals" and "Metal Hydride", Stuttgart, Germany (1988) ed. by R. Kirchheim, E. Fromm, E. Wicke: Z. Phys. Chemie N.F. **163–164** (1989)

7.13   *Metal-Hydrogen Systems: Fundamentals and Applications*, Proc. Int'l. Symp. Banff, Alberta, Canada (1990), ed. F.D. Manchester: J. Less-Comm. Met. **172–174** (1990)

7.14   *Metal-Hydrogen Systems: Fundamentals and Applications*, Proc. Int'l. Symp. Uppsala, Sweden (1992), ed. by D. Noréus, S. Rundqvist, E. Wicke: Z. Phys. Chem. **179–183** (1992)

7.15   *Hydrogen Energy Progress VII*, Proc. 7th World Hydrogen Energy Conf., Moscow, USSR, (1988), ed. by T.N. Veziroglu, A.N. Protsenko, Vol. I,II,III (Pergamon, New York 1988)

7.16   C.J. Winter: Int'l J. Hydrogen Energy **12**, 521 (1987)

7.17   H. Quadflieg: Int'l J. Hydrogen Energy **13**, 363 (1988)

7.18   J.O'M. Bockris, J.C. Wass: "About the real economics of massive hydrogen production at 2010 AD, in *Hydrogen Energy Progress VII*, **1**, 101 (Pergamon, New York 1988)

7.19   M.A. DeLuchi: Int'l J. Hydrogen Energy **14**, 81 (1989)

7.20   O. Bernauer: Development of hydrogen-hydride technology in the FRG, in *Hydrogen Energy Progress VII*, **1**, 181 (Pergamon, New York 1988)

7.21   A.N. Podgorny: Int'l J. Hydrogen Energy **14**, 599 (1989)

7.22   H. Tagawa, T. Otha: Japan hydrogen energy and technology program in *Hydrogen Energy Progress VII*, **1**, 153 (Pergamon, New York 1988)

7.23   A. Percheron-Guegan, J.M. Welter: Preparation of intermetallics and hydrides, in *Hydrogen in Intermetallic Compounds I*, ed. by L. Schlapbach, Topics Appl. Phys., Vol. **63** (Springer, Berlin, Heidelberg 1988) pp. 1–48

7.24   F.D. Manchester, D. Khatamian: Mechanisms for activation of intermetallic hydrogen absorbers, in [Ref. 7.8, pp. 261–296]

7.25   T.B. Flanagan, W.A. Oates: Thermodynamic of intermetallic compound hydrogen systems, in *Hydrogen in Intermetallic Compounds I*, ed. by L. Schlapbach, Topics Appl. Phys., Vol. **63** (Springer, Berlin, Heidelberg 1988) pp. 49–85

7.26   W.A. Oates, T.B. Flanagan: Mat. Res. Bull. **19**, 1397 (1984)

7.27   T.B. Flanagan, W.A. Oates: J. Less-Comm. Met. **100**, 299 (1984)

7.28   P. Dantzer, F. Marcelet: J. Phys. E **18**, 536 (1985)

7.29   D.H. Ryan, J.M.D. Coey: J. Phys. E **19**, 693 (1986)

7.30   H. Uchida, K. Terao, Y.C. Huang: Z. Phys. Chemie N.F. **164**, 1275 (1989)

7.31   H. Uchida, A. Hisano, K. Terao, N. Sato, A. Nagashima: J. Less-Common Met. **172–174**, 1018 (1991)

7.32   R. Griessen, T. Riesterer: Heat of formation models, in *Hydrogen in Intermetallic Compounds I*, ed. by L. Schlapbach, Topics Appl. Phys., Vol. **63** (Springer, Berlin, Heidelberg 1988) pp. 219–284

7.33   O.J. Kleppa, P. Dantzer, M.E. Melnichak: J. Chem. Phys. **61**, 4048 (1974)

7.34   P. Dantzer, O.J. Kleppa: J. Solid State Chem. **24**, 1 (1978)

7.35   G. Boureau, O.J. Kleppa, P. Dantzer: J. Chem. Phys. **64**, 5241 (1976)

7.36 G. Boureau, O.J. Kleppa: J. Chem. Phys. **65**, 3915 (1976)
7.37 P. Dantzer: J. Phys. Chem. Solid **44**, 913 (1983)
7.38 P. Dantzer, O.J. Kleppa: J. Solid State Chem. **35**, 34 (1980)
7.39 P. Dantzer, O.J. Kleppa: J. Chem. Phys. **73**, 5259 (1980)
7.40 C. Picard, O.J. Kleppa: High Temperature Sci. **12**, 89 (1980)
7.41 P.S. Rudman: Int'l J. Hydrogen Energy **3**, 431 (1978)
7.42 W. Luo, T.B. Flanagan: J. Phase Equilibria **15**, 20 (1994)
7.43 T.B. Flanagan, J.D. Clewley, T. Kuji, C.N. Park, D.H. Everett: J. Chem. Soc. Faraday Transl. **82**, 2589 (1986)
7.44 W.R.Mc. Kinnon: J. Less-Common Met. **91**, 293 (1983)
7.45 S. Quian, D.O. Northwood: Int'l J. Hydrogen Energy **13**, 25 (1988)
7.46 A.L. Shilov, N.T. Kuznetsov: J. Less-Common Met. **152**, 275 (1989)
7.47 B. Baranovski: J. Alloys and Compounds **200**, 87 (1993)
7.48 R. Balasubramaniam: Acta Metall. Mater. **41**, 3341 (1993)
7.49 K.H.J. Buschow, H.H. van Mal: J. Less-Common Met. **29**, 203 (1972)
7.50 C.N. Park, T.B. Flanagan: Ber. Bunsenges. Phys. Chem. **89**, 1300 (1985)
7.51 P. Dantzer, M. Pons, A. Guillot: Z. Phys. Chem. **183**, 205 (1994)
7.52 P.D. Goodell, G. Sandrock, E.L. Huston: J.Less-Common Met. **73**, 135 (1980)
7.53 Y. Josephy, M. Ron: Z. Phys. Chemie N.F. **147**, 233 (1986)
7.54 M. Ron, Y. Josephy: J. Less-Common Met. **131**, 51 (1987)
7.55 M. Groll, W. Supper, R. Werner: Z. Phys. Chemie N.F. **164**, 1485 (1989)
7.56 A. Percheron-Guegan, C. Lartique, J.C. Achard, P. Germi, F. Tasset: J. Less-Common Met. **74**, 1 (1980)
7.57 P. Dantzer, E. Orgaz, V.K. Sinha: Z. Phys. Chemie N.F. **163**, 141 (1989)
7.58 A.L. Shilov, M.E. Kost, N.T. Kuznetsov: J. Less-Common Met. **144**, 23 (1988)
7.59 G. Sandrock, P.D. Goodell: J. Less-Common Met. **73**, 161 (1980)
7.60 P.D. Goodell: J. Less-Common Met. **89**, 45 (1983)
7.61 F.G. Eisenberg, P.D. Goodell: J. Less-Common Met. **89**, 55 (1983)
7.62 G. Sandrock, P.D. Goodell: J. Less-Common Met. **104**, 159 (1984)
7.63 L. Schlapbach, J.C. Achard, A. Percheron-Guegan: Proc. 3rd Congr. Int'l Hydrogène et Matériaux (1982) A6
7.64 R. Suzuki, J. Ohno, H. Gondho: J. Less-Common Met. **104**, 199 (1984)
7.65 P. Selvam, B. Viswanathan, C.S. Swamy, V. Srinivasan: Z. Phys. Chemie N.F. **164**, 1199 (1989)
7.66 J.I. Han, J.Y. Lee: J. Less-Common Met. **152**, 319 (1989)
7.67 J.I. Han, J.Y. Lee: J. Less-Common Met. **152**, 329 (1989)
7.68 H. Wenzl, K.H. Klatt, P. Meuffels, K. Papathanassopoulos: J. Less Common Met. **89**, 489 (1983)
7.69 I.P. Jain, Y.K. Vijay, L.K. Malhotra, K.S. Uppadhyay: Int'l J. Hydrogen Energy **13**, 15 (1988)
7.70 A. Krozer, B. Kasemo: J. Less-Common Met. **160**, 323 (1990)
7.71 H. Shirai, H. Sakaguchi, G. Adachi: J. Less-Common Met. **159**, L17 (1990)
7.72 H. Sakaguchi, H. Seri, G. Adachi: J. Phys. Chem. **94**, 5313 (1990)
7.73 H.G. Wultz, E. Fromm: J. Less-Common Met. **118**, 315 (1986)
7.74 X.L. Wang, S. Suda: J. Alloys and Compounds **194**, 73 (1993)
7.75 P. Dantzer: J. Less-Common Met. **131**, 349 (1987)
7.76 G. Sandrock, P.D. Goodell, E.L. Huston, P.M. Golben: Z. Phys. Chemie N.F. **164**, 1285 (1989)
7.77 R.L. Cohen, K.W. West, R.H.J. Buschow: Solid State Commun. **25**, 2937 (1978)
7.78 R.L. Cohen, K.W. West, J.H. Wernick: J. Less-Common Met. **70**, 229 (1980)
7.79 R.L. Cohen, K.W. West, J.H. Wernick: J. Less-Common Met. **73**, 273 (1980)
7.80 R.L. Cohen, K.W. West: J. Less-Common Met. **95**, 17 (1983)
7.81 K.H.J. Buschow: Mat. Res. Bul. **19**, 935 (1984)
7.82 J. Bonnet, P. Dantzer, B. Dexpert, J.N. Esteva, R. Karnatak: J. Less-Common Met. **130**, 491 (1987)
7.83 J.M. Park, J.Y. Lee: Mat. Res. Bull. **22**, 455 (1987)
7.84 J.I. Han, J.Y. Lee: Int'l J. Hydrogen Energy **13**, 577 (1988)

7.85 T. Gamo, Y. Moriwaki, N. Yanagihara, T. Iwaki: J. Less-Common Met. **89**, 495 (1983)
7.86 Y.G. Kim, J.Y. Lee: J. Less-Common Met. **144**, 331 (1988)
7.87 Q.D. Wang, O.J. Wu, N. Qiu: Z. Phys. Chemie N.F. **164**, 1305 (1989)
7.88 D. Chandra, S. Bagchi, S.W. Lambert, W.N. Cathey, F.E. Lynch, R.C. Bowmann: J. Alloys and Compounds **199**, 93 (1993)
7.89 S. Ono, K. Nomura, E. Akiba, H. Uruno: J. Less-Common Met. **113**, 113 (1985)
7.90 R.C. Bowmann, C.H. Luo, C.C. Ahn, C.K. Witham, B. Fultz: J. Alloys and Compounds **217**, 185 (1995)
7.91 J.J. Reilly: *Proc. Int'l Symp on Hydrides for Energy Storage,* Geilo, (1977) (Pergamon, New York 1978) p. 301
7.92 H.J. Ahn, S.M. Lee, J.Y. Lee: J. Less-Common Met. **142**, 253 (1988)
7.93 J.Y. Lee, S.S. Park: Z. Phys. Chemie N.F. **164**, 1337 (1989)
7.94 M. Ron, Y. Josephy: Z. Phys. Chemie N.F. **164**, 1343 (1989)
7.95 E. Bershadsky, Y. Josephy, M. Ron: J. Less-Common Met. **172–174**, 1036 (1991)
7.96 P.H.L. Notten, R.E.F. Einerhand, J.L.C. Daams: J. Alloys and Compounds **210**, 221 (1994)
7.97 P.H.L. Notten, J.L.C. Daams, R.E.F. Einerhand: J. Alloys and Compounds **210**, 233 (1994)
7.98 P.D. Goodell, R.S. Rudman: J. Less-Common Met. **89**, 117 (1983)
7.99 P. Dantzer, E. Orgaz: J Less-Common Met. **147**, 27 (1989)
7.100 M.H. Mintz, J. Bloch: Prog. Solid State Chem. **16**, 163 (1985)
7.101 P.S. Rudman: J. Less-Common Met. **89**, 93 (1983)
7.102 S. Yagi, D. Kunii: Chem. Eng. Jpn: **19**, 500 (1955)
7.103 Y. Josephy, M. Ron: J. Less-Common Met. **147**, 227 (1989)
7.104 J.T. Koh, A.J. Goudy, P. Huang, G. Zhou: J. Less Common Met. **153**, 89 (1989)
7.105 J.I. Han, J.Y. Lee: Int'l J. Hydrogen Energy **14**, 181 (1989)
7.106 X.L. Wang, S. Suda: Z. Phys. Chemie N.F. **164**, 1235 (1989)
7.107 X.L. Wang, S. Suda: J. Less-Common Met. **159**, 83 (1990)
7.108 X.L. Wang, S. Suda: J. Less-Common Met. **159**, 109 (1990)
7.109 C. Bayane, E. Sciora, N. Gérard, M. Bouchdoug: Thermochimica Acta **224**, 193 (1993)
7.110 X.L. Wang, S. Suda: J. Less-Common Met. **172–174**, 969 (1991)
7.111 O. Bernauer, J. Töpler, D. Noreus, R. Hempelman, D. Richter: Int'l J. Hydrogen Energy **14**, 187 (1989)
7.112 A. Zarynow, A.J. Goudy, R.B. Schweibenz, K.R. Clay: J. Less-Common Met. **172–174**, 1009 (1991)
7.113 K.R. Clay, A.J. Goudy, R.B. Schweibenz, A. Zarynow: J. Less-Common Met. **166**, 153 (1990)
7.114 H. Uchida, M. Ozawa: Z. Phys. Chemie N.F. **147**, 77 (1986)
7.115 Y. Ohtani, S. Hashimoto, H. Uchida: J. Less-Common Met. **172–174**, 841 (1991)
7.116 J.J. Reilly, Y. Josephy, J.R. Johnson: Z. Phys. Chemie N.F. **164**, 1241 (1989)
7.117 Z. Gavra, J.R. Johnson, J.J. Reilly: J. Less-Common Met. **172–174**, 107 (1991)
7.118 S. Suda, N. Kobayashi, R. Yoshida, Y. Ishido, S. Ono: J. Less-Common Met. **74**, 127 (1980)
7.119 S. Suda, Y. Komazaki, E. Morishita, N. Takemoto: J. Less-Common Met. **89**, 325 (1983)
7.120 D. Kunii, J.M. Smith : A.I.Ch.E.J. **6**, 71 (1960)
7.121 S. Yagi, D. Kunii : A.I.Ch.E.J. **3**, 373 (1957)
7.122 S. Yagi, D. Kunii : A.I.Ch.E.J. **6**, 97 (1960)
7.123 E. Tsotsas, H. Martin: Chem. Eng. Process **22**, 19 (1987)
7.124 S. Suda, Y. Komazaki, N. Robayashi: J. Less-Common Met. **89**, 317 (1983)
7.125 M. Nagel, Y. Komazaki, S. Suda: J. Less-Common Met. **120**, 35 (1986)
7.126 S. Suissa, I. Jacob, Z. Hadari: J. Less-Common Met. **104**, 287 (1984)
7.127 A. Kempf, W.R.B. Martin: Int. J. Hydrogen Energy **11**, 107 (1986)
7.128 M. Pons, P. Dantzer, J.J. Guilleminot: Int. J. Heat Mass Transfer **36**, 2635 (1993)
7.129 M. Pons, P. Dantzer: Z. Phys. Chem. **183**, 213 (1994)

7.130 M. Pons, P. Dantzer: J. Less-Common Met. **172–174**, 1147 (1991)
7.131 M. Pons, P. Dantzer: Z. Phys. Chem. **183**, 225 (1994)
7.132 M. Devillers, M. Sirch, S. Bredendiek-Kämper, R.D. Penzhorn: Chem. Mat. **2**, 255 (1990)
7.133 E.H. Kisi, E. Gray: J. Alloys and Compounds **217**, 112 (1995)
7.134 E. Gray, C.E. Gray, C.E. Buckley, E.H. Kisi: J. Alloys and Compounds **215**, 201 (1994)
7.135 W. Supper, M. Groll, U. Mayer: J. Less-Common Met. **104**, 279 (1984)
7.136 M. Ron, D.M. Gruen, M.H. Mendelsohn, I. Sheft: J. Less-Common Met. **74**, 445 (1980)
7.137 E. Tüscher, P. Weinzierl, O.J. Eder: Int'l J. Hydrogen Energy **8**, 199 (1983)
7.138 Q.D. Wang, J. Wu, H. Gao: Z. Phys. Chemie N.F. **164**, 1367 (1989)
7.139 J. Töpler, O. Bernauer, H. Buchner: J. Less-Common Met. **74**, 385 (1980)
7.140 H. Ishikawa, K. Oguro, A. Kato, H. Suzuki, E. Ishii: J. Less-Common Met. **107**, 105 (1985)
7.141 H. Ishikawa, K. Oguro, A. Kato, H. Suzuki, E. Ishii: J. Less-Common Met. **120**, 123 (1986)
7.142 H. Ishikawa, K. Oguro, A. Kato, H. Suzuki, E. Ishii, T. Okada, S. Sakamoto: Z. Phys. Chemie N.F. **164**, 1409 (1989)
7.143 Y. Josephy, Y. Eisenberg, S. Perez, A. Bendavid, M. Ron: J. Less-Common Met. **104**, 297 (1984)
7.144 E. Bershadsky, Y. Josephy, M. Ron: Z. Phys. Chemie N.F. **164**, 1373 (1989)
7.145 E. Bershadsky, Y. Josephy, M. Ron: J. Less-Common Met. **153**, 65 (1989)
7.146 G. Anevi, L. Jansson, D. Lewis: J. Less-Common Met. **104**, 341 (1984)
7.147 N.P. Kherani, W.T. Shmayda, A.G. Heics: Z. Phys. Chemie N.F. **164**, 1421 (1989)
7.148 H. Aoki, H. Mitsui: Investigation of annular Metal hydride Reaction beds: Characteristics of reaction and heat transfer. Reports from Toyota Central Research & Development Labs. Inc. Japan (1988)
7.149 H. Aoki, H. Mitsui: Investigation of tubular heat exchangers with internal longitudinal fins for metals hydride reactor beds. Reports from Toyota Central Research & Development Labs. Inc. Japan (1988)
7.150 T. Ogawa, R. Ohnishi, T. Misawa: J. Less-Common Met. **138**, 143 (1988)
7.151 H. Uchida, T. Ebisawa, K. Terao, N. Hosoda, Y.C. Huang: J. Less-Common Met. **131**, 365 (1987)
7.152 J.R. Johnson, J.J. Reilly: Z. Phys. Chemie N.F. **147**, 263 (1986)
7.153 J.S. Noh, R.K. Agarwal, J.A. Schwarz: Int'l J. Hydrogen Energy **12**, 693 (1987)
7.154 R. Wiswall: In [Ref. 7.7, pp 230–233]
7.155 H. Suzuki, Y. Osumi, A. Kato, K. Oguro, M. Nakane: J. Less-Common Met **89**, 545 (1983)
7.156 S. Ono, Y. Ishido, E. Akiba, K. Jindo, Y. Sawada, I. Kitagawa, T. Kakutani: Proc. 5th World Hydrogen Energy Conf. V **3**, 1291–1301 (1984)
7.157 O. Bernauer, C. Halene: J. Less-Common Met. **131**, 213 (1987)
7.158 ERGENICS: Catalogue for hydrogen storage unit ST–l, ST–45, ST–90, US patent 4,396,114
7.159 HWT: Catalogue N° HUT 8602 E, laboratory hydride storage units
7.160 E.L. Huston, G.D. Sandrock: J. Less-Common Met. **74**, 435 (1980)
7.161 O. Bernauer: Int'l J. Hydrogen Energy **13**, 181 (1988)
7.162 K.C. Hofman, W.E. Winsche, R.H. Wiswall, J.J. Reilly, T.V. Sheehan, C.H. Waide: Metal hydrides as a source of fuel for vehicular propulsion, SAE technical paper series N°690232 (1969)
7.163 S. Furuhama: Int'l J. Hydrogen Energy **14**, 907 (1989)
7.164 J. Töpler, K. Feucht: Z. Phys. Chemie N.F. **164**, 1451 (1989)
7.165 F.E. Lynch: J. Less-Common Met. **172–174**, 943 (1991)
7.166 D. Davidson, M. Fairlie, A.E. Stuart: Int'l J. Hydrogen Energy **11**, 39 (1986)
7.167 S. Orimo, M. Tabata, H. Fujii, K. Yamamoto, S. Tanioka, T. Ogasawara, Y. Tsushio: J. Alloys and Compounds **203**, 61 (1994)
7.168 S. Orimo, M. Tabata, H. Fujii: J. Alloys and Compounds **210**, 37 (1994)

7.169 SAES: Milano, Italy

7.170 M.H. Mendelsohn, D.M. Gruen: J. Less-Common Met. **74**, 449 (1989)

7.171 K. Watanabe, M. Matsuyama, K. Ashida, H. Miyake: J. Vac. Sci. Technol. A **7**, 2725 (1989)

7.172 C.D. Cann, E.E. Sexton, A.A. Bahurmuz, A.J. White, G.A. Ledoux: J. Less-Common Met. **172–174**, 1297 (1991)

7.173 Ergenics: Catalogue HY-Pure, Gas Purifiers

7.174 HWT: Catalogue N° HUT 8604 E, gas purification systems

7.175 E. Peres Da Silva: Industrial prototypes of hydrogen compressor based on metallic hydrides technology, in Proc. 8th World Hydrogen Energy Conf. (1990), ed. by T.N. Veziroglu

7.176 J.J. Reilly, R.H. Wiswall: Hydrogen storage and purification systems, Rep. BNL 17136 (Brookhaven Nat Lab., Upton, New York, 1972)

7.177 J.J. Sheridan, F.G. Eisenberg, E.J. Greskovich, G.D. Sandrock, E.L. Huston: J. Less-Common Met. **89**, 447 (1983)

7.178 P.S. Rudman, G.D. Sandrock, P.D. Goodell: J. Less-Common Met. **89**, 437 (1983)

7.179 Q.D. Wang, J. Wu, C.P. Chen, Y. Zhou: J. Less-Common Met. **131**, 321 (1987)

7.180 H. Takeda, J. Satou, Y. Nishimura, T. Kogi, Y. Wakisaka: Develoment of generator cooling hydrogen purity improvement system using hydrogen absorbing alloy Presented at JSME-ASME Int'l Conf. on Power Eng. Tokyo (1993)

7.181 R.L. Swart, J.T. Tinge, Z. Meindersma: Z. Phys. Chemie N.F. **164**, 1435 (1989)

7.182 E.D. Snijder, G.F. Versteeg, W.P.M. van Swaaij: J. Chem. Eng. data **39**, 405 (1994)

7.183 H. Albrecht, U. Kuhnes, W. Asel: J. Less-Common Met. **172–174**, 1157 (1991)

7.184 C.L. Huffine: Fabrication of hydrides, in [Ref. 7.6, pp. 675–747]

7.185 R. Lässer, T. Schober: Some aspects of tritium storage in hydrogen storage materials, in [Ref. 7.8, pp. 40–75]

7.186 G. Sandrock, J.J. Murray, M.L. Post, J.B. Taylor: Mater. Res. Bull. **17**, 887 (1982)

7.187 Z.W. Wei: Int'l J. Hydrogen Energy **12**, 337 (1987)

7.188 B.M. Andreev, G.H. Sicking: Ber. Bunsenges. Phys. Chem. **91**, 177 (1987)

7.189 B. Jungblut, G. Sicking: Z. Phys. Chemie N.F. **164**, 1177 (1989)

7.190 M. Karas, E.P. Magomedbekov, G.H. Sicking: J. Less-Common Met. **159**, 307 (1990)

7.191 I.A. Fischer: J. Less-Common Met. **172–174**, 1320 (1991)

7.192 E. Gluekhauf, G.P. Kitt: Chromatographic separation of hydrogen isotopes, in *Vapor Phase Chromatography* (Butterworths, London 1957)

7.193 T. Altridge: Chromatographic hydrogen isotope separation, U.S. Patent 4,276,060 (June 1981)

7.194 F. Botter, J. Gowman, J.L. Hemmerich, B. Hircq, R. Lässer, D. Leger, S. Tistchenko, M. Tschudin: Fusion Technology **14**, 562 (1988)

7.195 G.S. Nickols: J. Less-Common Met. **172–174**, 1338 (1991)

7.196 Y.W. Wong, F.B. Hill, Y.N.I. Chan: Separ. Sci. Technol. **15**, 423 (1980)

7.197 F.B. Hill, V. Grzetic: J. Less-Common Met. **89**, 399 (1983)

7.198 N. Mitsuishi, S. Fukada, H. Tokuda, T. Nawata, Y. Takai: In [7.13] Vol. 1, pp. 647–671

7.199 H. Sakaguchi, Y. Yagi, J. Shiokawa, G. Adachi: J. Less-Common Met. **149**, 185 (1989)

7.200 S. Fukada, K. Fuchinoue, M. Nishikawa: J. Alloys and Compounds **201**, 49 (1993)

7.201 V.V. Solovey: Metal hydride thermal power installations, in [7.13] Vol. **1**, pp. 1391–1399

7.202 J.J. Reilly, A. Holtz, R.H. Wiswall: Rev. Sci. Instrum. **42**, 1485 (1971)

7.203 Van Mal: "Thesis title"; Ph.D. Dissertation Technological University, Delft (1976)

7.204 M. Blondeau, M. Bonneton, M. Jannot: Revue générale de thermique, France, N° 257, 411 (1983)

7.205 P.M. Golben: Multi-stage hydride-hydrogen compressor, Proc. 18th Intersociety Energy Conversion Engineering Conf. (1983) pp. 1746–1753

7.206 Ergenics catalogue: Model 1.5-15-40: Hydride/Hydrogen, Four Stage Compressor, Model HC-6, Hydride/Hydrogen Electric Compressor

7.207 T. Kumano, B. Tada, Y. Tsuchida, Y. Kuraoka, T. Ishige, H. Baba: Z. Phys. Chemie N.F. **164**, 1509 (1989)

7.208 R.C. Bowman, E.L. Ryba, B.D. Freeman: Adv. in Cryogenic Eng. **39**, 1499 (1994)

7.209 L.C. Wade, R.C. Bowman, D.R. Gilkinson, P.H. Sywulka: Adv. in Cryogenic Eng. **39**, 1491 (1994)

7.210 R.C. Bowman, B.D. Freeman, E.L. Ryba, J.R. Phillips: Proc. Int'l Absorption Heat Pump Conf., ASME **31**, 265 (1993)

7.211 R.C. Bowman, D.R. Gilkinson, R.D. Snapp, G.C. Abell, B.D. Freeman, E.L. Ryba, L.A. Wade: Fabrication and testing of the metal hydride sorbent bed assembly for a periodic 10K sorption Cryocooler, presented at 8th Int'l Cryocooler Conf., Vail, Colorado, USA (1994)

7.212 P.M. Golben: Solar energy, hydrogen sponge, keys to water pump operation, in Design News, Fluid Power, ed. by F. Yeapple (1987)

7.213 D.M. Gruen, M.H. Mendelsohn, I. Sheft: Solar Energy **21**, 153 (1978)

7.214 R. Gorman, P.S. Moritz: Metal hydride solar heat pump and power system (HYCSOS), AIAA/ASERC Conf. on Solar Energy: Technology Status, Phoenix, (1978) pp. 1–6

7.215 P. Dantzer, E. Orgaz: J. Chem. Phys **85**, 2961 (1986)

7.216 P. Dantzer, E. Orgaz: Int'l J. Hydrogen Energy **11**, 2961 (1986)

7.217 E. Orgaz, P. Dantzer: J. Less-Common Met. **131**, 385 (1987)

7.218 N. Douss, F. Meunier, L.M. Sun: Ind. Eng. Chem. Res. **17**, 310 (1988)

7.219 P. Dantzer, F. Meunier: What materials to use in hydride chemical heat pumps, in [Ref. 7.8 pp. 1–17]

7.220 M. Balakumar, S. Srinivasa Murthy, M.V. Krishna Murthy: Heat Recovery Systems & CHP. **7**, 221 (1987)

7.221 S. Srinivasa Murthy, M.V. Krishna Murthy, M.V.C. Sastri: Two-stage metal hydride heat transformer: a thermodynamic study, in [7.13 Vol. II, pp. 1253–1265]

7.222 Nomura, Hideo, Iguchi, Kazuyuki: Heat pump type air conditioning unit utilizing a hydrogen storage alloy, Daikin Industries Ltd, Osaka, Japan Patent 63-29163(A), Feb. 6, 1988

7.223 Iwasaki, Masahide, Nakajima, Masao, Inokuma, Yoshihiko, Kawai, Shigemasa: High temperature steam generating heat pump, Sekisui Chemical Co. ltd Osaka, Nihon Kagaku Gijutsu Co. ltd. Osaka, Japan Patent 63-34461(A), Feb. 15, 1988

7.224 Ebato, Kazuo, Tamura, Keiji, Yoshida, Hiroshi, Yasunaga, Tomohiro: Heat exchanger unit for hydrogen storage alloy, Nippon Yakin Kogyo Co Ltd. Tokyo, Japan Patent 63-259, 300, Oct. 26, 1988

7.225 Kusakabe, Hiroshi: Air conditioner. Sanyo Electric Co. Ltd Moriguchi, Japan Patent 63-231, 155 Sept. 27, 1988

7.226 Mochizuki, Kaorou Sato, Tatsuo: Air conditioner, Toshiba Co. Kawasaki, Japan Patent 63-210, 566, Sept. 1, 1988

7.227 M. Ron, Y. Josephy: Z. Phys. Chemie N.F. **164**, 1475 (1989)

7.228 N. Nagel, Y. Komazaki, M. Uchida, S. Suda, Y. Matsubara: J. Less-Common Met. **104**, 307 (1984)

7.229 T.A. Argabright: Heat/Mass Flow Enhancement Design for a Metal Hydride Assembly, Final Report, 1985, available from Nat. Technical Information Service, U.S. Dept. of Commerce, Springfield, VA. 22161

7.230 M. Ron: J. Less-Common Met. **104**, 259 (1984)

7.231 M. Ron, Y. Eisenberg, Y. Josephy, M. Gutman, U. Navon: Gas-solid reaction heat exchanger for vehicle engine exhaust vaste heat recovery, SAE technical paper series N° 860588, Feb 24–28, 1986

7.232 M. Ron, Y. Josephy: Z. Phys. Chemie N.F. **147**, 241 (1986)

7.233 A. Yanoma, M. Yoneto, T. Nitta, T. Okuda: Design and operation of the commercial size chemical heat pump system using metal hydrides, in Proc. joint Conf. ASME-JSNE, Honolulu (1987) pp. 431–437

7.234 K. Nasako, T. Yonesaki, K. Satoh, T. Imoto, S. Fujitani, N. Hiro, T. Hirose, K. Fukushima, T. Saito, I. Yonezu: Z. Phys. Chem. **183**, 235 (1994)

7.235 R. Werner, N. Groll: Design aspects of metal hydride heat transformers, in Proc. Int'l Congress on High Performances Heat Pumps, ed. by B. Spinner (Perpignan, France, 1988) pp. 320–329
7.236 R. Werner, M. Groll: J. Less-Common Met. **172–174**, 1122 (1991)
7.237 S. Suda, Y. Komazaki: J. Less-Common Met. **172–174**, 1130 (1991)
7.238 E. Tuscher, P. Weizierl, O.J. Eder: J. Less-Common Met. **95**, 171 (1983)
7.239 M. Nagel, Y. Komazaki, S. Suda: J. Less-Common Met. **120**, 35 (1986)
7.240 M. Nagel, Y. Komazaki, Y. Matsubara, S. Suda: J. Less-Common Met. **123**, 47 (1986)
7.241 H. Bjurström, Y. Komazaki, S. Suda: J. Less-Common Met. **131**, 225 (1986)
7.242 H. Bjurström, D. Lewis, S. Suda: Simulation of periodic heat pumps based on metal hydrides, in Proc. of Workshop, Absorption Heat Pumps, held in London (1988), ed. by P. Zegers, J. Miriam C.E.C. pp. 110–120
7.243 H. Bjurström, S. Suda: Int'l J. Hydrogen Energy **14**, 19 (1989)
7.244 M. Gambini: Int'l J. Hydrogen Energy **14**, 821 (1989)
7.245 D.W. Sun, S.J. Deng: J. Less-Common Met. **141**, 37 (1988)
7.246 D.W. Sun, S.J. Deng: J. Less-Common Met. **155**, 271 (1989)
7.247 B. Bogdanovic, B. Spliethoff: Int'l J. Hydrogen Energy **12**, 863 (1987)
7.248 Ali T. Raissi, D.K. Slattery, M.J. Axelrod, R. Zida: Synthesis and characterisation of nickel-doped magnesium hydride in Proc 8th World Hydrogen Energy Conf. (1990) ed. by T.N. Veziroglu
7.249 B. Bogdanovic, B. Spliethoff, A. Ritter: Z. Phys Chemie N.F. **164**, 1497 (1989)
7.250 M. Wierse, R. Werner, M. Groll: J. Less-Common Met. **172–174**, 1111 (1991)
7.251 J.M. Welter: J. Less-Common Met. **104**, 251 (1984)
7.252 D.E. Warren, K.A. Faughnan, R.A. Fellows, J.W. Godden, B.M. Seck: J. Less-Common Met. **104**, 375 (1984)
7.253 F.A. Lewis: *The Palladium Hydrogen System*, (Academic, London 1967)
7.254 M.A. Gutjahr, H. Buchner, K.D. Beccu, H. Säufferer: Power Sources N°4, 79 (1973), ed. by D.H. Collins
7.255 M. Earl, J. Dunlop: COMSAT Tech. Rev. **3**, 437 (1973)
7.256 E.W. Justi, H.H. Ewe, A.W. Kalberlak, N.M. Saridakis, M.H. Schaeffer: Energy Conversion **10**, 183 (1970)
7.257 H. Ewe, E.W. Justi, K. Stephan: Energy Conversion **13**, 109 (1973)
7.258 F.G. Will: U.S. Patent N°3874928 (1975)
7.259 A. Percheron-Guegan, J.C. Achard, J. Loriers, M. Bonnemay, G. Bronoel, J. Sarradin, L. Schlapbach: Nickel base alloys and their electrochemical applications, French patent N°7516160 (May 1975)
7.260 A. Percheron et al: Extension of the french Patent 7516160, N° 7706138 (March 1977)
7.261 A. Percheron-Guegan, J.C. Achard, J. Sarradin, G. Bronoel: Electrode materials based on lanthanum and nickel and electrochemical uses of such materials. US patent 4107405 (August 1978)
7.262 G. Bronoel, J. Sarradin, M. Bonnemay, A. Percheron-Guegan, J.C. Achard, L. Schlapbach: Int'l J. Hydrogen Energy **1**, 251 (1976)
7.263 A. Percheron-Guegan, F. Briaucourt, H. Diaz, J.C. Achard, J. Sarradin, G. Bronoel: Substitutions in LaNi$_5$ compound-comparative study of related hydrides in electrolytic and solid gas reactions, in Proc. 12th Rare Earth Res. Conf., ed. by Lundin, Vail Colorado (1976) pp. 300–308
7.264 G. Bronoel, J. Sarradin, A. Percheron-Guegan, J.C. Achard: Mater. Res. Bull. **13**, 1265 (1978)
7.265 J.J.G. Willems: Philips J. Res. **39**, Suppl. N°l (1984)
7.266 J.J.G. Willems, K.H. Buschow: J. Less-Common Met. **129**, 13 (1987)
7.267 H. Tamura, C. Iwakura, T. Kitamura: J. Less-Common Met. **89**, 567 (1983)
7.268 T. Sakai, T. Hazama, H. Miyamura, N. Kuriyama, A. Kato, H. Ishikawa: J. Less-Common Met. **172–174**, 1175 (1991)
7.269 P.H.L. Notten, R.E.F. Einerhand: Adv. Mat. **3**, 343 (1991)

7.270 P.H.L. Notten: Rechargeable nickel-metalhydride batteries: a successful new concept, in NATO-ASI series, eds. F. Grandjean, G.J. Long and K.H.J. Buschow (Kluwer, Dordrecht) V281 (1995) pp. 151–195

7.271 T. Sakai, A. Yuasa, H. Ishikawa, H. Myamura, N. Kuriyama: J. Less-Common Met. **172–174**, 1194 (1991)

7.272 D.E. Hall, J.M. Sarver, D.O. Gothard: Int. J. Hydrogen Energy **13**, 547 (1988)

7.273 H. Uchida, T. Ebisawa, Y.C. Huang: A new treatment for the pulverization of hydrogen storage alloys, Int'l Symp. on Hydrogen Produced for Renewable Energy, Florida, USA (1985)

7.274 Y. Matsumara, L. Sugiara, H. Uchida: Z. Phys. Chemie N.F. **164**, 1545 (1989)

7.275 T. Sakai, A. Takagi, N. Kuriyama, H. Miyamura: J. Less-Common Met. **172–174**, 1185 (1991)

7.276 A.N. Boonstra, G.J.M. Lippits, T.N.M. Bernards: J. Less-Common Met. **155**, 119 (1989)

7.277 A.H. Boonstra, T.N.M. Bernards: J. Less-Common Met. **161**, 245 (1990)

7.278 A.H. Boonstra, T.N.M. Bernards: J. Less-Common Met. **161**, 355 (1990)

7.279 L. Schlapbach: J. Phys. F **10**, 2477 (1980)

7.280 T. Sakai, K. Oguro, H. Miyamura, N. Kuriyama, A. Kato, H. Ishikawa, C. Iwakura: J. Less-Common Met. **161**, 193 (1990)

7.281 Y.Q. Lei, Z.P. Li, Y.M. Wu, J. Wu, Q.D. Wang: Multicomponent mischmetal-nickel rechargeable hydride electrodes, in Proc. Int'l Renewable Energy Conf. (IREC), Honolulu (1988) pp. 391–399

7.282 M. Matsuoka, T. Kohno, C. Iwakura: Electrochemica Acta **38**, 787 (1993)

7.283 H. Sawa, K. Ohseki, M. Otha, H. Nakano, S. Wakao: Z. Phys. Chemie N.F. **164**, 1521 (1989)

7.284 H. Sawa, M. Otha, H. Nakano, S. Wakao: Z. Phys. Chemie N.F. **164**, 1527 (1989)

7.285 Y. Moriwaki, T. Gamo, H. Seri, T. Iwaki: J. Less-Common Met. **172–174**, 1211 (1991)

7.286 D.N. Ryan, F. Dumais, B. Patel, J. Kycia, J.O. Ström-Olsel: J. Less-Common Met. **172–174**, 1246 (1991)

7.287 Y. Sakamoto, K. Kuruma, Y. Naritomi: J. Appl. Electrochem. **24**, 38 (1994)

7.288 C. Poinsignon, M. Forestier, M. Anne, D. Fruchart, S. Miraglia, A. Rouault, J. Pannetier: Z. Phys. Chemie N.F. **164**, 1515 (1989)

7.289 N. Kuriyama, T. Sakai, H. Miyamura, H. Ishikawa: Solid State Ionics **53**, 688 (1992)

7.290 B.Y. Liaw, R.A. Huggins: Z. Phys. Chemie N.F. **164**, 1533 (1989)

7.291 J. Kumar, S. Saxena: Electrochemical investigation of hydrogen absorption in titanium based intermetallics: in [Ref. 7.13, pp. 629–646]

7.292 C. Jordy, A. Percheron-Guegan, J. Bouet, P. Sanchez, J. Leonardi: J. Less-Common Met. **172–174**, 1236 (1991)

7.293 Y. Sato, A. Togami, K. Ogino, N. Ishii, K. Kobayakawa: Denki Kagaku **61**, 1429 (1993)

7.294 M. Yamashita, H. Higuchi, H. Takemura, K. Okuno: Denki Kagaku **61**, 729 (1993)

7.295 T. Ikeya, K. Kumai, T. Iwahori: J. Electrochem. Soc: **140**, 3082 (1993)

7.296 F. Meli, A. Züttel, L. Schlapbach: J. Alloys and Compounds **202**, 81 (1993)

7.297 A. Züttel, F. Meli, L. Schlapbach: J. Alloys and Compounds **203**, 235 (1994)

7.298 L. Schlapbach, F. Meli, A. Züttel: The surface of intermetallic hydride electrodes, in Proc. 186th ECS meeting, Miami, USA (1994)

7.299 A. Züttel, F. Meli, L. Schlapbach: J. Alloys and Compounds **200**, 157 (1993)

7.300 V.N. Zhuravleva, A.G. Pshenichnikov, Y.U.G. Chirkov, K.V. Shnepelev: Mass transfer in the hydrogen-absorbing metals grain. Hydrogen Energy Progress VII, 445, VI (1988)

7.301 M. Vitanen: J. Electrochem. Soc **140**, 936 (1993)

7.302 D. Staschevski: Intl J. Hydrogen Energy **11**, 279 (1986)

7.303 K. Kordesch, K. Holz, P. Kalal, M. Reindel, H. Steininger: Int'l J. Hydrogen Energy **13**, 475 (1988)

7.304 M.J. Rosso, O.J. Adlhart, J.A. Marmolejo: A fuel cell energy storage for space station extravehicular activity, 18th Intersociety Conf. on Environmental Systems, San Francisco (1988) SAE technical paper series N°881105
7.305 O.J. Adlhart, M.J. Rosso, J.A. Marmolejo: A fuel cell energy storage system concept for the space station freedom extravehicular mobility unit, Inter. Congress and Exposition, Detroit (1989) SAE technical paper series N°891582 (1989)
7.306 O.J. Adlhart, M.J. Rosso, J.A. Marmolejo: Design and performance of an air-cooled ion exchange membrane fuel cell, 33th Int'l Power Sources Symp. (1988) pp. 13–16
7.307 I.R. Harris: J. Less Common Met. **131**, 245 (1987)
7.308 S. Suda, G. Sandrock: Z. für Phys. Chem. **183**, 149 (1994)

Subject Index

activation 282
activation energy 11, 23, 62, 63, 65, 66, 70, 71, 73, 74, 82, 98, 122, 128, 231
  apparent 265
  distribution of 142
activation energy of diffusion 250
adiabatic approximation 14, 18, 77
adsorbed hydrogen 222
adsorption 230
  effective energy 247
  equilibria 230
  kinetics 230
  oxygen 253
  thermally activated 246
advanced neutron source (ANS) 157
AES 258
Ag 59
ageing/cycling, see cycling/ageing
Al 60, 61
alloys
  binary 73
  composition 250
  intermetallic 269
anelasticity 76, 82, 86
anharmonic corrections 23
anharmonic terms 23
anisotropy, in diffusion 70
Arrhenius behavior 60, 63
Arrhenius presentation 82, 86
Arrhenius relation 65, 70
asymmetry energy 44
asymmetry parameter 96
atom diffusion 109
Au 59

background gradient 115
band structure 117
barrier fluctuation effect 15
barrier shaking 22
batteries 325
  charge/discharge 326, 329
BPP
  formulation 99
  functions 99

  models 102, 137
  spectral densities 100, 109
bcc lattice 57
bcc metals 63, 68, 70
binding energy, apparent 243
Born-Mayer potential 190, 191, 203, 204
Born-Oppenheimer approximation 5, 6, 13, 36, 38, 77, 178, 186, 199
Born-von Karman model 201
boron 263
bound biphonons 205
Bragg scattering 166
Bravais lattice 159, 167
  cubic 57
Brillouin zone 161, 167
brittle failure 220
brittle-to-ductile transition, temperature 265
bubbles 226
  bubble growth 227
  hydrogen-filled 229
  water bubbles 227
bulk properties 3
Burgers vectors 234

cavity nucleation 215
Ce-D systems 123
Ce-H systems 123
chemical potential 54, 170, 259
chemical selectivity 305
  chromatographic process 311
  getters evaporable, bulk 305, 309
  isotopic 310
  pressure temperature swing technique 312
  purification 306
  separation 308
Chudley-Elliott model 160, 161, 162, 164, 166, 178, 179, 180
cleavage 220
cleavage crack 236
cleavage failure 228
cleavage fracture 268
cluster variation method 163

Topics in Applied Physics   Founded by Helmut K. V. Lotsch

Springer
and the
environment

 Springer